Hydrogenated Dilute Nitride Semiconductors

Hydrogenated Dilute Nitride Semiconductors

Theory, Properties, and Applications

edited by

Gianluca Ciatto

CRC Press
Taylor & Francis Group
Boca Raton London New York

CRC Press is an imprint of the
Taylor & Francis Group, an **informa** business

First published 2015 by Pan Stanford Publishing Pte. Ltd.

Published 2022 by CRC Press
Taylor & Francis Group
6000 Broken Sound Parkway NW, Suite 300
Boca Raton, FL 33487-2742

ISBN 978-981-4463-45-4 (hbk)
ISBN 978-981-4463-46-1 (ebk)

British Library Cataloguing-in-Publication Data
A catalogue record for this book is available from the British Library.

Hydrogenated Dilute Nitride Semiconductors: Theory, Properties, and Applications

Contents

Preface

Dilute nitride semiconductors are an example of the failure of "linear" models for the physical properties of solids: the addition of a few percent of N to GaAs causes changes in the physical properties that are opposite to those expected, assuming a linear interpolation of the GaAs and GaN properties. Even more surprisingly, the effect of nitrogen incorporation can be eliminated by exposure of the sample to a hydrogen flow. This book addresses in detail the modifications of the electronic structure and optical and structural properties induced by atomic hydrogen irradiation in technologically relevant dilute nitride semiconductors. The discussion of the experimental results from several techniques is enriched by state-of-the-art theoretical studies aimed at clarifying the origin of hydrogenation effects, which resides in the formation of specific nitrogen–hydrogen complexes. The nonlinear behavior of nitrogen and the passivation effect of hydrogen in dilute nitrides open the way to the manufactory of a new class of nanostructures with in-plane variation of the optical band gap.

The idea of this book emerged as an extension of an invited talk I gave in 2011 at the International Conference on Materials for Advanced Technologies (ICMAT 2011) in Singapore. The early outline of a monograph on the structural effects of hydrogenation was expanded in order to cover all aspects related to hydrogenated dilute nitride semiconductors, depending on the consideration that none of the books available on the topic of dilute nitrides had focused on the effects of hydrogenation. We estimated, indeed, that the huge number of recent published results on the theory, characterization, and nanomanipulation of hydrogenated dilute nitrides deserved collection in the form of a free-standing review volume.

With the present book, which is born out of several exchanges and collaborations between the different authors, we intend to tell the complete story of the amazing effects of hydrogen irradiation, from their first observation to the discovery of their physical origin and potential technology transfer. Our work primarily aims to guide graduate students and young scientists into the field but should be also of interest for more experienced scientists in research laboratories and academia. Moreover, we believe that the wide range and complementarity of the experimental techniques applied in the research here presented could inspire similar approaches in other fields of semiconductor science and condensed matter.

The book begins with an introductory chapter giving an overview of the unusual electronic structure and properties of dilute nitrides. This is followed by two chapters that present the hydrogenation technique and its effects on electronic properties and defects. The fourth chapter provides the theoretical basis of the mechanism of H-induced N passivation in dilute nitrides. This is followed by three chapters devoted to the effects of hydrogenation on the structure, addressed by different characterization techniques that allowed us to unveil the detailed structure of the specific N–H defect ruling the physical properties of these alloys. The last chapter describes how spatially selective hydrogenation of dilute nitride semiconductors can be used for the fabrication of a new class of site-controlled micro- and nanostructures with technological applications in nanophotonics and nanoelectronics.

It is a pleasure to acknowledge all the authors for the substantial time and efforts dedicated to the preparation of the different chapters. I appreciated very much their care and motivation toward a topic for which we share the same passion and enthusiasm. I would like to express many thanks to Stanford Chong for the invitation to develop the concepts of my talk into a book and to all Pan Stanford Publishing staff, in particular to Sarabjeet Garcha, Ritesh Kumar, and Archana Ziradkar, for their invaluable help while preparing the manuscript.

Gianluca Ciatto
Synchrotron SOLEIL
L'Orme des Merisiers, Saint-Aubin
BP 48, F-91192 Gif sur Yvette CEDEX, France
January 2015

Chapter 1

An Overview of Dilute Nitrides Theory and Properties

Eoin P. O'Reilly
Tyndall National Institute and Department of Physics, University College Cork, Lee Maltings, Cork, Ireland
eoin.oreilly@tyndall.ie

Dilute nitride alloys are a novel set of semiconductor alloys, of interest both from a fundamental perspective and for a wide range of potential applications. Their properties are markedly different from those of conventional alloys such as GaAs. This chapter presents an overview of the unusual electronic structure of dilute nitride alloys. We first describe that the band-anticrossing model provides a good basis to understand the electronic properties of alloys such as GaNAs and GaInNAs. We then apply the band-anticrossing model, with suitable modifications, to analyze the key features of the electronic structure. Finally we discuss the consequences of the unusual electronic structure for the observed transport and optical properties and for selected device applications.

Hydrogenated Dilute Nitride Semiconductors: Theory, Properties, and Applications
Edited by Gianluca Ciatto

ISBN 978-981-4463-45-4 (Hardcover), 978-981-4463-46-1 (eBook)
www.panstanford.com

1.1 Introduction

This book reflects the considerable recent interest in and the fascination of $Ga_{1-y}In_yN_xAs_{1-x}$ and related alloys.[a] When a small fraction of arsenic atoms in GaAs is replaced by nitrogen the energy gap initially decreases rapidly, at about 0.1 eV per % of N for $x < 0.03$ (Kondow et al., 1996, 1997). This behavior is markedly different from conventional semiconductors and is of interest both from a fundamental perspective and also because of its significant potential device applications. As an example, the strong bowing opens the possibility of using GaInNAs to get optical emission on a GaAs substrate at the technologically important wavelengths of 1.3 μm and 1.55 μm, considerably expanding the capabilities of GaAs for optoelectronics (Riechert et al., 2002; Bank et al., 2006, 2004; Broderick et al., 2012).

Our understanding of conventional III–V alloys has been built up through a range of approaches. Much progress is based on the use and application of relatively simple models, such as effective mass theory and the envelope function method (Bastard, 1990; Burt, 1999) to describe electronic states in quantum wells (QWs) and heterostructures. These simple and well-established models are underpinned and informed by more detailed and fundamental theoretical calculations, as well as by comparison with a wide range of experimental data.

It is well established that when a single N atom replaces an As atom in GaAs, it forms a resonant defect level, about 130 meV above the conduction band edge (CBE) of GaAs at low temperatures (Wolford et al., 1984; Liu et al., 1990b). This defect level arises because of the large difference in electronegativity and atomic size between N and As (Vogl, 1984; Hjalmarson et al., 1980; Lindsay and O'Reilly, 2003). A major breakthrough was achieved for dilute nitride alloys with the demonstration by Walukiewicz et al. (using hydrostatic pressure techniques; (Shan et al., 1999a)) that the reduction in energy gap in $Ga(In)N_xAs_{1-x}$ is due to a band-anticrossing (BAC) interaction between the CBE and a band of higher-lying localized nitrogen resonant states.

---

[a]For a review see (Buyanova and Chen, 2004; Henini, 2005; Erol, 2008).

Dilute nitride alloys have many novel features. Nitrogen is so strongly different from arsenic that adding hydrogen can lead to the formation of N–H complexes, which remove the resonant defect levels above the CBE and therefore eliminate the band-gap bowing (Polimeni et al., 2001a; Baldassarri Höger von Högersthal et al., 2001; Polimeni et al., 2002). This ability to reversibly engineer the electronic properties of dilute nitride alloys by hydrogenation and subsequent thermal annealing opens up many interesting possibilities, which form the major theme of this book. The hydrogenation and thermal annealing of GaNAs and related alloys effectively allow us to reversibly change the material electronic, optical, and transport properties between those of GaAs and of the very unusual alloy GaNAs. Given that the electronic properties of GaAs are in general well understood, and that the major theme of this book concerns this ability to undertake alchemy—reversibly interconverting GaAs and GaNAs—it is therefore of considerable value to first review the electronic structure of dilute nitride alloys and the consequences of the electronic structure for the alloy transport and optical properties.

Given the role of N defect levels, it is important to choose an appropriate level of detail when analyzing the band structure of dilute nitride alloys. We illustrate this here by focusing primarily on the electronic structure of Ga(In)NAs alloys—a very similar analysis can also be applied to describe other alloys such as Ga(In)NSb and Ga(In)NP (O'Reilly et al., 2009). We begin by showing in the next section that the two-level BAC model provides a good description of many properties of Ga(In)NAs. We then consider in Section 1.3 a number of examples where it is important to go beyond the two-level BAC model for GaNAs and show that it is possible to introduce a modified anticrossing model to account quantitatively for the electronic properties in these cases. The modified anticrossing model treats explicitly the full distribution of N states in the alloy. Various experimental studies show that this distribution of N levels plays a key role in determining some of the more detailed electronic properties of dilute nitride alloys. For instance, accurate measurements of the electron effective mass, m_e^* (Hai et al., 2000; Baldassarri Höger von Högersthal et al., 2003; Masia et al., 2003, 2006), and gyromagnetic factor, g_e^* (Pettinari et al.,

2006), in GaN_xAs_{1-x} as a function of x, reveal strong deviations from the smooth composition dependence predicted on the basis of a two-level BAC model (see Chapter 2). These deviations are well accounted for by considering hybridization of the CBE with states due to specific nitrogen clusters.

Adding nitrogen strongly perturbs the band structure of Ga(In)As and related alloys. Alloy scattering should therefore be considerably stronger than in conventional III–V semiconductors (Skierbiszewski, 2002; Fahy and O'Reilly, 2003; Fahy et al., 2006). We consider in Section 1.4 the consequences of the strong band-gap bowing for electron mobility in dilute nitride semiconductors. We have already described that there is a wide distribution of N resonant-state energies in GaN_xAs_{1-x}, associated with N–N nearest-neighbor pairs and clusters (Liu et al., 1990b; Masia et al., 2006). A significant number of these defect levels are calculated to be close to the CBE (Lindsay and O'Reilly, 2004b). We describe how inclusion of scattering due to this distribution of states can largely account for the low mobility of the order of 100 cm^2/Vs in dilute nitride alloys. We show that there is a fundamental connection between the band-gap bowing and the n-type carrier scattering cross section in the ultradilute limit, and that this imposes general limits on the carrier mobility in such alloys (Lindsay and O'Reilly, 2004b; Fahy and O'Reilly, 2003; Fahy et al., 2006). Using the BAC model, the carrier mobility for GaNAs is estimated to be $\sim$1000 cm^2/Vs for a N concentration of 1%, just under an order of magnitude lower than in GaAs and comparable to the highest measured mobility in high-quality GaInNAs samples at these N concentrations. This calculated mobility of $\sim$1000 cm^2/Vs is still approximately an order of magnitude higher than that found in most GaNAs samples.

The rapid reduction in energy gap with increasing N composition also has a significant potential benefit for optical devices, allowing GaAs-based telecom lasers (Broderick et al., 2012), as well as band-gap optimization in multijunction solar cells (Jones et al., 2011). We present in Section 1.5.1 theoretical calculations that show that the gain and loss characteristics of ideal dilute nitride lasers should be at least as good as those of conventional InP-based telecom lasers. This theoretical analysis is followed in Section 1.5.2 by a review of the substantial progress that was made in the development of GaInNAs telecom lasers. This included the demonstration of 1.3 μm edge-emitting and vertical-cavity lasers (Riechert et al., 2002) as well as devices emitting at 1.5 μm and beyond (Bank et al., 2006), with characteristics comparable to conventional InP-based devices.

Overall, dilute nitride alloys are a novel set of semiconductor alloys, of interest both from a fundamental perspective and for a wide range of potential applications (Buyanova and Chen, 2004; Henini, 2005; Erol, 2008). We conclude in Section 1.6 how their properties are markedly different from those of conventional alloys such as GaInAs. The ability to reversibly engineer their properties through hydrogenation and thermal annealing opens up a wide range of interesting possibilities, as described in more detail in the remainder of this book.

1.2 Band-Anticrossing Model of Dilute Nitrides

When a single N atom replaces an As atom in GaAs, it forms a resonant defect level above the CBE of GaAs (Wolford et al., 1984; Liu et al., 1990b). This defect level arises because of the large difference in electronegativity and atomic size between N and As. The BAC model introduced by Walukiewicz et al. explains the extreme band-gap bowing observed in $Ga_{1-y}In_yN_xAs_{1-x}$ in terms of an interaction between two levels, one at energy E_N associated with these localized N impurity states ψ_N and the other at energy E_c associated with the extended CBE state ψ_{c0} of the GaInAs matrix, with the two states linked by a matrix element V_{Nc} describing the interaction between them (Shan et al., 1999b). The conduction band (CB) dispersion of $GaInN_xAs_{1-x}$ is then given in the BAC model by the lower eigenvalue of the determinant:

$$\begin{vmatrix} E_N - E & V_{Nc} \\ V_{Nc} & E_c + \dfrac{\hbar^2 k^2}{2m_c^*} - E \end{vmatrix} \qquad (1.1)$$

where m_c^* is an appropriately chosen band-edge effective mass for the Ga(In)As host matrix material (Tomić et al., 2004). From Eq. 1.1, the CBE energy, E_-, is given by

$$E_- = \frac{E_\mathrm{N} + E_\mathrm{c}}{2} - \sqrt{\left(\frac{E_\mathrm{N} - E_\mathrm{c}}{2}\right)^2 + V_\mathrm{Nc}^2} \qquad (1.2)$$

and the alloy CBE wavefunction ψ_- can be found from Eq. 1.1 as $\psi_- = \alpha_\mathrm{c}\psi_{c0} + \alpha_\mathrm{N}\psi_\mathrm{N}$, where $\alpha_\mathrm{c}^2 + \alpha_\mathrm{N}^2 = 1$. The fractional Γ character, $f_{\Gamma c} = \alpha_\mathrm{c}^2$ of the E_- state provides a useful measure of how much the N-related states perturb the CBE wavefunction (Lindsay and O'Reilly, 2004b; Shan et al., 1999b). Since the interaction V_Nc between the N state and the host matrix CBE increases with N composition x as $x^{1/2}$ (Lindsay and O'Reilly, 1999), the Γ character $f_{\Gamma c}$ of the CBE state smoothly decreases with x in the BAC model, and the alloy electron mass m_e^* in a simple $\mathbf{k}\cdot\mathbf{p}$ model will then vary approximately as (Lindsay and O'Reilly, 2004b)

$$m_e^* = \frac{m_\mathrm{c}^* E_g(x)}{f_{\Gamma c} f_{\Gamma v} E_{g0}} \qquad (1.3)$$

When hydrostatic pressure is applied, the N resonant defect state at energy E_N is observed to have a lower pressure coefficient ($\sim$4 meV/kbar) than the GaAs CBE state at energy E_c (11 meV/kbar) (Wolford et al., 1984; Liu et al., 1990b). Because the E_- state is a linear combination of these two states, its measured pressure coefficient of 6–8 meV/kbar (Shan et al., 1999a) is intermediate between that of E_N and E_c. Likewise, the E_N level has a weaker variation with temperature than the GaAs E_c level, and so the energy gap of GaN_xAs_{1-x} also has a weaker temperature dependence than that of GaAs (Suemune et al., 2000).

A resonant feature associated with the upper eigenvalue, E_+, has also been observed in photoreflectance measurements (Shan et al., 1999a; Perkins et al., 1999; Klar et al., 2000), appearing in GaN_xAs_{1-x} for $x >\sim 0.2\%$ and remaining a relatively sharp feature until $x \sim 3\%$, beyond which composition it broadens and weakens when the resonant state becomes degenerate with the L-related CB levels (Lindsay and O'Reilly, 2004a). The data points in Fig. 1.1 show the measured variation of the E_- and E_+ energy levels with N composition x at room temperature in GaN_xAs_{1-x}; these measured values are very well fitted using the BAC model, as illustrated by the solid lines in Fig. 1.1, for which we use $E_\mathrm{N} = 1.67$ eV, $E_\mathrm{c} = 1.42 - 3.5\mathrm{x}$ eV and $V_\mathrm{Nc} = \beta \mathrm{x}^{1/2}$, with $\beta = 2.3$ eV (Klar et al., 2002).

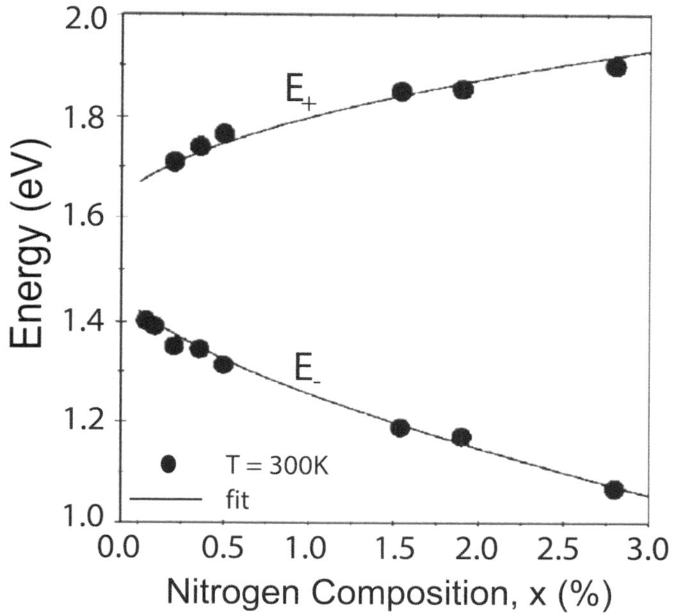

Figure 1.1 The E_- and E_+ data obtained by fitting the photoreflectance spectra of GaN_xAs_{1-x}. The solid lines are a fit to the two-level BAC model of Eq. 1.2. Adapted from (Klar et al., 2000) with permission of the American Institute of Physics.

Figure 1.2 shows the CB dispersion calculated using the two-level BAC model. The BAC interaction not only reduces the energy gap but also is predicted from Eqs. 1.1–1.3 to lead to an increased electron effective mass at the CB minimum and to a strongly nonparabolic CB dispersion. Direct evidence for this dispersion has been provided by magnetotunneling spectroscopy of GaNAs QW structures (Endicott et al., 2003). Photoreflectance measurements of GaNAs QW samples have also shown that strong band nonparabolicity is required to account for the QW excited state transition energies across a wide range of samples and as a function of hydrostatic pressure (Tomić et al., 2004; Klar et al., 2002).

Despite the wide success of the two-level BAC model, there are several sets of experimental data that it fails to explain. These include the observed composition dependence of the CBE effective

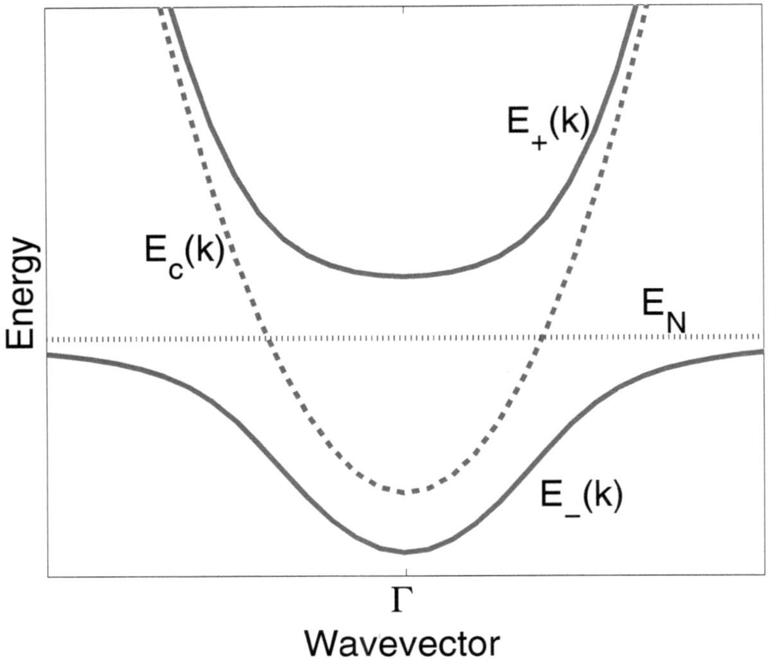

Figure 1.2 Solid lines show CB dispersion calculated using the two-level BAC model. Dashed line: Host matrix CB dispersion. Dotted line: N resonant defect energy level.

mass m_e^* in Ga(In)NAs alloys. The BAC model predicts, through Eq. 1.3, an enhancement of the CBE mass compared to GaAs. In practice, Eq. 1.3 provides a good estimate of the measured mass in GaNAs at very low N compositions (Masia et al., 2003) ($x < 0.05\%$) and also in indium-containing samples (Baldassarri Höger von Högersthal et al., 2003). However, it significantly underestimates the mass in GaN_xAs_{1-x} for $x > 0.1\%$ (Hai et al., 2000; Baldassarri Höger von Högersthal et al., 2003; Masia et al., 2003, 2006). In addition, the electron gyromagnetic ratio g_e^* in GaN_xAs_{1-x} has an unexpectedly large and nonmonotonic variation with x (Pettinari et al., 2006). These measured mass and gyromagnetic ratio values provide a stringent test of any model describing the electronic structure of GaN_xAs_{1-x} and related alloys.

The two-level BAC model has one further significant failing, giving an unphysical dispersion for the lower (E_-) band at large wavevector **k**. Because the BAC model effectively assumes an infinite density of N states per unit volume, this gives rise to an infinite density of states as the lower (E_-) band asymptotically approaches the nitrogen level (E_N) in Fig. 1.2. This problem has been addressed by treating the band structure using an Anderson many-impurity model and the Green's function approach (Vaughan and Ridley, 2007a,b). In the Anderson model, the Hamiltonian can be written as

$$H = H_0 + V \tag{1.4}$$

where H_0 is a sum of two terms describing the energies of extended and localized states, labeled by the wavevector **k** and position vector **j**, respectively, as follows:

$$H_0 = \sum_k E_k b_k^+ b_k + \sum_j E_j b_j^+ b_j \tag{1.5}$$

where E_k and E_j are the energies of the host matrix CB states and of the N resonant states and b_n^+ and b_n denote creation and annihilation operators, respectively. The second term in Eq. 1.4 represents an interaction between the extended and localized states, equivalent to the off-diagonal matrix element in Eq. 1.1. Vaughan and Ridley have shown using the Green's function approach that this model, which includes a finite density of N states per unit volume, retains the enhanced effective mass and also strong nonparabolicity of the BAC model but without the unphysical density of states due to the infinite number of N states in the two-level BAC model.

The linear combination of isolated nitrogen states (LCINS) model, which we introduce in the next section, closely resembles the many-impurity Anderson model. The reader is referred to Vaughan and Ridley for further detail of the Green's function approach (Vaughan and Ridley, 2007a,b). We will continue in later sections to compare experimental results with the two-level BAC model, while bearing in mind that this model has an unphysical aspect associated with the calculated band dispersion at large wavevector **k**.

1.3 Conduction Band States in GaNAs

The BAC model omits much of the detail of the band structure. Experimental studies show a range of resonant defect levels due to the formation of N complexes above the CBE in ultradilute nitride GaAs:N alloys (Liu et al., 1990b; Polimeni et al., 2008a) and below the CBE in GaNP (Zhang et al., 2000). Similar states are found in empirical pseudopotential (Kent and Zunger, 2001b; Kent et al., 2002) and tight-binding (TB) (O'Reilly et al., 2009; Lindsay and O'Reilly, 2004b) studies of N complexes. Such calculations support many aspects of the BAC model but also provide additional insight into the role of disorder and nitrogen clustering in GaNAs alloys. A full description of this distribution of N defect states is key to understanding the detail of the band structure in dilute nitride alloys.

A range of theoretical techniques are possible to investigate the isolated resonant state ψ_{N} and the wider distribution of N states in dilute N alloys. We have found it particularly fruitful to use an accurate sp^3s^* TB Hamiltonian that we have developed to describe the electronic structure of $\mathrm{GaInN}_x\mathrm{As}_{1-x}$ (O'Reilly et al., 2002) of $\mathrm{GaN}_x\mathrm{P}_{1-x}$ (Harris et al., 2008) and of $\mathrm{GaInN}_x\mathrm{Sb}_{1-x}$ (Lindsay et al., 2008). This TB Hamiltonian accounts well for the observed experimental data and also gives results in good agreement with pseudopotential calculations (Kent and Zunger, 2001b; Mattila et al., 1999; Bellaiche et al., 1996; Wei and Zunger, 1996).

We have used the TB Hamiltionian to investigate the nitrogen resonant state ψ_{N0} associated with an isolated N atom (Lindsay and O'Reilly, 2001). By comparing the calculated CBE states ψ_- and ψ_{c0} in large supercells ($\mathrm{Ga}_{864}\mathrm{N}_1\mathrm{As}_{863}$ and $\mathrm{Ga}_{864}\mathrm{As}_{864}$, respectively), we can derive ψ_{N0}. In the BAC model, ψ_- is a linear combination of ψ_{c0} and ψ_{N0}, with ψ_{N0} then given by

$$\psi_{\mathrm{N0}} = \frac{\psi_- - \alpha\psi_{c0}}{\sqrt{1-\alpha^2}} \tag{1.6}$$

where $\alpha = <\psi_-|\psi_{c0}>$.

We find for ordered supercell calculations that ψ_{N0} is highly localized, with over 50% of its probability density on the N site and the four neighboring Ga atoms. Further calculations show that we can also associate localized resonant states with N pairs and clusters, whose role becomes increasingly important with increasing N composition, x (Lindsay and O'Reilly, 2004b).

To investigate the role of disorder, we have extended the TB and two-level model to disordered GaN_xAs_{1-x} supercells. We found for $Ga_LN_MAs_{L-M}$ supercells containing $\sim$1000 atoms that the energy and wavefunction of the CBE state can be well represented as an **LCINS** and the GaAs CBE wavefunction, ψ_{c0} (LCINS method) (Lindsay and O'Reilly, 2004a,b; O'Reilly et al., 2004). For any given $Ga_LN_MAs_{L-M}$ supercell, we associate a localized resonant state ψ_{Ni} with each of the M nitrogen atoms ($i = 1, \ldots. M$) and a wavefunction ψ_{c0} with the (unperturbed) CBE.

Because N introduces such a strong perturbation, the result of an individual calculation on a 1000-atom supercell depends strongly on the statistical distribution of the N atoms, including, for example, the number of N–N pairs in the supercell and the presence or otherwise of larger and less common N clusters (O'Reilly et al., 2004). Figure 1.3 demonstrates the range of the nitrogen cluster state energies ε_l and the strength of their interactions $V_l = \langle\psi_{Nl}|H|\psi_{c0}\rangle$ with the CBE. To ensure a realistic distribution of N pair and cluster levels, we extend the LCINS approach in Fig. 1.3 to study ultralarge supercells containing a random distribution of up to $M = 8000$–10000 N atoms, with the composition x determined by the size of the supercell considered. The large values of M ensure minimal statistical variation between different random supercells.

Figure 1.3 shows the calculated distribution of the N state energies ε_l and their interaction V_l with the CBE state ψ_{c0} for $x =$ 0.21%, 0.5%, and 1.2% respectively, plotted as $V_{\mathrm{N}}(E)$, where $V_{\mathrm{N}}(E)$ is given by

$$V_{\mathrm{N}}(E) = \sum |V_l|^2 T(E - \varepsilon_l), \tag{1.7}$$

where $T(E)$ is a Gaussian function with 1 meV broadening. The GaN_xAs_{1-x} CBE, E_-, is shown by the vertical arrow for each value of x. For very low N composition ($x \sim 0.2\%$), most of the interaction $V_{\mathrm{N}}(E)$ arises from states that lie close to the isolated N resonant-level energy ($E_{\mathrm{N}} = 1.706$ eV at 0 K in our calculations). A small but important feature due to N–N pairs is observed at about 1.48 eV, with another weak feature at 1.63 eV, due to second-neighbor N (220) pairs on opposite corners of a cubic unit cell face. The energy

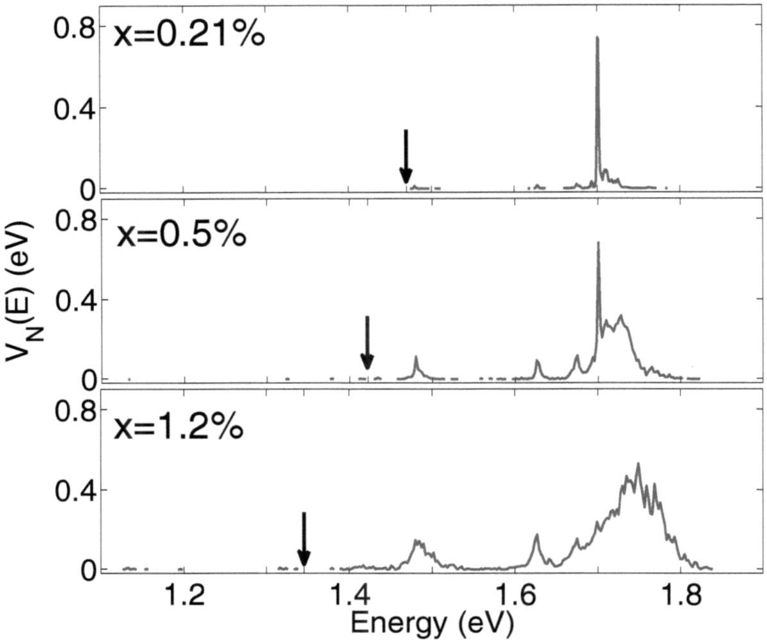

Figure 1.3 Calculated distribution of N cluster state energies, ε_l, weighted by the square of their interactions, $|V_l|^2$, with the CBE state for three bulk GaN_xAs_{1-x} compositions, with $x = 0.21\%$, 0.5%, and 1.2%, respectively. The vertical arrows show the GaN_xAs_{1-x} CBE in each case.

spectrum of the N cluster states broadens considerably at higher N compositions, and a small number of N states start to be observed below the N–N pair states due to the random formation of a small proportion of larger N clusters in the supercell.

Using a modified **k**·**p** model, the CBE effective mass m_e^* is from Eq. 1.3 approximately proportional to the energy gap and inversely proportional to $f_{\Gamma c}$ and the valence band (VB) fractional Γ character, $f_{\Gamma v}$. We calculated the fractional Γ character, $f_{\Gamma c}$, by first solving for the interaction between the GaAs CBE state ψ_{c0} and the full distribution of N states of Eq. 1.17, with energy levels, ε_l, and LCINS eigenstates, ψ_l (Lindsay and O'Reilly, 2004b). We were then able to determine the energy spectrum, $G_\Gamma(E)$, of these mixed CBE–N states with energies E_i and eigenstates φ_i by projecting onto the GaAs CBE

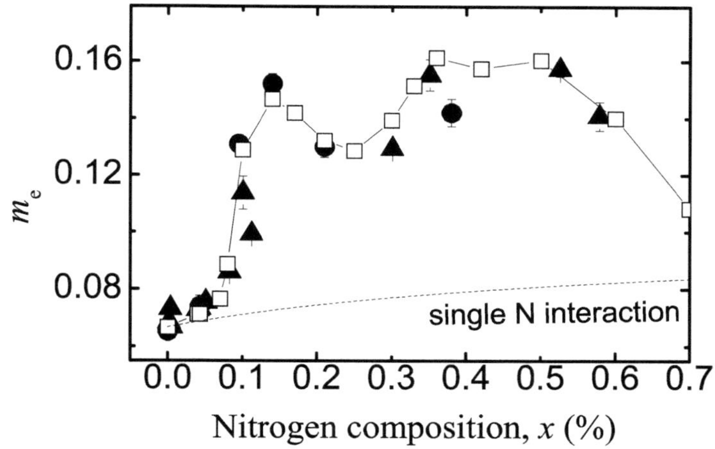

Figure 1.4 Measured values of the electron effective mass as a function of the N concentration. Several untreated (filled circles) and H-irradiated (filled triangles) samples are considered. Error bars indicate the uncertainty on the mass values (in some cases the uncertainty is within the symbol size). Open squares are the calculated effective mass values, and the solid line is a guide to the eye. The dotted line shows the predicted variation of effective mass based on the two-level band anticrossing model of Eq. 1.1. Adapted from (O'Reilly et al., 2009) with permission of the Institute of Physics.

state, ψ_{c0}:

$$G_\Gamma(E) = \sum_i | < \varphi_i | \psi_{c0} > |^2 \delta(E - E_i) \tag{1.8}$$

E_- is then defined as the state with the largest Γ character $f_{\Gamma c} = | < \varphi_i | \psi_{c0} > |^2$ in this spectrum.

The filled data points in Fig. 1.4 show the low-temperature electron effective mass determined by magnetophotoluminescence for a wide range of GaN_xAs_{1-x} samples (Masia et al., 2006). The dotted line shows the predicted variation of the band-edge effective mass in bulk GaN_xAs_{1-x} using the two-level BAC model of Eq. 1.1 (Shan et al., 1999a). This model significantly underestimates the measured mass, even for x as low as 0.1%. The open symbols show the low-temperature mass calculated for selected compositions, x, using the LCINS model and Eq. 1.3 (Masia et al., 2006).

The calculated and experimental data are in remarkable agreement with each other, confirming that hybridization between the

CBE and nitrogen cluster states causes the observed enhancement of effective mass values. The mass $m^*{}_e$ initially increases with composition x, as the band edge passes through the N–N pair states between $x = 0.1\%$ and 0.2%. For $x > 0.2\%$, we might expect a decrease in m_e^* toward the two-level BAC value, as the CBE E_- level passes through the N pair states. However, the density of other low-energy N-related states starts to increase markedly with increasing x. We attribute the states located at about 1.42–1.44 eV as due to C_S triplets, a noncoplanar chain of N-Ga-N-Ga-N atoms, whose density initially increases quickly, approximately as x^3 (Masia et al., 2006; Polimeni et al., 2008a). For $x = 0.36\%$ and 0.50%, the E_- level interacts strongly with these C_S-related levels, with a weaker interaction also occurring with the N–N pair levels whose distance in energy from the E_- level increases with x. These combined interactions lead to a further reduction in $f_{\Gamma c}$, consistent with the additional, sudden increase in m_e observed experimentally in Fig. 1.4 in this concentration range.

We note that the measurements in Fig. 1.4 provide a very interesting example of how hydrogenation can be used to probe the band structure of dilute nitride alloys. The filled circular data points were measured on dilute nitride samples of the given composition. The filled triangular data points show measurements taken on partially hydrogenated samples, where the effective N composition was determined on the basis of the measured energy gap of the sample. Hydrogenation in this case allowed the same sample to be used to determine the mass for several different effective compositions, x, rather than needing a large number of distinct samples for the measurements.

These detailed studies of GaN_xAs_{1-x} and further detailed studies both of GaN_xAs_{1-x} and of $Ga_{1-y}In_yN_xAs_{1-x}$ reveal the importance of the distribution of N levels and in particular the importance of the relative energy of the different defect levels compared to the alloy CBE energy. The BAC model has been widely applied to investigate the CB dispersion in Ga(In)NAs, being used, for example, both to interpret excited state transition energies in Ga(In)NAs QW structures (Tomić et al., 2004; Klar et al., 2002) and also to analyze the gain characteristics of GaInNAs/GaAs QW laser devices (Hofmann et al., 2001, 2002; Tomić et al., 2003). We expect that cluster states will disrupt the BAC dispersion at higher energies. However, the CBE energy shifts down with respect to the N cluster states both with increasing temperature and with increasing indium content, and so we conclude that the conventional BAC model should provide an appropriate starting point to describe the lowest conduction states and hence to analyze gain and loss processes in GaInNAs-based 1.3 and 1.5 μm lasers.

1.4 Alloy Scattering and Transport in Dilute Nitride Alloys

Progress was initially slow to develop models to describe the transport and mobility properties of dilute nitride alloys. These properties are difficult to analyze, precisely because N introduces such a strong perturbation to the band structure of Ga(In)As. This must lead to strong alloy scattering. There is a well-established model (Harrison and Hauser, 1976), based on the Born approximation, to describe the relatively weak alloy scattering that occurs in conventional semiconductor alloys. This model is, however, entirely insufficient for extreme alloys such as GaNAs, underestimating the alloy scattering cross section by over 2 orders of magnitude (Fahy and O'Reilly, 2003). We therefore use the distorted Born wave approximation to treat the strong effects of N. We do so below, first using the two-level BAC model to show how strong scattering substantially limits the electron mobility in dilute nitride alloys, consistent with the maximum mobility observed experimentally of the order of 1000 cm^2/Vs (Volz et al., 2003). We then modify the analysis to include the broad distribution of N levels in a random alloy, showing that these dominate the scattering near the CBE, playing a crucial role in further limiting the mobility to values less than 200 cm^2/Vs, in good agreement with typical experimental measurements, as shown in Fig. 1.5.

We first calculate the scattering cross section for an isolated N impurity in GaAs using S-matrix theory (distorted Born wave approach). This approach was previously applied (Sankey et al., 1982) to successfully describe resonant scattering due to conventional impurities in GaAs (Fisher et al., 1987). For a sufficiently localized perturbation, ΔV_N, the total scattering cross section σ for an isolated impurity is given by

$$\sigma = 4\pi \left(\frac{m^*}{2\pi\hbar^2} \right)^2 \ | < \psi_{c1} | \Delta V_{\mathrm{N}} | \psi_{c0} > |^2 \, \Omega^2, \qquad (1.9)$$

where m^* is the electron effective mass at the band edge and Ω is the volume of the region in which the wavefunctions are normalized. The state ψ_{c0} is the Γ-point CB Bloch wavefunction (in the absence of the N atom) and ψ_{c1} is the exact band-edge state in the presence of the N atom.

Consider a perfect crystal for which the electron Hamiltonian is H_0 and the CBE state has wavefunction ψ_{c0} and energy E_{c0}. When we introduce a single N atom into a large volume Ω of the otherwise perfect lattice, the new Hamiltonian, $H_1 = H_0 + \Delta V_{\mathrm{N}}$, leads to a modified band-edge state ψ_{c1} with energy E_{c1}. We can therefore rewrite the scattering matrix element of Eq. 1.9 as

$$< \psi_{c1} | \Delta V_{\mathrm{N}} | \psi_{c0} > = < \psi_{c1} | H_1 - H_0 | \psi_{c0} > = (E_{c1} - E_{c0}) < \psi_{c1} | \psi_{c0} > . \qquad (1.10)$$

Because $< \psi_{c1} | \psi_{c0} > \rightarrow 1$ for sufficiently large Ω, we derive that at low impurity concentrations

$$\Omega < \psi_{c1} | \Delta V_{\mathrm{N}} | \psi_{c0} > = \frac{dE_{\mathrm{c}}}{dn}, \qquad (1.11)$$

where E_{c} is the CBE energy and n is the number of impurities per unit volume. Substituting Eq. 1.11 in Eq. 1.9, and noting that n is related to the concentration x by $n = 4x/a_0^3$, where a_0 is the GaAs unit cell dimension, the scattering cross section for an isolated impurity is then given by

$$\sigma = \frac{\pi}{4} \left(\frac{m^*}{2\pi\hbar^2} \right)^2 \left[\frac{dE_{\mathrm{c}}}{dx} \right]^2 a_0^6. \qquad (1.12)$$

This result is key: it establishes a fundamental connection between the composition dependence of the CBE energy and the n-type carrier scattering cross section in the ultradilute limit for semiconductor alloys, imposing general limits on the carrier mobility in such alloys.

We can show this by extending the isolated N result of Eq. 1.12 to the case of a dilute nitride alloy, GaN_xAs_{1-x}. The mean free path l of carriers depends in an independent scattering model on the scattering cross section σ for a single defect and the number of

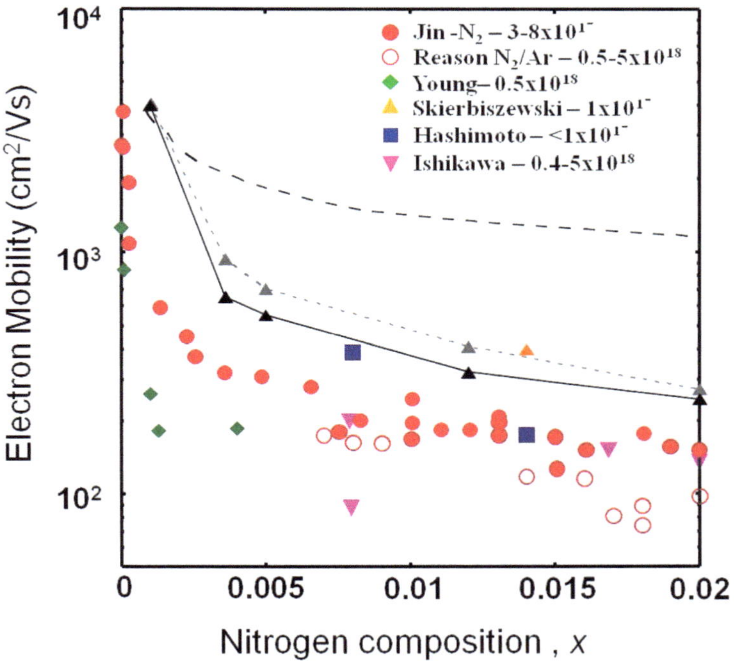

Figure 1.5 Data points. Measured variation of electron mobility with N composition x in $GaAs_{1-x}N_x$ (from (Jin, 2010)). The uppermost dashed line shows the calculated mobility, assuming scattering by isolated N atoms only (Eq. 1.13), while the lower lines show the calculated mobility, assuming a distribution of N states (Fahy et al., 2006) and a carrier density of $1 \times 10^{17} cm^{-3}$ (gray triangles; short dashed line) and 1×10^{18} cm^{-3} (black triangles; solid line).

defects n per unit volume as $l^{-1} = n\sigma$. Assuming such a classical model and the values of m^* and dE_c/dx at $x = 0$, we estimate for a N content of 1% a mean-free path of only 15 nm. The mobility μ is related to the mean free path l in a nondegenerate semiconductor as $\mu = e\tau/m^*$, with the scattering time $\tau = l/\bar{u}$, where $\bar{u}$ is the mean electron velocity. Setting $\bar{u}^2 = 3kT/m^*$, where T is the temperature, we estimate that the mobility μ is given by

$$\mu^{-1} = \frac{\sqrt{3m^*kT}}{e} \pi \left(\frac{m^*}{2\pi\hbar^2}\right)^2 \left[\frac{dE_c}{dx}\right]^2 a_0^3 x. \tag{1.13}$$

The uppermost dashed line in Fig. 1.5 shows the estimated variation of the room-temperature electron mobility with x in GaN_xAs_{1-x}, calculated allowing both m^* and dE_c/dx to vary with x on the basis of the two-level model of Eq. 1.1. The electron mobility is estimated to be of the order of 1000 cm^2/Vs when $x = 1\%$, of similar magnitude to the highest values observed to date in dilute nitride alloys (Volz et al., 2003), but larger than that found in many samples, where $\mu \approx 100$–400 cm^2/Vs (Jin, 2010, and references therein).

We saw earlier that the CBE effective mass was strongly modified due to the interaction of the band-edge state with N cluster states with which it is approximately degenerate in energy. These states tend also to reduce the mobility. We can extend the expression for mobility in Eq. 1.13 to include the influence of cluster states by assuming that all N states act as independent scattering centers, with the mobility then given by

$$\mu^{-1} = \frac{\sqrt{3m^*kT}}{e}\pi\left(\frac{m^*}{2\pi\hbar^2}\right)^2\sum_i\left[\frac{dE_c}{dx_i}\right]^2 a_0^3\,x_i. \qquad (1.14)$$

where x_i is the concentration and dE_c/dx_i is the rate of change of the CBE energy due to the ith type of cluster state.

The two lower lines with triangles in Fig. 1.5 show the calculated variation of room-temperature mobility in GaN_xAs_{1-x} when we include the effect of cluster states. The cluster states dominate the band-edge scattering with increasing x, reducing the estimated mobility by a factor of about 5 to 200 cm^2/Vs for $x \approx 2\%$, comparable to the experimental values. We note that factors omitted in the calculation here, including film quality and composition fluctuations, may also contribute to limiting the mobility in actual samples. The intrinsic alloy-scattering-limited mobility should be larger in GaInNAs samples due to the weaker band-gap bowing observed in indium-containing samples (Klar et al., 2002).

The intrinsically low electron mobilities in dilute nitride alloys have significant consequences for potential device applications. The low electron mobility, combined with the short nonradiative lifetimes observed to date, limit the electron diffusion lengths achievable in GaInNAs-based solar cells. Further efforts are leading to increased nonradiative lifetimes but are unlikely to see significant further improvements in the alloy-scattering-limited mobility. The inclusion of a GaInNAs layer in a multijunction solar cell can provide an absorbing layer with an energy gap of the order of 0.9 eV to 1 eV. Despite the relatively low electron mobility in GaInNAs, material of sufficient quality has been demonstrated to allow the recent realization of a multijunction cell with a record efficiency of 43.5%, a notable improvement[a] on the previous record value of 42.3% (Jones et al., 2011). In other applications, the low electron mobility may allow reduced surface recombination in mesa structures containing GaInNAs QWs, of benefit for ultracompact photonic device applications.

1.5 Dilute Nitride Lasers

The exponential growth in optical telecommunications and the internet has been underpinned by the development of semiconductor lasers emitting at 1.3 μm and 1.55 μm, wavelengths at which, respectively, dispersion is zero and losses are minimized in standard optical fibers (Zory, 1993; Kapon, 1999). The lasers designed to operate at these wavelengths are based primarily on quaternary InGaAsP and InGaAlAs alloy structures grown on InP substrates. Despite their widespread application, there are several significant drawbacks associated with these devices, mostly associated with the constraints of growing on InP substrates. The development of telecom lasers grown on GaAs substrates could bring several advantages. Firstly, because GaAs is a more robust material, growth can be carried out on larger substrates. Secondly, better optical confinement can be achieved in GaAs heterostructures because of the larger refractive index difference between GaAs and AlGaAs compared to that between InP and the quaternary alloys. Vertical-cavity surface emitting lasers (VCSELs) can therefore be grown monolithically on GaAs but only with great difficulty on InP. In addition, because AlGaAs has a considerably larger energy gap than InP and can be grown lattice-matched on GaAs, it should be possible

---

[a]http://optics.org/news/2/4/22.

to achieve much better electrical confinement in a telecom laser based on GaAs.

Considerable advantage could therefore be gained if high-quality and reliable telecom lasers could be developed on GaAs. This is very difficult to achieve using conventional QW structures. Highly efficient lasers using strained InGaAs QW structures are very well established for emission around 1 μm. However, too much In is needed, and there is therefore too large a lattice mismatch relative to GaAs to achieve reliable lasers emitting at 1.3 μm and beyond (Coleman, 1993). There was therefore considerable interest generated when it was shown that replacing a small fraction of As atoms by N in Ga(In)As leads to a rapid reduction in the energy gap, as shown earlier in Fig. 1.1. This opens up the possibility of achieving longer-wavelength emission on a GaAs substrate.

The major driver toward the development of dilute nitride III–V compounds was undoubtedly the realization by Kondow et al. (Kondow et al., 1996, 1997) that these alloys offer a viable route to achieve GaAs-based QW laser diodes emitting in the 1.3 μm and 1.5 μm optical windows. These GaInNAs/GaAs QW structures also provide the benefits associated with compressively strained devices (Adams, 1986; O'Reilly, 1989). In addition, the inclusion of N in the GaInNAs layers increases the CB offset, leading to improved electron confinement when compared with conventional InGaAsP 1.3 μm lasers (Kondow et al., 1996, 1997). Both edge-emitting and VCSEL structures have been reported with impressive characteristics, as discussed further below.

Given the significant differences in the CB structure of GaInNAs compared to conventional III–V semiconductors, it is worthwhile to elucidate the influence of N not only on the electronic structure but also on the gain characteristics of ideal dilute nitride lasers. We summarize below that the incorporation of N degrades properties such as transparency and threshold carrier density compared to that of the best GaInAs/GaAs structures (Tomić et al., 2003; Tomić and O'Reilly, 2003). Nevertheless, the overall characteristics of an ideal GaInNAs/GaAs laser are still expected to be at least as good as those of conventional InP-based telecom lasers.

1.5.1 *Theoretical Analysis of Dilute Nitride Lasers*

We present here a theoretical overview of the consequences of the electronic structure of dilute nitride alloys for laser emission. We summarize the influence of N incorporation on the calculated dipole matrix elements, electron effective mass, and gain versus carrier density and radiative current density, as well as the effect of N on the primary loss mechanisms of defect-related and Auger recombination. We then compare the theoretically estimated laser characteristics of GaInNAs-based and conventional InGaAsP-based 1.3 μm lasers.

The solid lines in Fig. 1.6 show the calculated band dispersion for an ideal 7 nm wide $In_{0.36}Ga_{0.64}N_{0.02}As_{0.98}$/GaAs QW, calculated using a 10-band **k·p** Hamiltonian that includes the CB, heavy-hole (HH), light-hole (LH), and spin-split-off (SO) states from the conventional 8-band **k·p** model, as well as two additional spin-degenerate N-related states (Tomić et al., 2003). The dashed line shows the calculated band dispersion of the equivalent N-free 7 nm wide $In_{0.36}Ga_{0.64}As$/GaAs QW structure. An increased band-edge effective mass and strong nonparabolicity are clearly visible in the CB dispersion in the GaInNAs QW due to the interaction between the lowest conduction states and the N resonant levels.

The coupling between the N level and the CBE modifies the CB wavefunctions and reduces the interband optical transition matrix element $|M_{e-h}|^2$ compared to a conventional N-free alloy. The band-edge, zone-center, transverse electric (TE) matrix element is calculated to decrease by $\sim$30% due to the incorporation of N. The N–CB coupling therefore leads to an increased CBE effective mass. We calculate that the band-edge density-of-states effective mass in the first CB increases to $0.060m_0$ in the structure with 2% N compared to a value of $0.046m_0$ in the N-free case.

The reduction in $|M_{e-h}|^2$ and the consequent increase in the CBE effective mass, m_e^*, causes the product $m_e^* \times |M_{e-h}|^2$ to stay approximately constant, as would be expected from **k·p** theory: the dominant contribution to the bulk CB inverse effective mass, m_e^{*-1}, is directly proportional to $|M_{e-h}|^2$. The band structure presented in Fig. 1.6 was used to calculate the variation of material gain with temperature T and as a function of carrier density n and of radiative

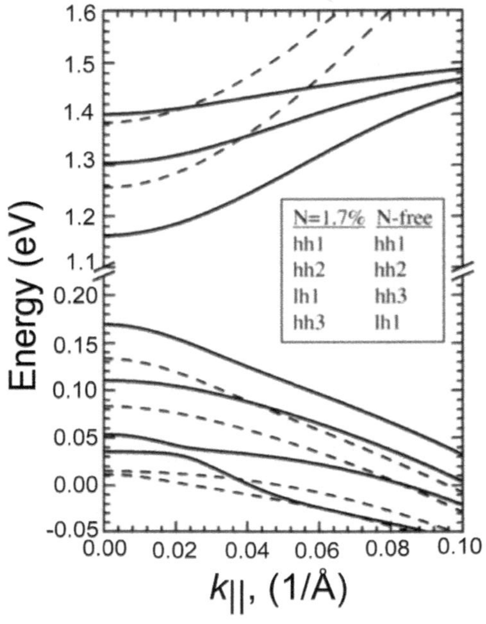

Figure 1.6 Solid lines: Band structure of an $In_{0.36}Ga_{0.64}N_{0.02}As_{0.98}/GaAs$ single-QW laser device calculated from a 10-band **k·p** Hamiltonian that was parametrized to fit the experimentally observed QW transitions at $k_{||} = 0$ (Tomić et al., 2003). Dashed lines: Calculated band structure of an equivalent N-free single-QW device. Copyright © 2003 IEEE. Reprinted, with permission, from Tomić et al., 2003.

current density J_{rad}, both in an $In_{0.36}Ga_{0.64}N_{0.02}As_{0.98}/GaAs$ structure and in an equivalent N-free structure (Tomić and O'Reilly, 2002). The increase in the CB effective mass leads to an increase in the carrier concentration at transparency for the $In_{0.36}Ga_{0.64}N_{0.02}As_{0.98}/GaAs$ QW laser structure and a decrease in the separation between the conduction and VB quasi-Fermi energies, $E_{\text{Fc}} - E_{\text{Fv}}$, for a fixed carrier concentration. As a consequence the peak gain decreases at a fixed carrier density in the $In_{0.36}Ga_{0.64}N_{0.02}As_{0.98}/GaAs$ structure when compared to the N-free case, as evidenced in Fig. 1.7a. There is, however, a much weaker variation in the peak gain versus radiative current density, as shown in Fig. 1.7b. This weak variation reflects the fact that for a fixed

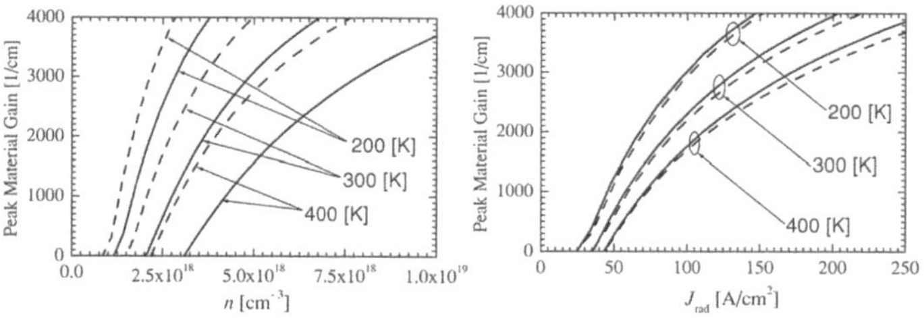

Figure 1.7 Calculated variation of the peak material gain as a function of (a) carrier density and (b) radiative current density, in the same N-containing (solid lines) and N-free (dashed lines) single-QW laser device of Fig. 1.6. Copyright © 2003 IEEE. Reprinted, with permission, from Tomić et al., 2003.

quasi-Fermi level separation, the radiative current J_{rad} in a QW laser is approximately proportional to $m_r^* \times |M_{e-h}|^2$, where m_r^* is the band-edge reduced effective mass. Because the VB mass, m_v^*, is always larger than the CB mass, m_e^*, the reduced mass m_r^* is determined primarily by m_e^*. We saw above that $m_e^* \times |M_{e-h}|^2$ is approximately constant, thus accounting for the calculated weak variation in J_{rad}.

Figure 1.7 and related calculations suggest that the optimal GaInNAs/GaAs QW laser device should contain minimal nitrogen, ideally being N free! This, however, is not possible because of the excessively large strain required to achieve 1.3 μm emission in a GaInAs QW. It can be estimated, using the strain-thickness criteria in Refs. (Zory, 1993; Kapon, 1999), that one needs $y > 1.5\%$ and $x < 39\%$ to achieve 1.3 μm emission in a pseudomorphic $Ga_{1-x}In_xN_yAs_{1-y}$/GaAs QW structure. The N content can be further reduced by adding more In to narrower QWs and also by growing tensile-strained layers above and below the QW, as demonstrated by Tansu et al., who achieved 1.3 μm emission with $y = 0.5\%$, a QW width of 6 nm, and utilizing strain-compensating $GaAs_{0.85}P_{0.15}$ tensile layers in the barrier region (Tansu et al., 2002; Tansu and Mawst, 2002).

Comparing our theoretical results on GaInNAs/GaAs (Tomić et al., 2003) with the previous theoretical analysis of 1.3 μm InGaAsP/InP and InGaAlAs/InP structures (Seki et al., 1994; Pan and Chyi, 1996), we find that the GaInNAs material has a higher differential gain of $\sim 0.8 \times 10^{-15}$ cm^{-1} compared to $\sim 0.6 \times 10^{-15}$

cm^{-1} in InP-based devices for the same amount of compressive strain (~1.5%), optical gain (~1300 cm^{-1}), or well thickness (~7 nm), despite the significantly larger inhomogeneous broadening (18 meV) that we assumed in GaInNAs compared to the InP-based devices. Stronger optical confinement is possible in 1.3 μm GaInNAs-/GaAs-based lasers compared to InP-based devices, because of the larger refractive index step achievable through use of AlGaAs cladding layers. This reduces the number of QWs required in an edge-emitting device, while also opening up the possibility of GaInNAs-based 1.3 μm and 1.5 μm VCSELs. The theoretical analysis presented here therefore confirms the potential of GaInNAs-based lasers both for edge- and surface-emitting laser applications in the telecommunication wavelength range.

1.5.2 *Experimental Analysis of Dilute Nitride Lasers*

Nonradiative recombination processes make a significant contribution to the threshold current in conventional InP-based lasers. This leads to a strong temperature dependence of the threshold current in such devices (Phillips et al., 1999). A number of factors can contribute to this temperature dependence, such as carrier leakage due to weak electron confinement (Chen et al., 1983; Barrau et al., 1992) or intervalence band absorption (Adams, 1986; Seki et al., 1994), but the dominant loss mechanism in InP-based lasers at threshold is generally regarded to be Auger recombination (Phillips et al., 1999; Dutta and Nelson, 1981). Given the excellent radiative characteristics of ideal GaInNAs lasers, it is of considerable interest therefore to also investigate the relative importance of radiative and nonradiative recombination paths in dilute nitride lasers compared to their conventional InP-based counterparts.

We begin our discussion of GaInNAs lasers with a description of the quantitative experimental analysis of recombination pathways in 1.3 μm GaInNAs/GaAs lasers carried out in (Fehse et al., 2002), highlighting the importance of defect-related and Auger recombination in such devices. We then provide an interpretation of the observed device characteristics in terms of detailed theoretical calculations of the effects of N on Auger recombination rates in GaInNAs/GaAs systems (Andreev and O'Reilly, 2004). This is then followed by a discussion of the observed threshold characteristics of GaInNAs lasers emitting across the full wavelength range from 1.2 μm to 1.6 μm.

Under the assumptions of charge neutrality in the active region (electron density n = hole density p) and negligible carrier leakage, the current density in a single-QW laser can be written in the Boltzmann approximation as

$$J = J_{\text{mono}} + J_{\text{rad}} + J_{\text{Auger}} = e(An + Bn^2 + Cn^3) \qquad (1.15)$$

where J_{mono}, J_{rad}, and J_{Auger} refer, respectively, to the current densities due to monomolecular/defect-related ($\propto n$), radiative ($\propto n^2$), and Auger ($\propto n^3$) recombination, with A, B, and C the monomolecular, radiative, and Auger recombination coefficients, respectively. In each case the recombination rate varies in the Boltzmann approximation as carrier density n to the power of the number of carriers involved—for example, because radiative recombination is a bimolecular process, it varies as n^2.

It is possible from Eq. 1.15 to express the total current density J over a limited current range as $J \propto n^z$, with $1 \leq z \leq 3$ (Phillips et al., 1999; Higashi et al., 1999) and with the value of z varying depending on whether the dominant contribution to the current density is due to monomolecular ($z = 1$), radiative ($z = 2$), or Auger ($z = 3$) recombination. It is also possible to measure the integrated spontaneous emission (SE) rate L from a semiconductor laser by etching a window in the laser substrate and then collecting the SE emitted through the window. Because L is proportional to Bn^2, this allows the total current density J to be related to the integrated SE L as $J \propto L^{z/2}$, which can be rewritten as $\ln J \propto z \ln L^{1/2}$. This then enables z to be quantified experimentally for a given laser by plotting $\ln J$ against $\ln L^{1/2}$, with the slope of the log-log plot giving the value for z.

Such an analysis was applied in (Fehse et al., 2002) to the study of two 1.3 μm GaInNAs/GaAs laser structures. By careful analysis of the data measured over a wide temperature range for a single-QW laser, it was shown that at low temperatures ($T <\sim 200$ K) defect-related recombination constituted ~70% of the total threshold current in the device considered and contributed ~55% of the total current at room temperature. A full analysis

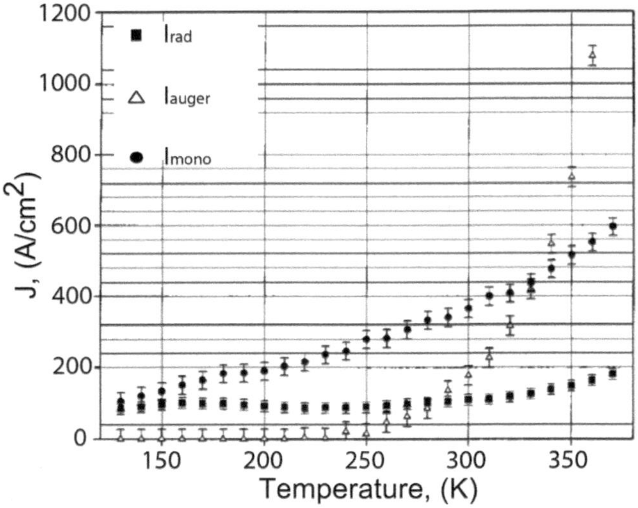

Figure 1.8 Experimentally measured variation of the monomolecular (I_{mono}), radiative (I_{rad}), and Auger (I_{Auger}) current densities at threshold as a function of temperature for a $Ga_{0.64}In_{0.36}N_{0.017}As_{0.983}$ single-QW laser device, as determined from the analysis outlined in the text. Copyright © 2002 IEEE. Reprinted, with permission, from Fehse et al., 2002.

indicated defect-related, radiative, and Auger currents constituted, respectively, ~55%, 20%, and 25% of the laser threshold current density J_{th} at room temperature in the device considered. The measured variation of each of the monomolecular, radiative, and Auger current paths in this single-QW device are shown in Fig. 1.8. It can be seen that the Auger contribution to the threshold current increases rapidly at higher temperatures, becoming the dominant current path in the given device above $T = 350$ K. This is confirmed by the measured value of $z_{th} = 2.8$ at $T = 370$ K. Comparing this to the value of $z_{th} = 2.9$ for the 1.3 μm InGaAsP/InP device of (Phillips et al., 1999) shows that the GaInNAs/GaAs material system does not differ significantly in its Auger-dominated high-temperature characteristics from conventional InP-based lasers.

A detailed theoretical study of the effects of N on a GaInNAs/GaAs QW structure (Andreev and O'Reilly, 2004) confirmed that the contribution of Auger recombination is relatively unchanged in a GaInNAs laser compared to that observed in conventional InP-based lasers. On the basis of this experimental (Fehse et al., 2002) and theoretical (Andreev and O'Reilly, 2004) evidence it is clear that Auger recombination is then as important in 1.3 μm GaInNAs lasers as in their InP-based counterparts.

By increasing the N content, it is possible to extend the emission of GaInNAs-based lasers to longer wavelengths. The first emission beyond 1.5 μm with GaInNAs-/GaAs-based devices was reported in 2000 by Fischer et al. using a 5% N and a 38% In double-QW structure (Fischer et al., 2000). The threshold current density of these initial devices was very large, with $J_{th} \approx 60$ kA cm^{-2} at $\lambda = 1.51$ μm. The development of GaInNAs 1.5 μm devices and the reduction in the reported threshold current was then rapid, with further improvements achieved by introducing Sb to give GaInNAsSb QW systems (Ungaro et al., 1999; Yang et al., 1999, 2001; Li et al., 2003). The development of the GaInNAs material system was largely driven by Infineon. In 2004 it reported long-wavelength devices at $\lambda = 1.4$ μm with $J_{th} = 690$ A cm^{-2} and at $\lambda = 1.43$ μm with $J_{th} = 1090$ A cm^{-2} (Averbeck et al., 2004). At the time these values were the lowest values for Sb-free devices emitting beyond 1.3 μm. By 2005 they had extended the emission to 1.51 μm with a GaInNAs single QW and GaAs barriers (Averbeck et al., 2005). This device had a record-low threshold current density for any GaAs-based device at this wavelength with $J_{th} = 780$ A cm^{-2}.

The development of the GaInNAsSb system was largely driven by the Harris group at Stanford, who extended the emission wavelength to 1.55 μm in 2005 (Bank et al., 2005). With further developments of its device design and optimization of its growth the Stanford group produced devices with a record-low threshold current density at 1.55 μm of 579 A cm^{-2} (Bank et al., 2006, 2004) using GaInNAsSb and GaNAs barriers in 2006. These devices remain the benchmark for 1.55 μm emission using the GaInNAs(Sb) material system. The Harris group led the use of antimony in the growth of GaInNAs-active regions. The general consensus is that antimony acts as a reactive surfactant enhancing the incorporation of N (Massies and Grandjean, 1993). In addition, it allows higher growth temperatures, enabling improved optical quality, because of the formation of fewer point defects (Bank, 2006).

In summary, the rapid reduction in the energy gap with increasing x in GaN_xAs_{1-x} has opened up the possibility of achieving longer-wavelength emission on GaAs. Theoretical calculations have shown that the gain and loss characteristics of ideal dilute nitride lasers should be at least as good as those of conventional InP-based telecom lasers. Substantial progress was made in the development of GaInNAs telecom lasers, leading to the demonstration of 1.3 μm edge-emitting and vertical-cavity lasers (Riechert et al., 2002), as well as devices emitting at 1.5 μm and beyond, with characteristics comparable to conventional InP-based devices. Overall very good characteristics have been demonstrated by dilute nitride lasers, but the advantages of using dilute nitrides, for example, in VCSEL structures have not been sufficiently pronounced to displace the well-tested incumbent technology based on InP.

We note that the lack of uptake of dilute nitride-based lasers was at least in part driven by economic uncertainty in the telecom industry during the mid-2000s, where some of the most successful commercial producers of devices left the sector. This led to a large loss of momentum in the development of the technology. However, dilute nitrides are making a resurgence, for example, as the 1 eV junction in multijunction solar cells, which currently hold the efficiency record of 43.5% (Jones et al., 2011)[a] and for use in III–V alloys for direct integration of lasers with silicon (Liebich et al., 2011), both of which are proving to be promising applications of this interesting class of materials.

1.6 Summary and Conclusions

Dilute nitride alloys are of great interest both from a fundamental perspective and also for a wide range of potential applications. We have summarized in this chapter how the CB structure of dilute nitride alloys is markedly different from conventional alloys. It can be well explained using a BAC model that describes the electronic structure in terms of the interaction between N-related resonant defect states and the host matrix material CB states. The

[a]http://optics.org/news/2/4/22.

BAC model provides a very good basis to understand not just the electronic structure but also the transport and optical properties of these alloys. Overall, nitrogen is so strongly different from arsenic that adding H can lead to the formation of N–H complexes, which remove the resonant defect levels above the CBE and therefore eliminate the band-gap bowing. This allows us to reversibly engineer the electronic properties of dilute nitride alloys by hydrogenation and subsequent thermal annealing, opening up many interesting possibilities, as discussed in detail in the remainder of this book.

Acknowledgments

I am grateful to many colleagues with whom I have had the pleasure to interact on dilute nitride alloys, and I particularly thank Alf Adams, Mario Capizzi, Stephen Fahy, Peter Klar, Andrew Lindsay, Antonio Polimeni, Henning Riechert, Stephen Sweeney, and Stanko Tomić for their major contributions to the work described here, as well as Masoud Seifikar and Muhammad Usman for their help in preparing this chapter. Our work at Tyndall was supported by Science Foundation Ireland (00/PI.1/C072, 06/IN.1/I90, 10/IN.1/I2994).

Chapter 2

Effects of Hydrogenation on the Electronic Properties of Dilute Nitrides

Giorgio Pettinari,[a] Antonio Polimeni, and Mario Capizzi
Dipartimento di Fisica, Sapienza Università di Roma,
Piazzale A. Moro 2, 00185 Roma, Italy
giorgio.pettinari@cnr.it

This chapter reviews the effects that hydrogen incorporation produces on the electronic properties of dilute nitrides, as determined by optical spectroscopy experiments. In the first instance, H brings the band-gap energy of Ga(AsN) and related materials back to that of the N-free host material. This dramatic finding is due to the *electronic passivation* of N atoms by H atoms following the formation of specific N–H complexes. These complexes can be progressively and controllably deactivated by either thermal, electron, or optical treatments. Nitrogen passivation involves the entire band structure of dilute nitrides and causes a tunable variation of the transport (electron mass) and spin (electron gyromagnetic factor) properties. The application of magnetic fields and hydrostatic pressures combined with the effects of hydrogenation permits to unveil an electronic hierarchy between different aggregates of N atoms

[a]National Research Council (CNR), Institute for Photonics and Nanotechnologies (IFN-CNR), via Cineto Romano 42, 00156 Roma, Italy

Hydrogenated Dilute Nitride Semiconductors: Theory, Properties, and Applications
Edited by Gianluca Ciatto

ISBN 978-981-4463-45-4 (Hardcover), 978-981-4463-46-1 (eBook)
www.panstanford.com

(N clusters [NCs]) that contribute to a different extent to determine the puzzling compositional dependence of the physical properties of these fascinating materials.

2.1 Introduction

Hydrogen is an impurity common to most phases of growth and processing of semiconductor materials and to the related mass production of electronic and optoelectronic devices (Pankove and Johnson, 1991). The importance of H-induced effects in semiconductors stems from the high chemical reactivity and diffusivity of this element (Weiringen and Warmholtz, 1956). Indeed, H tends to bind to dangling bonds associated with point and extended defects and thus cleans up the band gap from the energy levels associated with those defects (Pankove et al., 1983; Chevallier et al., 1985). This process, usually referred to as *defect passivation*, leads to a large modification of conductivity in most semiconductors and removes often the detrimental consequences associated with crystal imperfections (Northrup, 2002; Matsubara et al., 2010; Polimeni et al., 1994). In particular, the formation of complexes made of H atoms and impurities has been observed in a vast number of elemental, compound, and alloyed semiconductors (Pankove and Johnson, 1991). Interestingly, hydrogen itself may act as a dopant in narrow-gap semiconductors and in III nitrides (Pettinari et al., 2008; De Luca et al., 2012; Kozlova et al., 2013). It also turns conductivity from *p*-type into *n*-type in diamond (Teukam et al., 2003), neutralizes the native *p*-type conductivity in Ga(AsBi) alloys (Pettinari et al., 2012a), and transforms graphene from a highly conductive semimetal into an insulator (Elias et al., 2009).

Probably, the most impressive effects of hydrogen incorporation into a semiconductor have been observed in dilute nitrides (see Chapter 1) where small percentages of nitrogen atoms ($x \sim 1\%$) replace the group V elements in III–V compounds such as arsenides, phosphides, and antimonides (Buyanova and Chen, 2004; Henini, 2005; Erol, 2008; O'Reilly et al., 2009). Despite the small amount of substituted atoms, N alloying perturbs dramatically the electronic properties of the host crystal, leading to a counterintuitive and

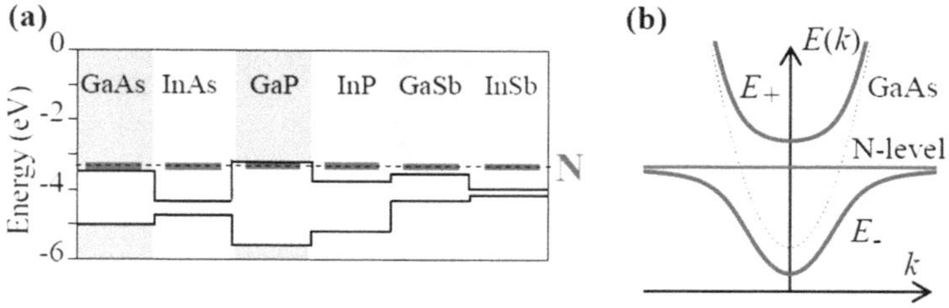

Figure 2.1 (a) Band lineups and position of the N-level (horizontal thick dash) for a range of III–V compounds. For each material, the upper and lower lines indicate the position of the conduction band (CB) minimum and the valence band maximum, respectively. Band lineups have been calculated considering the host electron affinity as in Ref. (Van de Walle and Neugebauer, 2003). (b) Sketch of the energy–wavevector dispersion curves (thick solid lines) assuming a simple two-level interaction between a single-N level (horizontal solid line) and the parabolic conduction band of GaAs (dotted line).

large decrease of the band-gap energy (O'Reilly et al., 2009; Kent and Zunger, 2001a,b; Tisch et al., 2002), a nonmonotonic variation of the electron effective mass (Masia et al., 2006; Allison et al., 2008; Dannecker et al., 2010) and gyromagnetic factor (Pettinari et al., 2006; Ivchenko et al., 2010; Kunold et al., 2011), and a softened response to external perturbations, such as hydrostatic pressure (Klar et al., 2000; Weinstein et al., 2003; Pettinari et al., 2007) and temperature (Polimeni et al., 2001b; Luo et al., 2001). Summarizing the results presented in Chapter 1, all these effects are ascribable to the formation of N-related localized states that are resonant with (or close to) the extended states of the conduction band (CB) of the host lattice, as exemplified in Fig. 2.1a (Van de Walle and Neugebauer, 2003) showing the relative alignment of the band extrema with respect to a single-N level, E_N, in different semiconductors (O'Reilly et al., 2009; Kent and Zunger, 2001a; Shan et al., 1999a; Endicott et al., 2003; Lindsay et al., 2008). The quantum interaction between these two sets of states leads to a distortion of the energy–wavevector dispersion curve and to the formation of a fully developed energy gap in the CB of the host crystal, as shown in Fig. 2.1b for GaAs. This determines a drastic reduction of the band-gap energy, as displayed for many different

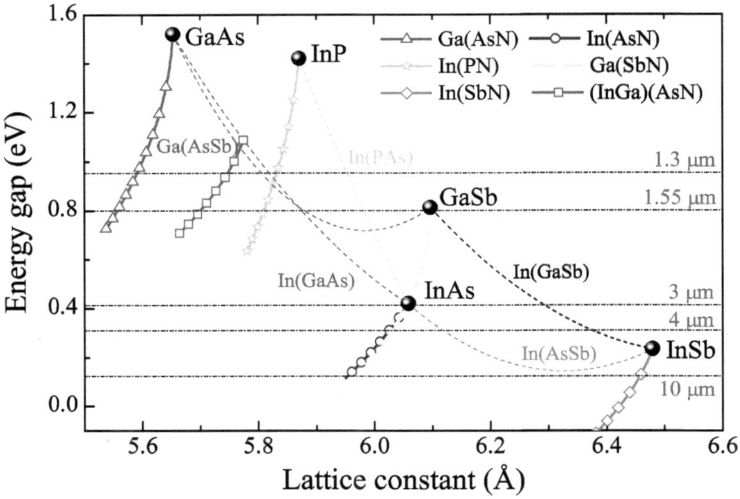

Figure 2.2 Band-gap energy (at low temperature) versus lattice constant for different III–N–V dilute nitride alloys. Horizontal dotted lines highlight the region of interest for telecommunications through optical fiber systems (1.3 μm and 1.55 μm) and environmental sensing or medical diagnostics (3–4 μm and 10 μm).

material systems in Fig. 2.2. Furthermore, N alloying permits a tuning of the band-gap energy down to wavelength regions of interest for telecommunications and an extension of the frequency range exploitable for solar cells (Carrère et al., 2006; Volz et al., 2007). Also, it provides a means for modifying the effective mass and group velocity of conduction electrons with a potential for novel electronic devices in the terahertz region (Ignatov et al., 2006; Spasov et al., 2008).

This chapter illustrates the effects of hydrogen incorporation on the electronic properties of different dilute nitride semiconductors. In the next section, we describe briefly the main experimental methods we employed. In the third section, we present the modifications induced by H on the band-gap energy of different materials over a range of nitrogen concentrations (x) spanning from the very dilute ($x < 0.1\%$) to the concentrated ($x > 1\%$) limit. We discuss next the microscopic mechanism responsible for these effects, the spatial distribution of H atoms in the crystal, and its relationship with the measured optical properties. The fifth section discusses the stability of the modifications induced by H with respect to thermal annealing and irradiation of the hydrogenated samples with photons. Finally, we show the variations that the formation of N–H complexes exert on the CB of Ga(AsN) as probed by two fundamental band structure parameters, namely, the electron effective mass and the gyromagnetic factor.

2.2 Experimental Methods

2.2.1 *Hydrogenation*

Sample hydrogenation was performed by ion beam irradiation from a Kaufman source with the samples held at high temperature (usually 300°C) during the whole process (Pankove and Johnson, 1991; Trotta et al., 2012b). The ion energy was set at 100 eV, and the ion current density varied from less than 0.1 μA/cm^2 up to more than 100 μA/cm^2, depending on the target H dose to be irradiated on the sample. The impinging H dose, d_{H}, may vary from $d_{\mathrm{H}} = 10^{14}$ ions/cm^2 to 10^{20} ions/cm^2. By using secondary ion mass spectrometry (SIMS), it was found that less than 1% of the impinging H ions are actually retained in the crystal (most of the ions are reflected by the sample surface and/or repelled by surface electric fields). It should be noted that H ions move into the crystal by thermal diffusion, while the initial ion acceleration is needed to overcome the surface barrier (Trotta et al., 2008, 2009a). The lowest temperature at which H ions sizably diffuse in the sample was found to be 170°C in dilute nitrides. Figure 2.3 sketches the experimental apparatus for hydrogen implantation in semiconductors.

2.2.2 *Optical Characterization*

The electronic properties of the samples were investigated by continuous-wave photoluminescence (PL) using a duplicated Nd:YVO_4 laser ($\lambda = 532$ nm) for optical excitation. The luminescence was spectrally analyzed by a 0.75 m focal length monochromator and detected by a liquid-nitrogen-cooled Si charge-coupled device

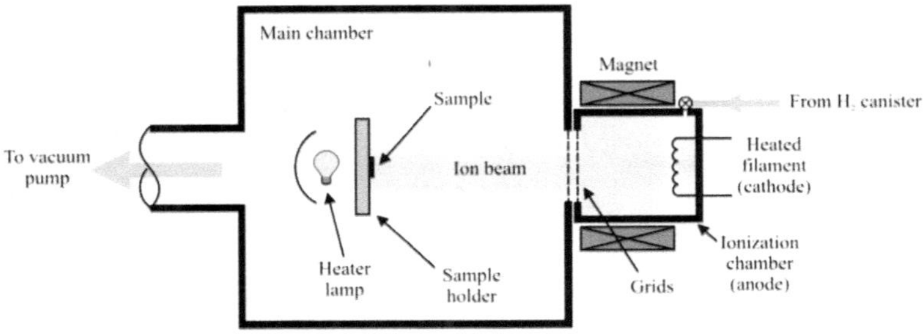

Figure 2.3 Schematic illustration of a Kaufman source. Molecular hydrogen is pumped in the ionization chamber, where it is ionized and/or dissociated into *protons* by the electrons emitted from a hot tungsten filament (cathode) toward the walls of the chamber (anode). A magnet spirals the electrons' trajectories, enhancing the cross section for the hydrogen ionization process. Hydrogen ions (H^+) are then accelerated by a set of grids held at high voltage, so that an ion beam is generated, enters the main chamber where it has a Gaussian profile (sigma ~25 mm for a 100 eV beam energy), and hits the sample, which is electrically grounded. A very low pressure ($\sim 5 \times 10^{-7}$ mbar) is made by a turbomolecular pump before opening the H flux, and a vacuum of $\sim 3 \times 10^{-4}$ mbar is maintained in the main chamber during H irradiation.

(CCD) or InGaAs array detector. A Ti-sapphire tunable laser was used as the excitation source for photoluminescence excitation (PLE) measurements. In this case, the luminescence signal was dispersed by a double 0.75 m focal length monochromator and detected by a photomultiplier tube with a Cs-activated GaAs photocathode operating in a single-photon counting mode.

2.2.3 *Structural Characterization*

The hydrogen profile along the growth direction of the samples was determined by SIMS measurements. A Cs^+ primary beam ($E_{Cs} = 1$ keV) at oblique incidence (55°) was used in a CAMECA Sc-ultramass spectrometer, where $^{2}H^-$ and $^{75}As^-$ secondary ions were collected with a depth resolution of 2–3 nm/decade. The sputtering time was converted to an in-depth scale by measuring the obtained crater depths by a mechanical stylus profilometer with a resolution of a few nanometers (Trotta et al., 2008, 2009a).

2.2.4 *Sample Growth*

All samples were grown by molecular beam epitaxy or metal-organic vapor-phase epitaxy. Here, we focus mainly on Ga(AsN) alloys. These samples were grown at 500°C on top of a 500 nm thick GaAs buffer layer deposited at 600°C on a (001) semi-insulating GaAs substrate. Different nitrogen concentrations (x) and Ga(AsN) layer thicknesses (t) were employed depending on the specific application. We will also present results on (InGa)(AsN)/GaAs quantum wells (QWs) (with different well widths, L) and Ga(PN) alloys (Xin and Tu, 2000; Fischer et al., 2003).

2.3 Hydrogen-Induced Band-Gap Tuning

We focus on two dilute nitride systems, Ga(AsN)/GaAs and Ga(PN)/GaP. Although these materials emit in very different wavelength ranges, they share quite similar electronic properties, as apparent in the following figures. Figure 2.4a,b shows the PL spectra of two Ga(AsN) samples having very different N concentrations and irradiated at different H doses (Bissiri et al., 2002b; Polimeni et al., 2002; Klar et al., 2003).

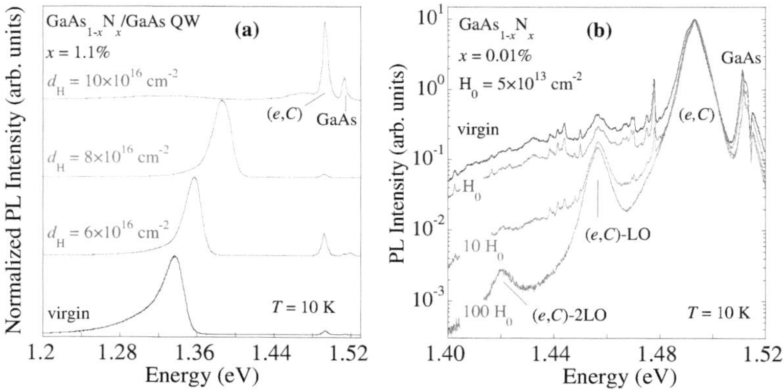

Figure 2.4 (a) PL spectra at $T = 10$ K of a Ga(AsN)/GaAs QW with width $L = 6$ nm treated with increasing hydrogen doses d_H. (b) The same for a Ga(AsN) epilayer (thickness $t = 110$ nm) in the N very dilute limit.

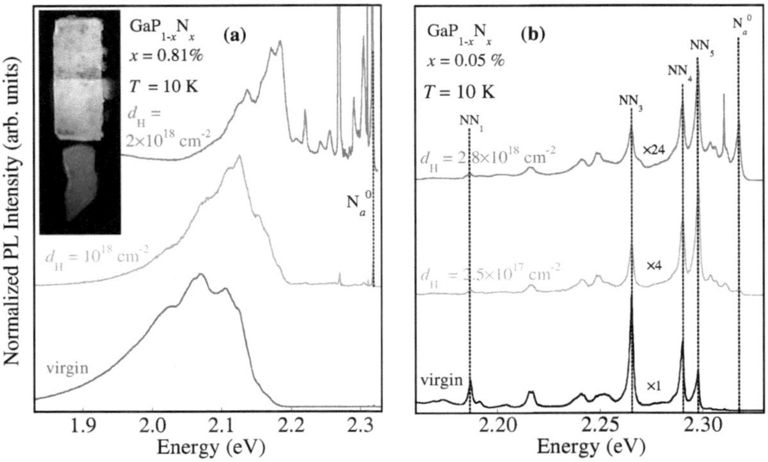

Figure 2.5 (a) PL spectra at $T = 10$ K of a Ga(PN) epilayer (thickness $t = 250$ nm) treated with increasing hydrogen doses. The inset is a photographic image of the samples, excited with the 458 nm line of a Ar^+ laser, to which the spectra of the panel refer. (b) The same as in panel (a) in the N dilute limit ($x = 0.05\%$).

Figure 2.4a refers to a Ga(AsN)/GaAs QW with $L = 6$ nm and $x = 1.1\%$. Clearly, with increasing d_H the PL peak, which is due to exciton recombination in the well, blue-shifts until it reaches the energy position expected for a N-free material. Figure 2.4b displays the PL spectra of a Ga(AsN) epilayer ($t = 110$ nm) in the very dilute concentration limit, $x < 0.1\%$. In this limit, the band-gap energy is slightly affected by N incorporation, while the crystal emission is characterized by a series of very sharp lines (linewidth ~0.5 meV) due to radiative recombination of excitons localized on N atom aggregates (such as N pairs (NP), triplets, or higher-order clusters); see the topmost spectrum. These lines are superimposed on a broad band also related to N alloying. Hydrogenation leads to a progressive and finally complete quenching of the N-related lines as well as of the broad underlying band.

Figure 2.5a shows the PL spectra of a Ga(PN) epilayer ($x = 0.81\%$ and $t = 250$ nm) irradiated with different H doses. The center of mass of the emission spectra *blue-shifts* with increasing

d_H because of the combined action of two effects. First, the energy gap of the material reopens toward that of N-free GaP, leaving uncovered N-related levels, which were resonant with the CB before hydrogenation. In addition, hydrogen apparently passivates *deeper* NCs (emitting at lower energies) much more efficiently, thus leading to a transfer of spectral weight from low-energy to high-energy PL lines (Polimeni et al., 2003a). The effects of H on NC states can be better appreciated in Fig. 2.5b that shows the PL spectra of a Ga(PN) sample with nitrogen concentration $x = 0.05\%$ (namely, in the dilute concentration limit) and $t = 250$ nm before and after irradiation with different H doses. Therein, the emission lines due to different NCs are spectrally well resolved (Felici et al., 2005), and the evolution of their intensity with increasing d_H can be followed in detail. It is found that deeper NN_i lines (index i denotes the order of separation of NPs, NN, in the lattice) disappear earlier from the PL spectrum, thus suggesting a stronger interaction between hydrogen atoms and the NCs responsible for these lines. This behavior can be accounted for by a greater attractive potential exerted by NCs having deeper energy levels on the incoming hydrogen ions.

PLE measurements, too, show clearly a band gap reopening upon hydrogenation. This is shown in Fig. 2.6 for a Ga(AsN) epilayer having $x = 0.095\%$ and $t = 500$ nm. PL and PLE spectra are compared for each dose and both show a blue shift of the peak and absorption edge, respectively (Felici et al., 2006b).

The effects above described are of utmost importance for the fabrication of nanostructures based on dilute nitrides, as extensively reported in Chapter 8. In the next section, we will touch on the structural origin of the band-gap reopening, related to the formation of N–H complexes addressed in Chapters 4–7, and will describe how these complexes distribute in the crystal.

2.4 Nitrogen–Hydrogen Complexes and Their Spatial Distribution

Infrared absorption studies and X-ray absorption studies of Ga(AsN) epilayers treated with both hydrogen and deuterium, which will be described in detail in Chapters 5 and 6, respectively, revealed the

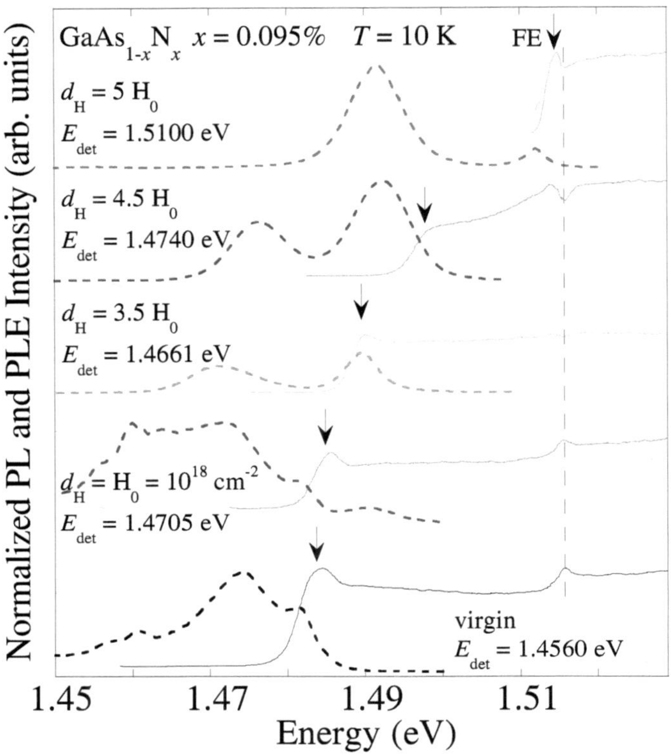

Figure 2.6 PL (dashed lines) and PLE (solid lines) spectra of a Ga(AsN) epilayer (thickness $t = 500$ nm) treated with increasing hydrogen doses. The vertical dashed line indicates the GaAs band-gap energy. The solid arrows indicate the free exciton (FE) energy positions as derived by PLE.

microstructure of the specific N–H defect causing the passivation of the electronic activity of N (Jiang et al., 2004; Kleekajai et al., 2008; Wen et al., 2010; Ciatto et al., 2005a). To date, such a defect is considered the most likely complex responsible for nitrogen passivation (Fowler et al., 2005; Du et al., 2005; Amore Bonapasta et al., 2007).

Figure 2.7a shows the SIMS depth profiles of deuterium for an $x = 0.4\%$ sample with thickness 180 nm (deuterium, rather than hydrogen, was used in order to increase sizably the SIMS sensitivity). The sample irradiation temperature was 300°C, and two different

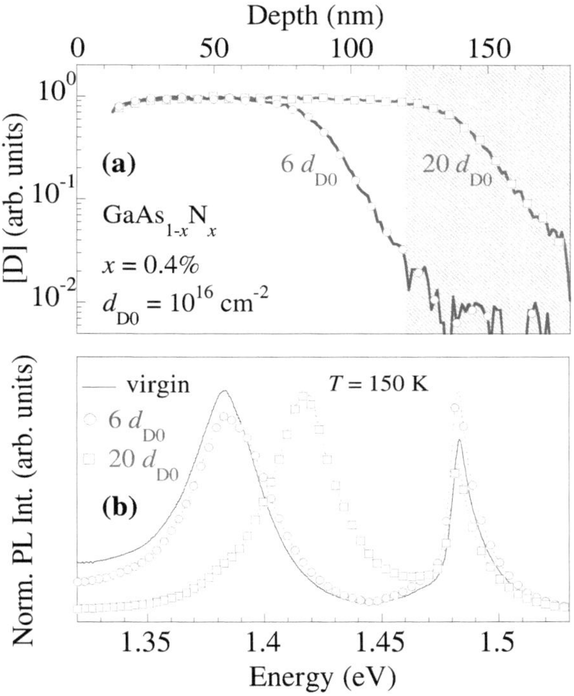

Figure 2.7 (a) SIMS depth profiles of deuterium and (b) PL spectra at 150 K in a Ga(AsN) sample ($x = 0.4\%$ and $t = 180$ nm) treated with different doses of deuterium: $d_D = 6 \times 10^{16}$ ions/cm^2 (circles) and $d_D = 2 \times 10^{17}$ ions/cm^2 (squares). The PL spectrum of a virgin sample is also reported in (b) (solid line). The PL bands peaked at about 1.38 eV correspond to carriers recombining in deuterium-free Ga(AsN). The PL peak at about 1.48 eV is due to carrier recombination in GaAs or fully passivated Ga(AsN) regions. The band at about 1.42 eV originates from carriers recombining in the sample region located at the deuterium forefront where a partial passivation of N atoms occurred, highlighted by the shaded region in (a).

impinging D-ion doses were employed (obtained by using the same ion current but different exposure times). This results in a nearly rigid shift of the D concentration trailing edge. In particular, we notice that the deuterium concentration falls by a factor of 10 within 30 nm. Remarkably, this quantity decreases considerably with deuteration temperature and is less than 5 nm/decade for $x \sim$

1% and $T_D = 200°C$, as detailed in Chapter 8 (Trotta et al., 2008, 2009a).

The steepness and shape of the D trailing edge reflect on the sample optical properties. In particular, in synergy with SIMS, PL can be usefully used to estimate the extent of the D-free Ga(AsN) thickness after irradiation. The PL spectra at 150 K of the two deuterated samples, whose SIMS is displayed in Fig. 2.7a, are shown in Fig. 2.7b along with the PL spectrum of the virgin Ga(AsN) sample. Notice that the *peak energy* of the Ga(AsN)-related emission is the same in the virgin and in the lower-dose irradiated sample (open squares), while the *weight* of the GaAs-related emission relative to that of the Ga(AsN)-related emission is slightly higher in the latter sample. In fact, carriers relax toward regions of the sample having the lowest band-gap energy before recombining radiatively. Therefore, in the sample deuterated at a lower dose no sizable carrier recombination occurs in the fully or partially deuterated regions closer to the sample surface; almost all luminescence originates from the *D-free* 50 nm thick Ga(AsN) layer (highlighted by a shaded area in Fig. 2.7a) at the Ga(AsN)/GaAs interface. In the sample irradiated at a higher D dose, the low-energy band due to Ga(AsN) is blue-shifted by about 40 meV because partial N passivation occurs at the interface between the Ga(AsN) and the GaAs buffer layers, where the SIMS profile shows a sizable D concentration; see Fig. 2.7a. Consequently, a fine-tuning of the effective N concentration in dilute nitrides can be attained in Ga(AsN) epilayers with small N concentrations (Polimeni et al., 2004) or in Ga(AsN), or (InGa)(AsN), QWs (Polimeni et al., 2001a; Baldassarri Höger von Högersthal et al., 2001); see, for example, Fig. 2.4a. In these latter heterostructures, one exploits the small width of the well (usually a few nanometers) in order to vary quite uniformly the density of electronically active N atoms.

2.5 Reversibility of the Hydrogen Effects

This paragraph describes different methodologies that can be employed in order to dissociate the N–H complexes responsible for the electronic passivation of N activity in dilute nitrides. This is an issue of high relevance since it addresses both the stability of the effects induced by hydrogen and the possibility to spatially modulate these effects after N–H complexes have been formed. Two different approaches are presented here, namely, thermal- and photon-induced N–H complex dissociation. The results of a third approach (i.e., electron-induced complex dissociation) are presented in Chapter 8.

2.5.1 *Thermal-Induced N–H Complex Dissociation*

Thermal annealings at temperatures, T_a, ranging from 200°C to 600°C for duration times, t_a, ranging from 1 hour to 50 hours were performed at 10^{-6} torr.

PL spectra of a Ga(AsN) sample with $x = 0.01\%$ hydrogenated at $d_H = 5\times10^{15}$ ions/cm^2 $= 100$ H_0 (see Fig. 2.4b for comparison) and annealed at $T_a = 330$°C are shown in Fig. 2.8a for various t_as. Quite remarkably, *thermal annealing fully restores the optical properties that the material exhibited before hydrogenation*. Most of the N lines as well as broad bands are gradually recovered for increasing t_as, until their intensity saturates to a value only a bit lower than that seen in the virgin sample. Analogously to what is observed in the case of hydrogenation of Ga(PN), a closer inspection reveals that the recovery rate of the N lines is considerably different from line to line. For instance, for $T_a = 330$°C, the PL intensity (I) of line "h" easily reaches a saturation value, while, for example, line "p" does not reappear at all. Other transitions, such as line "i" and "a," exhibit an intermediate behavior. This suggests the presence of different N–H complexes with different bond strengths (Bissiri et al., 2002a,b).

A quantitative analysis can be carried out if one assumes that N–H complexes dissociate irreversibly, namely, that H recapture is absent or negligible. The probability per unit time for the dissociation process of the N–H bonds is $p = \nu \cdot \exp[-E_D/(k_B T_a)]$ (Bergman et al., 1998), where the attempt frequency ν is set equal to 93 THz (3100 cm^{-1}), namely, the value of the N–H local vibrational mode (Jiang et al., 2004; Kleekajai et al., 2008; Wen et al., 2010). E_D is the activation energy for the dissociation of the complex formed between H and a given N-related cluster. In turns, the percent of surviving bonds is given by $n = \exp(-p \cdot t_a)$. Since the intensity of the

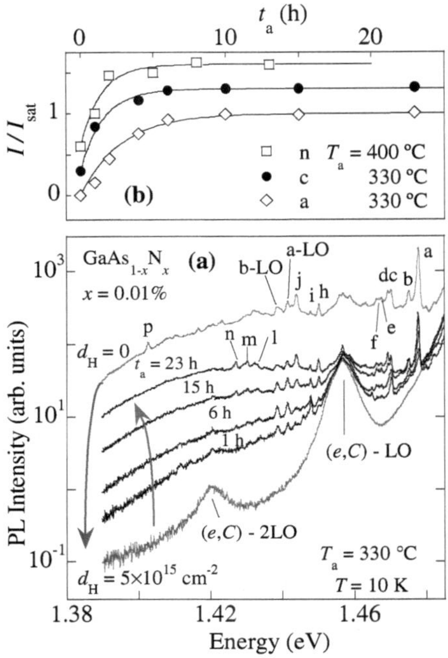

Figure 2.8 (a) PL spectra at $T = 10$ K of a Ga(AsN) epilayer ($x = 0.01\%$) for different postgrowth treatments. The topmost trace corresponds to a virgin sample ($d_{\mathrm{H}} = 0$), and the bottommost trace corresponds to a hydrogenated sample. Intermediate spectra have been recorded on the hydrogenated sample annealed at $T_{\mathrm{a}} = 330°$C for different annealing times, t_{a}. Transitions corresponding to NCs whose activation energies have been determined are labeled with letters. X-nLO is the phonon replica of an X line. (b) Dependence of the percentage of broken N–H bonds, I/I_{sat}, on the annealing time t_{a} for three of the PL lines shown in (a). The continuous lines are fits to the data meant to determine the activation energy (E_{D}) for the N–H bond dissociation ($E_{\mathrm{D}} = 2.12$ eV, 2.15 eV, and 2.36 eV for lines "c," "a," and "n," respectively); see text. Data has been offset along the vertical axis for clarity purposes.

PL sharp lines is proportional to the number of the corresponding NCs optically active, $n = [1 - I(T_{\mathrm{a}}, t_{\mathrm{a}})/I_{\mathrm{sat}}]$ for each PL line, where $I(T_{\mathrm{a}}, t_{\mathrm{a}})$ is the line intensity after an annealing time t_{a} at T_{a}, and I_{sat} is its saturation value for t_{a} (or T_{a}) $\rightarrow \infty$. The intensity of most of the lines emitting in the energy range 1.46–1.48 eV reaches a well-defined saturation value with E_{D} ranging between 2.00 eV and

2.25 eV; see full points and open diamonds in Fig. 2.8b, where the dependence of I/I_{sat} ($= 1-n$) on t_a is shown for a few representative cases.

The PL intensity of lines emitting below 1.46 eV does not saturate even for $t_a > 23$ hours at 330°C. It saturates, instead, for $T_a \geq$ 400°C, leading to E_D values varying between 2.28 eV and 2.40 eV, as shown in Fig. 2.8b for line "n." Such an increase in E_D for low-energy emitting states indicates that these centers are most likely NCs or chains with a large number of N atoms, where H can be caged more tightly than in small N aggregates. Hydrogen recapture by long N atom chains may result in a higher effective value of E_D, as well.

A correlation between the emission energy of the N aggregate, E_N, and the strength of the N–H bond is found. This relationship is shown in Fig. 2.9, where E_D is plotted versus the emission energy, E_N, for all lines. Deeper states (presumably clusters or chains with a large number of N atoms) are characterized by higher values of E_D.

We now turn to the alloy limit. Figure 2.10a shows the PL spectra at 150 K of an $In_{0.41}Ga_{0.59}As_{0.978}N_{0.022}$/GaAs QW ($L$ =

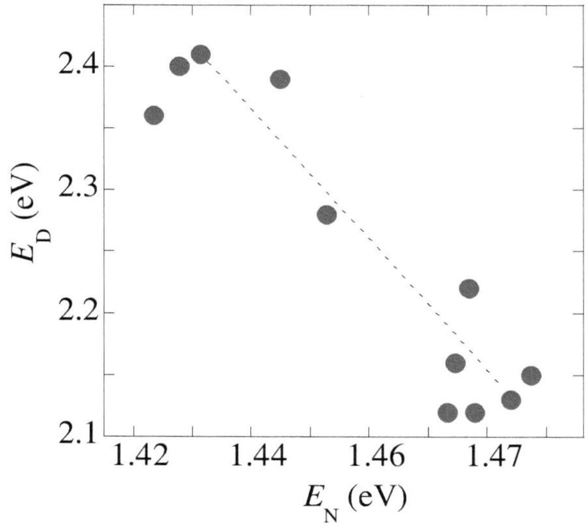

Figure 2.9 Dependence of the activation energy (E_D) versus the emission energy (E_N) for various NCs.

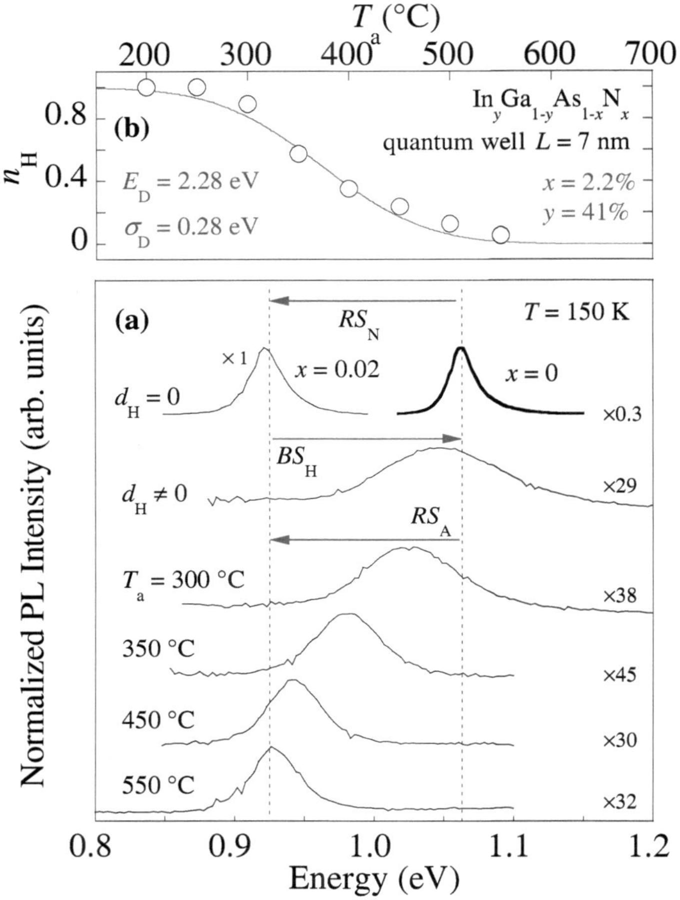

Figure 2.10 (a) Normalized PL spectra of an $In_{0.41}Ga_{0.59}As_{0.978}N_{0.022}$/GaAs QW hydrogenated and subsequently annealed for one hour at different annealing temperatures (T_a). (b) Percentage of N–H bonds (n_H) as a function of T_a, as determined by Eq. 2.1. Empty circles denote experimental values.

7.0 nm). The N-induced large red shift of the band gap (RS_N; see topmost spectrum) is counterbalanced by the blue shift due to H irradiation (BS_H, second spectrum from the top). Thermal annealing red-shifts the gap again (RS_A; see bottommost spectrum) until it reaches that of the virgin N-containing sample (Baldassarri Höger von Högersthal et al., 2001). The PL line width is broadened by the hydrogenation and reversibly narrowed by thermal annealing. The activation energy for the N–H bond dissociation can be derived assuming that the *effective* N concentration, $x_{\mathrm{eff}}(T_{\mathrm{a}},t_{\mathrm{a}})$, in a sample irradiated with H and thermally annealed at temperature T_{a} for time t_{a} can be deduced from $E_{\mathrm{p}}(x_{\mathrm{eff}} \neq 0) = E_{\mathrm{p}}(x = 0) - a \cdot x_{\mathrm{eff}}^{b}$. The two parameters a and b are found by a fit of this formula to the $In_yGa_{1-y}As_{1-x}N_x$, $y \approx 0.4$, band-gap values obtained at $T = 150$ K. This fit gives $a = 2.7$ eV and $b = 0.77$. If x is the as-grown N concentration and $x_{\mathrm{eff}}(0,0)$ is that of the N atoms electronically active after H irradiation and no annealing, the percentage of residual N–H bonds after annealing is given by $n = [x - x_{\mathrm{eff}}(T_{\mathrm{a}},t_{\mathrm{a}})]/[x - x_{\mathrm{eff}}(0,0)]$. Figure 2.10b shows the dependence of n on $T_{\mathrm{a}}(t_{\mathrm{a}} = 1\ \mathrm{h})$. A very good fit is obtained when we assume that a *distribution* of NCs contributes to the sample band-gap reduction. This gives rise to a Gaussian distribution for the N–H bond dissociation energies, $G(E - < E_{\mathrm{D}} >)$, with a mean value $< E_{\mathrm{D}} >$ and a standard deviation σ. In turn, this leads to the formula (Bissiri et al., 2002a)

$$n(t_a, T_a) = \int_0^{\infty} e^{-\upsilon\, t_a \cdot \exp(-E/k_{\mathrm{B}} T_a)}\, G(E - < E_D >)\, dE, \qquad (2.1)$$

whose fit to the data is shown by the solid line ($< E_{\mathrm{D}} >= 2.28$ eV and $\sigma = 0.28$ eV).

Similar results are obtained for Ga(PN), both in the dilute and in the alloy limit (Polimeni et al., 2003a). Figure 2.11a shows the PL spectra of a Ga(PN) sample with $x = 0.05\%$ before and after irradiation with different H doses. Therein, the emission lines due to different NCs are spectrally well resolved, and the evolution of their intensity with increasing d_{H} can be followed in detail. As for Ga(AsN), it is found that deeper NN_i lines disappear earlier from the PL spectrum, thus suggesting a stronger interaction between hydrogen atoms and the NCs responsible for these lines. This behavior can be accounted for by a greater attractive potential exerted by NCs having deeper energy levels on the incoming hydrogen ions. Also, in the alloy limit, the same kind of behavior of Ga(AsN) is found for Ga(PN); see Fig. 2.11b for $x = 0.6\%$. A quantitative analysis leads to results similar to those discussed above for arsenides.

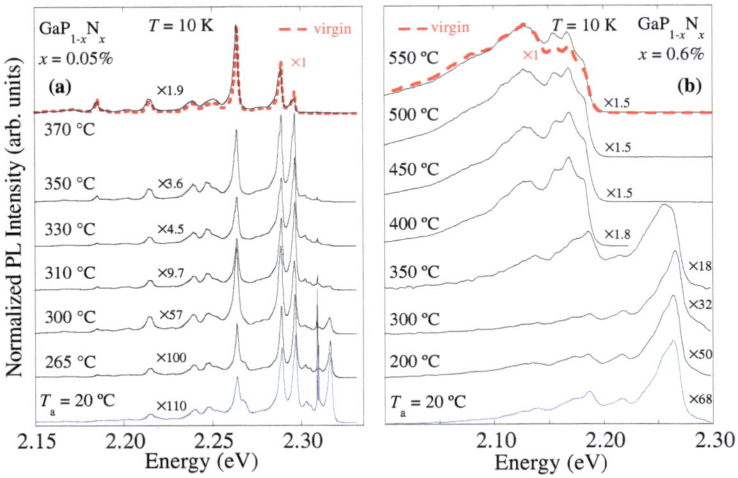

Figure 2.11 (a) Peak-normalized PL spectra ($T = 10$ K) of a Ga(PN) epilayer with $x = 0.05\%$ hydrogenated with a hydrogen dose $d_H = 8 \times 10^{18}$ ions/cm^2 (bottommost blue curve) and annealed for a time $t_a = 10$ minutes at different annealing temperatures T_a. The topmost red dashed curve is the PL spectrum of the virgin sample. (b) Same as for (a), but $x = 0.6\%$.

2.5.2 *Photon-Induced N–H Complex Dissociation*

The combination of hydrogenation and optical excitation with a focused laser beam can provide a successful tool for the tuning and spatial control of the optical properties of dilute nitride alloys. Indeed, in the following section we show that the N–H complex responsible for N-effect passivation can be dissociated, by means of a purely photonic effect, upon exposing the sample to a focused laser beam. This process differs from the thermal-induced dissociation described in the previous section and provides a means of controlling the optical properties of dilute nitrides on a micrometer scale (being limited only by the diffraction limit of the laser light) (Balakrishnan et al., 2012, 2011).

Figure 2.12a shows the room temperature ($T = 300$ K) microphotoluminescence (μPL) spectra and μPL map of a hydrogenated Ga(AsN) sample (H dose $= 6 \times 10^{16}$ ions/cm^2, [N] $= 0.9\%$) after exposure for a time $t_a = 15$ seconds to a focused laser beam (of diameter $d \sim 1$ μm and $\lambda = 515$ nm) with different values of laser

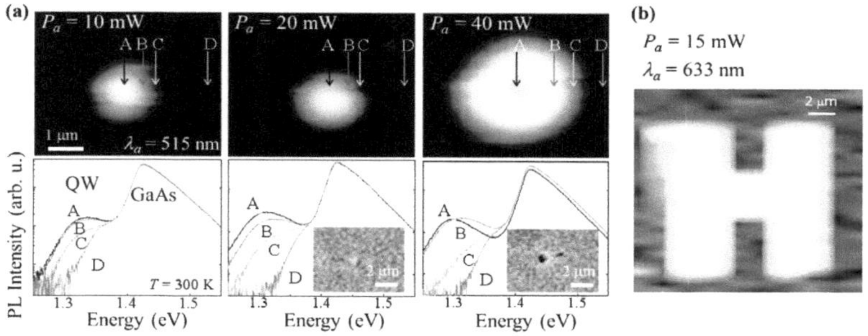

Figure 2.12 (a) Microphotoluminescence (μPL) maps at $T = 300$ K of a hydrogenated Ga(AsN) QW sample (H dose $= 6 \times 10^{16}$ ions/cm^2, [N] $= 0.9\%$) following the exposure for a time $t_a = 15$ seconds to a focused laser beam with $\lambda_a = 515$ nm and $P_a = 10$ mW, 20 mW, and 40 mW. The μPL spectra at positions A, B, C, and D are shown at each power on the bottom. The μPL maps are obtained by plotting the μPL intensity around $E = 1.30$ eV, corresponding to the peak energy of the QW emission in the virgin sample (see gray area in the μPL spectra) and were acquired with $P_{exc} = 1.2$ mW and $\lambda = 633$ nm. Insets are 1 kV scanning electron microscopy (SEM) images of the laser-annealed spots. (b) H-shaped laser-written ($P_a = 15$ mW, $t_a = 10$ seconds, $\lambda_a = 633$ nm) emitting area mapped at 300 K with a laser power of 0.1 mW.

power P_a. The μPL spectra were acquired at low laser power within and around the laser-exposed spot. It can be seen that the QW PL emission is peaked at lower energies in proximity to the spot center; see the μPL spectra at positions A, B, C, and D in Fig. 2.12a. The μPL map in the spectral range of the QW emission shows a light-emitting region with an approximately Gaussian intensity distribution and a full-width at half-maximum (FWHM) of 0.95 µm that corresponds closely to the laser spot diameter. With increasing laser power, the light-emitting area increases in size, showing spot-like (for $P_a \leq 30$ mW) or ring-like (for $P_a \geq 40$ mW) patterns, whose origin will be discussed later.

In the previous sections, we showed how hydrogen incorporation in dilute nitrides gradually restores the N-free band-gap energy (see, for example, Figs. 2.4a and 2.5a) and how H removal from the whole sample via thermal annealing recloses the band gap down to the pristine value (see, for example, Figs. 2.10a and 2.11b). Here, the hydrogen-induced neutralization effect is reversed *locally* by using a focused laser beam, which dissociates the N–H complexes and restores the electronic activity of nitrogen atoms only in the laser-treated region. This gives high versatility both on the size of the treated region—down to micrometer scale—and on its shape, as exemplified by the laser-written H-shaped emitting area shown in Fig. 2.12b (Balakrishnan et al., 2012, 2011).

We use the peak energy of the Ga(AsN) QW PL emission at different positions in each μPL map (Fig. 2.12a) to probe the spatial distribution of active N atoms and hence the percentage variation in the concentration of N–H complexes, $\Delta n_{\mathrm{N-H}}$. Our estimate of $\Delta n_{\mathrm{N-H}}$ uses the two-level band-anticrossing model of Ref. (Shan et al., 1999a), with the single-N level placed 0.23 eV above the CB minimum of GaAs at $T = 300$ K and an interaction parameter $V_{\mathrm{N}} = 2.7$ eV. The values of $\Delta n_{\mathrm{N-H}}$ at various laser powers and positions (X) in the QW plane are plotted in Fig. 2.13a. It can be clearly seen that with increasing P_{a}: (*i*) the minimum in the $\Delta n_{\mathrm{N-H}}$ profile becomes deeper; (*ii*) all N–H complexes in the Ga(AsN) QW can be dissociated at laser powers $P_{\mathrm{a}} \sim 40$ mW corresponding to $|\Delta n_{\mathrm{N-H}}| \sim 100\%$; and (*iii*) the FWHM of the $\Delta n_{\mathrm{N-H}}(X)$ curve ($d_{\mathrm{N-H}}$) increases above a minimum value corresponding to the diffraction-limited size of the laser spot diameter; see also Fig. 2.13c.

To probe the pure photonic origin of the laser-induced N–H complex dissociation, we estimate the contribution of laser-induced heating of the crystal lattice to the ionization of the N–H complex and corresponding $\Delta n_{\mathrm{N-H}}$ profiles (Balakrishnan et al., 2012). For $P_{\mathrm{a}} < 40$ mW, the calculated temperatures (see Fig. 2.13b) are well below those required ($T_{\mathrm{a}} > 250^{\circ}$C) for thermal dissociation of N–H complexes and outgassing of H from the sample, as measured by the thermal annealing studies reported in the previous section; see also Ref. (Bisognin et al., 2008). Thus, under these laser exposure conditions, the laser-induced local heating *cannot* be responsible for the ionization of the N–H bonds. Photon-induced N–H complex dissociation is also supported by laser-writing experiments at liquid helium temperatures ($T = 4.2$ K) and laser exposure conditions ($P_{\mathrm{a}} < 40$ mW) that correspond to calculated temperatures $T_{\mathrm{a}} < 25$ K (not shown here).

In contrast, for laser exposure with $P_{\mathrm{a}} \geq 40$ mW the temperature increases very steeply in the center of the laser spot up to values

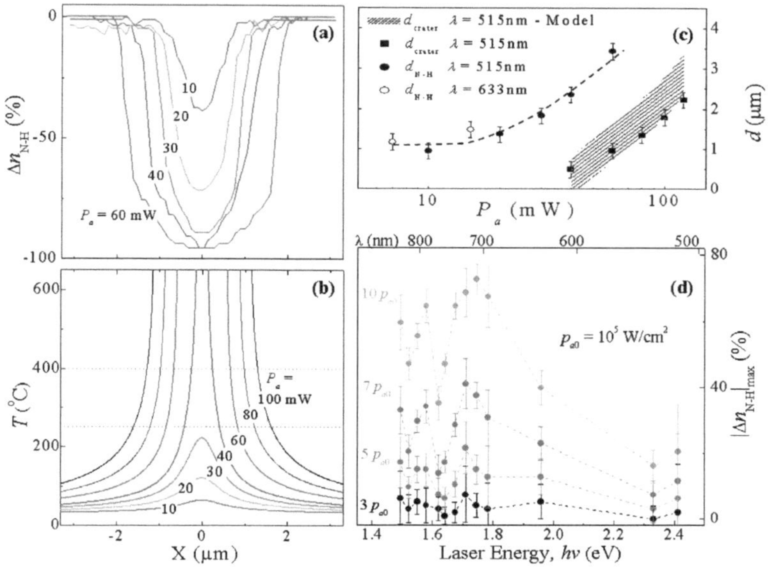

Figure 2.13 (a) Spatial dependence of the percentage variation in the concentration of N–H complexes (Δn_{N-H}) upon laser exposure at various $P_a(\lambda_a = 515$ nm, $t_a = 15$ seconds), as obtained by the μPL maps in Fig. 2.12. (b) Calculated spatial profile of laser-induced sample heating upon laser exposure at various $P_a(\lambda_a = 515$ nm, $t_a = 15$ seconds, beam waist 0.4 μm). Horizontal dotted lines highlight temperatures required for H outgassing ($T_a \geq 250°$C) and As desorption ($T_a \geq 400°$C). (c) Dependence of the FWHMs of the Δn_{N-H} profile (d_{N-H}) on P_a at $\lambda_a = 515$ nm (full points) and 633 nm (circles). The dashed line is a guide to the eye. The squares and shaded area describe the P_a dependence of d_{crater} (the diameter of the damaged surface area of the sample), as measured from the SEM study and estimated by an As desorption model, respectively. (d) Dependence of the value of the $|\Delta n_{N-H}(X)|$ profile at the center ($X = 0$) of the laser spot, $|\Delta n_{N-H}(X)|_{max}$, on the laser photon energy $h\nu$ (and wavelength, λ) at various power densities p_a for $t_a = 20$ seconds.

that can exceed those required for H outgassing ($T_a > 250°$C), As desorption ($T_a > 400°$C) (Kobayashi and Kobayashi, 1991), and GaAs melting ($T_a > 1200°$C). Thus, at high laser powers ($P_a \geq 40$ mW), the high temperatures ($T_a \geq 250°$C) induced around the laser spot can cause thermal dissociation of the complex, diffusion, and outgassing of H. This results in a broadening of the Δn_{N-H} profile to $d_{\mathrm{N-H}}$ values that exceed the diffraction-limited size of the laser beam (Fig. 2.13) and in an irreversible damage of the sample surface; see, for example, the darker spot in the SEM and μPL map for $P_{\mathrm{a}} = 40$ mW in Fig. 2.12a. No topographic changes can be revealed, instead, in samples exposed with lower laser powers ($P_{\mathrm{a}} < 40$ mW).

We investigated the possible mechanisms responsible for photon-induced N–H complex dissociation by studying the dependence of $|\Delta n_{\mathrm{N-H}}|$ on the laser power density p_{a} and photon wavelength λ (Fig. 2.13d) (Balakrishnan et al., 2012). Here the photon excitation is tuned over a range of energies $h\nu = 1.49 - 2.41$ eV ($\lambda = 830 - 515$ nm), below and above the dissociation energy $E_A = 1.77 - 1.89$ eV ($\lambda = 700 - 656$ nm) of the N–H complexes, as derived from thermal annealing studies of Ga(AsN) (Bisognin et al., 2008). From Fig. 2.13d, it can be seen that at each power the value of $\Delta n_{\mathrm{N-H}}$ at the center of laser-treated spot ($|\Delta n_{\mathrm{N-H}}|_{\mathrm{max}}$) tends to peak at $h\nu = 1.76$ eV ($\lambda \sim 700$ nm), close to the reported values of E_A (Bisognin et al., 2008). PLE measurements of both hydrogenated and virgin samples exclude the presence of specific resonances in the density of states of the crystal at the energy mentioned above. Thus the data in Fig. 2.13d suggests resonant photon absorption by the N–H complex itself, accompanied by its dissociation. Perhaps surprisingly, the phenomenon is also observed under laser photon energies that are significantly smaller than E_A. To explain this result, we envisage that the dissociation can be assisted by phonons and that the photogenerated electron–hole pairs can act to screen the N–H chemical bond, effectively reducing the dissociation energy of the complex. Similar mechanisms were invoked to explain the light-enhanced H diffusion in amorphous Si (Santos et al., 1991) and the photoinduced reactivation of neutralized donors in hydrogenated Si-doped GaAs (Loridant-Bernard et al., 1998).

The laser-induced reversibility of hydrogen effects reported above for Ga(AsN) has been observed also in Ga(PN) and (InGa)(AsN) alloys (Balakrishnan et al., 2012), thus indicating universal behavior of hydrogenated dilute nitride alloys. Therefore, if properly harnessed, the photoninduced N–H complex dissociation could be employed to laser-write the electronic properties of dilute nitrides (as we already demonstrated in Ref. (Balakrishnan et al., 2012)) and serve as the basis for the development of a new, inexpensive method for fast and flexible writing of dilute nitride nanostructures, complementary to those described in Chapter 8 but without the need of complex lithographic/etching steps. A very promising step in this direction is represented by the recent finding—reported by F. Intonti et al. in Ref. (Intonti et al., 2011)—that laser treatments can also be performed through the tip of a scanning near-field optical microscope (SNOM). In principle, this may allow for reducing the extent of the illuminated area to sizes compatible with the resolution of SNOM ($<$100 nm)—much below the diffraction limit spot size ($\sim$1 μm) achievable with a conventional optical microscope—thus making possible the realization of quantum-confined nanostructures by laser writing. Finally, photon-induced N–H complex dissociation retains the advantage to work at low temperatures, thus allowing us to manufacture nanostructures in a cryogenic environment and/or of temperature-sensitive samples.

2.6 Hydrogen-Induced Tuning of Electronic Properties

In the previous sections, we have described the main effects of hydrogen incorporation on the optical properties of dilute nitride alloys and reported on the reversibility of hydrogen effects through thermal and laser treatments. The introduction of hydrogen strongly modifies also other fundamental *electronic properties* of dilute nitride alloys, as will be shown in this section for the electron effective mass (Masia et al., 2006) and electron gyromagnetic factor (Pettinari et al., 2006) of Ga(AsN). In particular, the dependence of the exciton effective mass on hydrostatic pressure unveils an electronic hierarchy between different nitrogen complexes as regard the extent of the perturbation these complexes exert on the electronic properties of the GaAs host (Pettinari et al., 2007). Also, PL spectroscopy at high hydrostatic pressures discloses that different NP complexes show quite different rates of passivation by hydrogen, thus indicating strain relaxation as the key mechanism driving the interaction of hydrogen with N atoms in Ga(AsN) (Polimeni et al., 2008a).

2.6.1 *Electron Effective Mass in Ga(AsN)*

The compositional dependence of the electron effective mass in Ga(AsN) and its modification upon hydrogen incorporation have been investigated by means of magnetophotoluminescence (magneto-PL) spectroscopy (Masia et al., 2006). Indeed, the diamagnetic shift of a free-electron to bound-hole recombination, such as the free-electron to neutral-carbon-acceptor (e, C) transition observed in the investigated samples, is directly related to the electron effective mass at the CB minimum m_e^* (Rossi et al., 1970; Rühle and Göbel, 1976; Bimberg, 1978). Because of the dispersionless characteristic (i.e., infinite effective mass) of the C impurity level in **k** space, the energy shift of this level induced by an external magnetic field (B) is very small (~0.02 meV at 10 T in GaAs; see Ref. (Bimberg, 1978)). Therefore, the diamagnetic shift $\Delta E_{(e,C)}$ of the (e, C) recombination follows the Landau quantization of the CB minimum with a slope proportional to the inverse of the electron effective mass:

$$\Delta E_{(e,C)} = \frac{e\hbar}{2m_e^*} B. \tag{2.2}$$

Figure 2.14 displays the magneto-PL spectra of a series of Ga(AsN) samples both virgin and treated with different hydrogen doses. As shown previously (see Section 2.3 and Fig. 2.4a), hydrogen incorporation tunes the optical properties of the Ga(AsN) alloys by gradually restoring the band-gap energy of the N-free material. This results in an *effective* nitrogen concentration (hereinafter x), which for hydrogenated samples is derived by inverting the relation obtained by fitting the band-gap exciton energy of the virgin samples: $E_{\text{exc}}(x) = E_{\text{exc}}^{\text{GaAs}} - \alpha \cdot x^{\beta}$, where $E_{\text{exc}}^{\text{GaAs}} = 1.515$ eV is the FE energy in GaAs, $\alpha = 0.23$ eV, and $\beta = 0.86$.

Relatively low temperatures ($T \leq 30$ K) and low excitation power densities ($P_{\text{exc}} \leq 20$ W/cm^2) were used to enhance the contribution of the (e, C) transitions with respect to the transitions related to FEs and excitons trapped on different NCs (*NC*). The magneto-PL spectra highlight the different degree of localization that carriers involved in those different transitions exhibit. Indeed, the FE and (e, C) recombinations regard free carriers, and their PL peaks blue-shift with an increasing magnetic field, the former

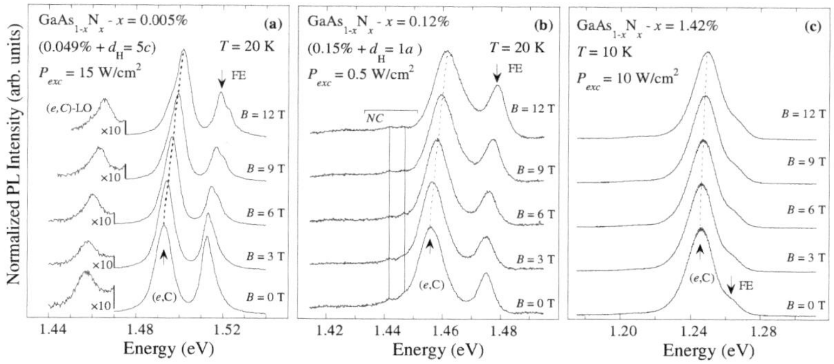

Figure 2.14 Peak-normalized PL spectra of Ga(AsN) samples with effective nitrogen concentration $x = 0.005\%$ (0.049% irradiated with a hydrogen dose $d_H = 5c = 5 \times 10^{18}$ ions/cm^2, panel [a]), $x = 0.12\%$ (0.15% irradiated with $d_H = 1a = 1 \times 10^{16}$ ions/cm^2, panel [b]), and $x = 1.42\%$ (virgin, panel [c]) under different magnetic field intensities. Downward (upward) arrows indicate FE ([e, C]) recombinations. The dotted line, roughly indicating the position of the (e, C) peak, is as a guide to the eye. Solid vertical lines show the energy position of NC states *NC*, which are not perturbed by the magnetic field.

showing a sublinear dependence on B (whose origin will be discussed later; see also Ref. (Pettinari et al., 2012b)), the latter exhibiting a linear blue shift at high fields (as discussed above; see Eq. 2.2 and Ref. (Pettinari et al., 2012b)). Instead, the *NC*-related transitions do not vary their energy with B (actually, they shift less than 0.2 meV at 12 T; see Fig. 2.14b). This indicates that the *NC* transitions involve carriers whose wavefunctions are strongly localized around nitrogen atoms and, therefore, are very weakly perturbed by the application of an external magnetic field.

The relative energy shift of the (e, C) transition as a function of B is shown in Fig. 2.15a for some characteristic nitrogen concentrations, including a 0.049% fully hydrogenated (H dose $= 5.0 \times 10^{18}$ ions/cm^2) sample (*effective* $x = 0.005\%$) whose shift matches that of a GaAs reference sample. The dependence of the measured electron effective mass m_e^* on the *effective* nitrogen concentration is shown in Fig. 2.15b, where full circles and diamonds—which refer respectively to virgin and hydrogenated samples—follow the same compositional dependence. After the first abrupt doubling for

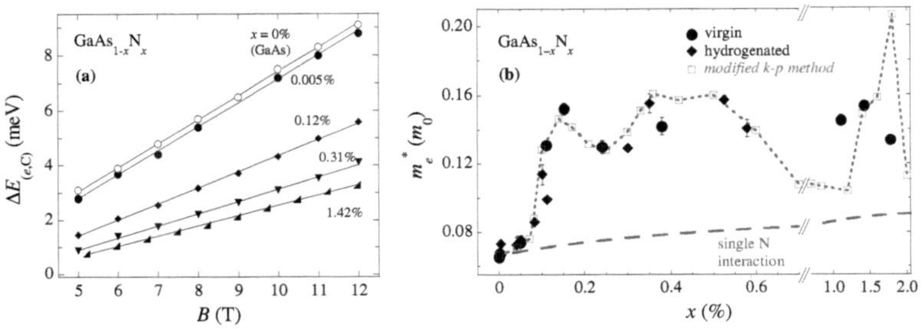

Figure 2.15 (a) Magnetic field–induced shift of the (e, C) recombination band for some Ga(AsN) samples (full symbols) with different nitrogen concentrations. The shift of the (e, C) recombination band for a GaAs reference sample (open circles) is also shown. Solid lines are linear fits to the data from which the value of m_e^* is obtained according to Eq. 2.2. (b) Dependence of the electron effective mass in Ga(AsN) on nitrogen concentration (full symbols). Circles and diamonds refer to virgin and hydrogenated samples, respectively. The dashed line is the theoretical expectation for a single-N band-anticrossing model; see Ref. (Shan et al., 1999a). The dotted line through the open squares shows the result of a *modified k·p* method, which takes into account the effects of the interaction of the CB minimum with multiple N states. An abscissa axis break at $x = 0.7\%$ is introduced for ease of comparison.

$x \sim 0.1\%$ ($m_e^* \sim 0.13$ m$_0$), m_e^* undergoes a second increase for $x \sim 0.35\%$, and finally it shows fluctuations around $m_e^* \sim 0.15$ m$_0$ for $0.38\% \leq x \leq 1.78\%$. This dependence displays a behavior very different from that expected for a conventional alloy, where the compositional dependencies of band structure parameters evolve monotonously. Furthermore, a simple interaction model taking into account the mixing between the CB minimum and a single-N level clearly fails to reproduce the experimental data; see the dashed line in the figure. Instead, the experimental results are remarkably well reproduced up to $x = 0.6\%$ by a *modified k·p method* (open squares). Within this approach, the interaction between the CB edge of Ga(AsN)—which red-shifts with increasing x—and the energy-pinned states due to NCs in the band gap is duly taken into account; see Refs. (O'Reilly et al., 2009; Masia et al., 2006; Lindsay and O'Reilly, 2004b) and Chapter 1. In turn, this explains the compositional dependence of the electron effective mass in terms of on- and off-resonance conditions, driven by the nitrogen concentration, between states having a different degree of localization.

The role of different NC states in determining the electronic properties of Ga(AsN) can be very effectively investigated by measurements under hydrostatic pressure, p. Indeed, the energy of N-related levels shifts with p at a rate ($\beta_p \sim$ 4–5 meV/kbar) that is about half of that of the CB edge (Klar et al., 2000; Weinstein et al., 2003; Pettinari et al., 2007; Ma et al., 2005). In turn, this allows a continuous tuning of the repulsion/hybridization degree between delocalized and localized states. In particular, the dependence of m_e^* on the energy distance between specific N complex levels and the CB minimum can be probed by performing pressure-dependent magneto-PL measurements on a sample with a given N concentration—where the statistical distribution of N clusters is "frozen-in." Figure 2.16a shows the PL spectra at $T = 90$ K of a Ga(AsN) sample with $x = 0.11\%$ for different hydrostatic pressures.

As p approaches 6 kbar, *NC* states (labeled Ci; $i = 0 \ldots 7$ in the figure), which are resonant with the CB continuum at ambient pressure, emerge into the band gap. In turn, these *NC* levels provide low-energy recombination channels competing with the FE recombination, which indeed loses intensity. The energy shift with p of the *NC* states depends slightly on the type of cluster considered and varies between 4.0 meV/kbar and 4.9 meV/kbar, in agreement with the literature (Klar et al., 2000; Weinstein et al., 2003; Ma et al., 2005). For FE we find $\beta_p = 10.0$ meV/kbar. Hereafter, we restrict our analysis to magneto-PL measurements taken at moderately high temperature (90 K) because we found that at low temperature (5 K) *NC* bands dominate the emission spectra as soon as $p > 1$ kbar (not shown here).

Figure 2.16b–e shows the magneto-PL spectra and the derived FE diamagnetic shift ΔE_{FE} as a function of the magnetic field for different applied pressures in two representative samples with $x =$ 0.11% and 0.24%. The exciton Hamiltonian in the presence of a magnetic field B is solved numerically by expanding the exciton wavefunction in terms of spherical harmonics (Pettinari et al., 2012b; Cabib et al., 1971, 1972). The only external parameter entering in the Hamiltonian is the exciton reduced mass (μ_{exc}) on which both the exciton energy and the exciton wavefunction size

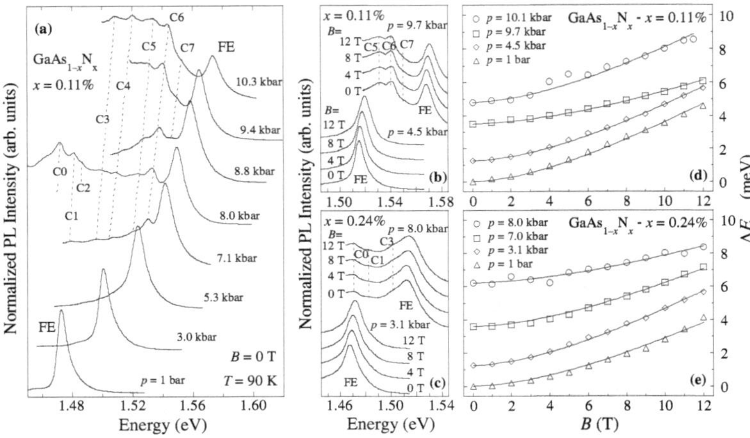

Figure 2.16 (a) PL spectra at $T = 90$ K and $B = 0$ T of a Ga(AsN) sample with $x = 0.11\%$ recorded at different hydrostatic pressures p. FE and Ci indicate the FE and NC-related recombinations, respectively. PL spectra for different magnetic fields B and different values of hydrostatic pressures are also shown for $x = 0.11\%$ (b) and $x = 0.24\%$ (c). Notice that in both cases the spectral weight of NC states increases with p. The dependence of the FE diamagnetic shift ($\Delta E_{\rm FE}$, open symbols) on the magnetic field is shown for different pressures for (d) $x = 0.11\%$ and (e) $x = 0.24\%$, respectively. The solid lines are a fit to the data by means of the model reported in Refs. (Pettinari et al., 2012b; Cabib et al., 1971, 1972), where the exciton reduced mass is the only fitting parameter. $\Delta E_{\rm FE}$ data is vertically shifted for clarity purposes.

depend. The solid lines in Fig. 2.16d–e are best fits to $\Delta E_{\rm FE}$ data. By this method, in a GaAs reference sample we derived $\mu_{\rm exc} = 0.054\ {\rm m}_0$, in excellent agreement with previous determinations. In particular, the data displayed in Fig. 2.16d–e shows that $\mu_{\rm exc}$—which decreases with increasing $\Delta E_{\rm FE}(B)$ curvature—and, hence, $m_{\rm e}^*$ do not depend monotonically on p.

Figure 2.17a,c shows the pressure dependence of the exciton effective mass for $x = 0.11\%$ and $x = 0.24\%$, respectively. At ambient pressure, in both samples $\mu_{\rm exc}$ is much greater than the value it has in GaAs ($0.054\ {\rm m}_0$) because of an already strong admixing between NP states and the CB edge; see Fig. 2.15b. For $x = 0.11\%$, $\mu_{\rm exc}$ slightly decreases up to ~5.6 kbar, except for a

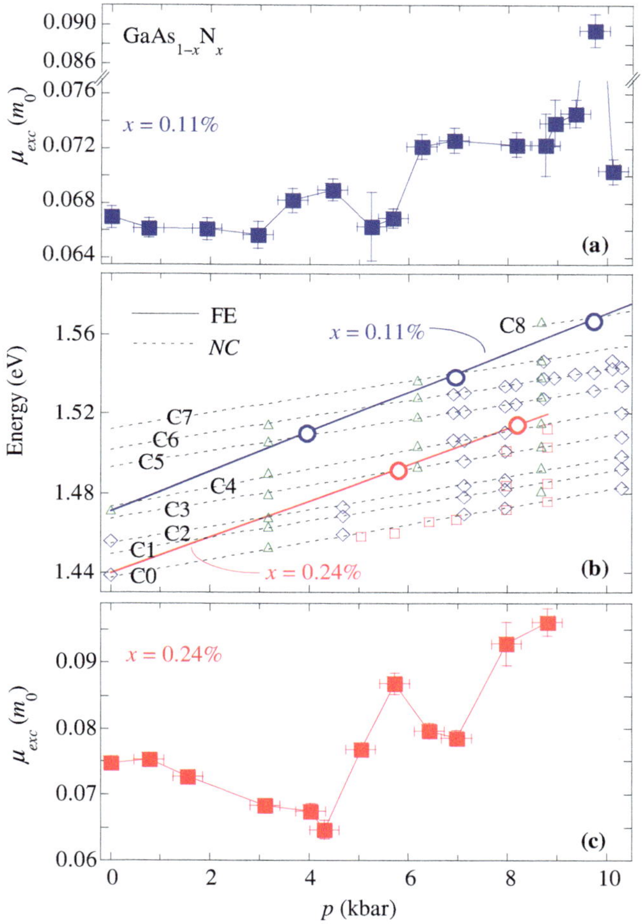

Figure 2.17 Dependence of the exciton effective mass (μ_{exc}) on hydrostatic pressure for Ga(AsN) samples with $x = 0.11\%$ (a) and $x = 0.24\%$ (c). FE (solid lines) and NC state (open symbols) energies at $T = 90$ K as a function of hydrostatic pressure are also reported for investigated samples (b). Different symbols in (b) indicate the energies of recombination bands related to NCs, as obtained by a PL analysis on different samples (Δ: from the $x = 0.049\%$ sample, $\diamond$: from the $x = 0.11\%$ sample, and $\square$: from the $x = 0.24\%$ sample). Dashed lines are linear fits to the *NC* data. Ci label different N clusters shown in Fig. 2.16. Open larger circles highlight pressure values related to a marked increase of the exciton effective mass, as shown in (a) and (c).

slight increase between 3.0 kbar and 5.0 kbar. Then, after an increase by ~10% at 6.3 kbar, μ_{exc} remains constant until $p \sim 8.7$ kbar, suddenly increases by ~30% at 9.7 kbar, and finally decreases at 10.1 kbar. For $x = 0.24\%$, μ_{exc} decreases with increasing p up to 4.2 kbar, then increases by about 30% at $p = 5.8$ kbar, displays a second minimum at 7.0 kbar, and finally gets over 0.095 m_0 at the highest applied pressure. The pressure dependence of the exciton effective mass shows marked differences on going from the 0.11% sample to the 0.24% sample. These differences are accounted for qualitatively by the fast evolution in kind and density of the NCs forming in the N concentration range $x = 0\%–0.5\%$, where Ga(AsN) exhibits strong and sudden changes in its physical properties (as detailed in the previous sections). To identify the clusters interacting more with the CB continuum states, the energy of the *NC* states measured in different samples is shown versus p in Fig. 2.17b by symbols and dashed lines. In the same figure, the energy of the band-gap exciton recombination, which is a few millielectron volts smaller than the CB edge energy, is shown by solid lines. For the $x = 0.11\%$ sample, the increases in μ_{exc} observed at $p \sim 4.0$ kbar, 6.3 kbar, and 9.7 kbar correspond to a crossing between the CB edge and the NCs (likely N triplets and chains) C5, C7, and C8, respectively, as evidenced by blue circles in Fig. 2.17b. Importantly, the different extent of the increases in μ_{exc} highlights quantitative differences between different N clusters. Moreover, the crossing with the C4 level at ~0.5 kbar produces no sizable change in the exciton mass, while a *minimum* in μ_{exc} is observed when the CB edge resonates with the C6 level at ~5.0 kbar. In the case of the $x = 0.24\%$ sample, the initial decrease in μ_{exc} for $p \leq 4.0$ kbar should be related to the CB minimum moving away from level C0, as well as to the absence of interaction with cluster levels C1 and C2. The two increases in μ_{exc} at $p \sim 5.8$ kbar and ~8.0 kbar can be attributed to a mixing between the CB edge and cluster levels C3 and C4, respectively, as evidenced by red circles in Fig. 2.17b. These results improve our understanding of the interplay between extended and localized states in Ga(AsN). First, each increase in μ_{exc} induced by pressure can be traced back to a crossing of the CB edge with a *NC* state, in agreement with the linear combination of isolated nitrogen states (LCINS) method and its results presented in Chapter 1. However, such a crossing appears as a *necessary but not sufficient condition* for having an increase in the exciton (and electron) effective mass, as observed for the crossing with the C1, C2, and C6 cluster levels. This suggests that the degree of mixing between the *NC* and CB states may depend on the symmetry of the NC wavefunction. In turns, this would justify the absence of a strong interaction with N clusters whose wavefunction has an odd symmetry.

Furthermore, the amount of level admixing should increase with the density of a given *NC* cluster. As an example, the strong interaction between the CB minimum and cluster C4 observed in the $x = 0.24\%$ sample is not noticeable for $x = 0.11\%$, likely because of the lower C4 density in the latter sample. In fact, for a random distribution, the density of NCs containing n nitrogen atoms varies as x^n—or faster if N atoms segregate, as observed in Ga(AsN) for $x \sim 5\%$—thus resulting in a relative increase of the population of higher-order clusters (e.g., N triplets and chains) faster than that of lower-order ones (e.g., NPs).

Despite the fact that the pressure-induced variations observed in the carrier effective mass of Ga(AsN) alloys are exceedingly larger than those observed in conventional semiconductors (e.g., GaAs and InP; see Ref. (Shantharama et al., 1984)), these variations are qualitatively consistent with a modified $k \cdot p$ method, which duly takes into account the interaction between the CB edge and the *NC* states (as described previously for the compositional dependence of the electron effective mass at ambient pressure; see Fig. 2.15b). In fact, this approach predicts a pressure-driven modulation of the fractional Γ character of the CB as a result of on- and off-resonance conditions occurring between the CB edge and specific NC levels; therefore, a highly nonmonotonic pressure-induced variation of m_e^* (and μ_{exc}) is expected; see Ref. (Lindsay and O'Reilly, 2004b). In particular, the selective character of the hybridization process contributes to identify which NCs perturb more the band structure of Ga(AsN) and to what extent the symmetry and relative population of NCs matter; see also the next section.

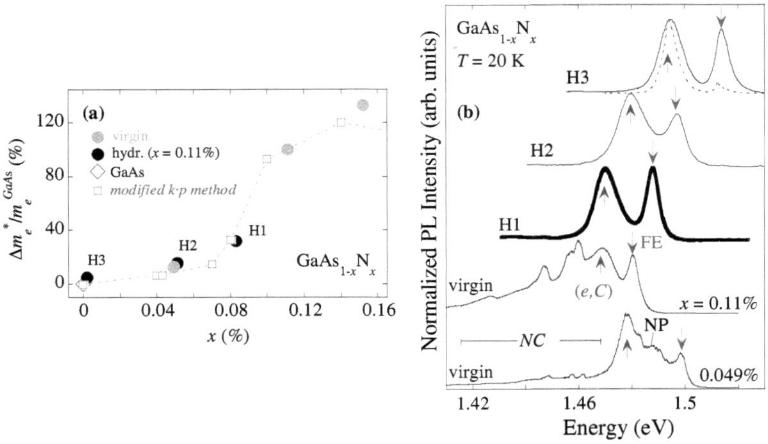

Figure 2.18 (a) Percentage variation of the electron effective mass in Ga(AsN) as a function of the N concentration, as deduced by Fig. 2.15b. Gray dots refer to virgin samples and black dots to the $x = 0.11\%$ sample hydrogenated with different hydrogen doses (H1 = 3.5×10^{18} ions/cm^2, H2 = 4.5×10^{18} ions/cm^2, and H3 = 5.0×10^{18} ions/cm^2). The error bars are always within the symbol size. The open squares connected by a dotted line are the results of a modified $k\cdot p$ method; see Refs. (Masia et al., 2006) and (Lindsay and O'Reilly, 2004b). The open diamond indicates the GaAs reference sample. The PL spectra of most of the samples reported in (a) are shown in (b) at low temperature (T = 20 K). free exciton and (e, C) indicate the FE and the free-electron to neutral-carbon recombination bands, respectively. NP is due to carrier recombination on a N pair. *NC* indicates a nearly continuum of states due to excitons recombining on various N clusters.

2.6.2 *Effective Mass Restoration upon Hydrogenation and Relative Importance of Different NC States*

We address now in more details the relative importance that different NCs may have in the sudden increase of the electron effective mass of Ga(AsN) for $x \sim 0.1\%$ (see Fig. 2.15b), as well as the role played by H in the gradual recovery of the GaAs electron effective mass.

Figure 2.18a shows the percentage variation with respect to the GaAs value of the electron effective mass ($\Delta m_e^*/m_e^{\mathrm{GaAs}}$) in the nitrogen concentration region around 0.1%. For clarity reasons, only few representative data points are shown. The abrupt "jump" occurring around $x \sim 0.1\%$ is extremely well defined by the three virgin samples having $x = 0.049\%$, 0.11%, and 0.15% (gray dots) and by the 0.11% sample irradiated with different H doses (black dots). The PL spectra of most of the investigated samples are shown in Fig. 2.18b. In the virgin $x = 0.049\%$ and 0.11% samples (bottommost curves), several lines due to carrier recombination on different NCs are present. In particular, narrow bands on the top of a long tail of states have been observed below 1.47 eV, with a spectral weight increasing with x. In addition, a band centered at 1.488 eV, indicated as NP, is also observed in the $x = 0.049\%$ sample.

In the hydrogenated samples, NC states seem to disappear *all at once* with a hydrogen irradiation dose H1 (3.5×10^{18} ions/cm^2), while the band gap reopens gradually with increasing hydrogen doses. In the hydrogenated samples, however, high-energy NC bands show up again once a sufficiently high pressure is applied, as in the case of the H1-hydrogenated sample highlighted with a thick line in Fig. 2.18b. Therein, the energy-pinned NP cluster state—which is resonant with the conduction continuum at ambient pressure—becomes visible in the emission spectra at high pressure, as shown in the following figure; see Fig. 2.19a (Polimeni et al., 2008a). In the H1 sample, the effective N concentration decreases only from 0.11% to 0.083%; however, the $\Delta m_e^*/m_e^{GaAs}$ value diminishes by as much as 50% with respect to the value measured in the virgin sample; see Fig. 2.18a. On the one hand, these findings highlight the resonant behavior of m_e^* on the N concentration; on the other hand, they indicate that some NC state ought to be *still* electronically active in the H1 sample. This is indeed shown in Fig. 2.19a, where the PL spectra of the H1 sample are displayed for different applied hydrostatic pressures. With increasing p, the FE and (e, C) recombination bands blue-shift at a rate of $\beta_p \sim 10$ meV/kbar. Interestingly, at $p = 4.4$ kbar, a narrow peak, labeled NP*, appears between the FE and (e, C) bands. At higher p values, this narrow emission gains intensity with respect to that of the FE and eventually dominates the spectrum at $p \geq 6.5$ kbar together with its transverse acoustic (TA) and longitudinal optical (LO) phonon replicas. For $p > 5$ kbar, other weaker recombination bands related

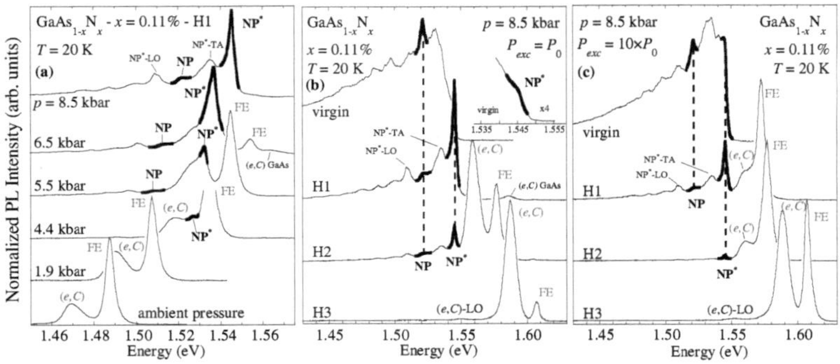

Figure 2.19 (a) PL spectra at low temperature ($T = 20$ K) of the Ga(AsN) sample ($x = 0.11\%$) hydrogenated with dose H1. The different spectra were recorded under different hydrostatic pressure p. (b, c) PL spectra at low temperature and high hydrostatic pressure ($p = 8.5$ kbar) of the virgin and H-irradiated $x = 0.11\%$ samples. (H1 $= 3.5 \times 10^{18}$ ions/cm^2, H2 $= 4.5 \times 10^{18}$ ions/cm^2, and H3 $= 5.0 \times 10^{18}$ ions/cm^2) measured at low (b) and high (c) excitation power density ($P_0 = 1.5$ W/cm^2). FE and (e, C) indicate the free exciton and the free-electron to neutral-carbon recombination bands, respectively. NP and NP* refer to carrier recombination on different N aggregates (see text), and NP*–TA/LO are the phonon replicas of the NP* band. The NP and NP* bands are highlighted by thick lines. Notice the much higher passivation rate by hydrogen of the NP band with respect to the NP* band, as evidenced by vertical dashed lines in (b) and (c). The inset in (b) is meant to better highlight the NP* contribution to the PL spectrum of the virgin sample.

to NCs are observed. In particular, the peak indicated as NP emerges in the 5.5 kbar to 8.5 kbar interval.

The NP* and NP lines have been observed in the virgin as well as in other hydrogenated samples (Pettinari et al., 2007). The energy shift with pressure of the NP and NP* lines ($\beta_p = 4.5$ meV/kbar and 4.3 meV/kbar, respectively) well matches the value reported for N cluster states and gives an extrapolated energy at an ambient pressure of 1.484 eV and 1.508 eV, respectively. The extrapolated energy value of the NP state is very close to that of the band indicated as NP in Fig. 2.18b for the $x = 0.049\%$, while 1.508 eV is the energy position *usually attributed to isolated NPs* (Liu et al., 1990a,b; Zhang and Mascarenhas, 2000; Francoeur et al., 2004; Karaiskaj et al., 2007).

The microscopic origin of the NP and NP* bands has been investigated by employing the LCINS method; see Ref. (Polimeni et al., 2008a) and Chapter 1. The attribution of the NP and NP* bands to, respectively, *isolated* NPs and *perturbed* N pairs (with a third N atom sitting on some neighboring sites) on the one hand contradicts the usual attribution of the 1.508 eV band, while on the other hand it can explain the puzzling electron compositional dependence of the electron effective mass shown in Figs. 2.15b and 2.18a. These two states share an almost identical electron charge density (49.4% and 48.9% for NP and NP*, respectively). Consequently, the magnitude of the interaction between the CB edge and a single NP or NP* state is almost identical, while the overall interaction depends on the fraction y of NP and z of NP* clusters as $2.84 \times y^{1/2}$ eV and $2.79 \times z^{1/2}$ eV, respectively. Despite these similarities, the crossing of these two states with the CB minimum affects very differently the value of m_e^* (see Fig. 2.18a), whose largest variation occurs when the CB minimum resonates with a level located at ~1.487 eV (NP). This energy corresponds to the band-gap energy of a sample with $x \sim 0.1\%$. Instead, m_e^* increases smoothly when the CB minimum passes through the 1.508 eV state (NP*), which corresponds to the band-gap energy of a sample with $x \sim 0.05\%$. These findings are consistent with the much larger density of NP states ($y \sim 6x^2$) with respect to the NP* states ($z \sim 12x^3$, for a specific NP* cluster with the third N atom sitting directly above one of the NP atoms). For a purely random distribution of N atoms, the ratio y/z of the density of the NP to the considered NP* cluster varies as $1/(2x)$ and is approximately equal to 500 at $x = 0.1\%$ and to 1000 at $x = 0.05\%$ (with a similar variation for other configurations of the perturbed NP). This result indicates a more important role of the isolated NP state with respect to the perturbed NP (NP*) state as far as the interaction with the CB minimum is concerned.

Another interesting property of *NC* states can be determined by means of optical measurements under hydrostatic pressure on the hydrogenated samples. As mentioned in Section 2.3, the introduction of a sufficiently high dose of hydrogen leads to a complete quenching of the *NC* related PL bands both in Ga(AsN) and in Ga(PN) alloys; see Fig. 2.4b and Fig. 2.5b. However, the hydrogen passivation of different cluster states occurs at different rates as, for example, for the NP and NP* states—the latter being passivated less efficiently than the former; see Fig. 2.19b,c. In particular, the intensity of the NP line decreases with increasing H doses faster than that of NP* line, despite the fact that the NP band is at lower energy than the NP* band and should be, therefore, thermally favored. This finding is independent of the level population, as verified by employing two different laser power densities. Moreover, at the highest H dose, the PL spectra are not distinguishable from that of GaAs; see also upper curves in Fig. 2.18b for a comparison at ambient pressure. This indicates that all NC states (at least those optically accessible at 8.5 kbar) are fully passivated, consistent with the complete recovery of the GaAs band-gap energy and electron effective mass; see Figs. 2.18b and 2.18a, respectively. Since both NP and NP* states have very similar wavefunction extents, the faster passivation rate of one kind of state cannot be accounted for in terms of electronic charge density arguments. Instead, one can envisage that the larger *cross section* for H capture by the *isolated* NPs responsible for the NP band originates from the greater amount of strain induced by these clusters (Polimeni et al., 2008a). In other words, the total system energy will decrease more when H goes on those NCs whose passivation causes a larger decrease in strain—as witnessed by the fact that H counteracts the native tensile strain of Ga(AsN); see Refs. (Bisognin et al., 2008, 2006) and Chapter 7. The validity of this hypothesis can be qualitatively assessed by rather simple calculations of the elastic energy in the crystal in terms of a parameterized valence force field (Polimeni et al., 2008a). The total elastic energy stored at equilibrium for a specific NP* cluster (with the third N atom directly above one of the NP atoms) within a $Ga_{500}N_3As_{497}$ supercell is 4.789 eV. The total elastic energy for an NP and an isolated N atom, respectively, within a $Ga_{500}N_2As_{498}$ and a $Ga_{500}N_1As_{499}$ supercell, is 3.397 eV (NP) + 1.545 eV (N) = 4.942 eV. The smaller energy for the NP* cluster is due to the presence of the third nitrogen atom (N3), which allows (*i*) the Ga–N bond lengths of NP and N3 to become slightly shorter, namely, closer to the natural Ga–N bond length, and (*ii*) the lattice to have a smaller surrounding tetrahedral distortion (angular distortion), that is, a smaller deviation from the ideal zincblende crystal. For these reasons, one may conclude that in the NP* cluster the local strain energy associated with the NP is smaller than that associated with the isolated NP pair (where N3 is at infinity). Accordingly, H will be preferably driven toward NP rather than toward NP* clusters. To complete the picture, calculations have shown that for those perturbed NPs that induce energy levels *below* that of the isolated NP level, the total elastic energy is, instead, *greater* than the total elastic energy of an isolated NP plus an isolated N atom. This result is consistent with the larger activation energy required for thermal debonding of H from these energy levels, as reported in Section 2.5.1; see Fig. 2.9 and Ref. (Bissiri et al., 2002a). Therefore, the pressure studies reveal that the *strain relaxation process is the key mechanism driving the interaction of hydrogen with nitrogen in dilute nitrides.*

2.6.3 *Electron Gyromagnetic Factor in Ga(AsN)*

The high sensitivity of the CB states to nitrogen incorporation and passivation, which induces the anomalous compositional behavior of electron effective mass of Ga(AsN), affects also another important electronic parameter, namely, the electron effective gyromagnetic factor, g_e^* (Pettinari et al., 2006).

Figures 2.20a–c show the PL spectra of three Ga(AsN) epilayers in the spectral region of the free-electron to neutral-carbon (e, C) recombination band for different values of the magnetic field intensity and for different circular polarization states of the emitted photons. The spectra were taken at low temperature ($T = 10 - 30$ K) and different laser excitation power densities ($P_{exc} \leq 20$ W/cm^2) in order to enhance the contribution arising from the (e, C) transition with respect to those related to NC states. The g_e^* value of GaAs, though temperature dependent, varies only slightly in this temperature range (Oestreich and Rühle, 1995). As previously shown, with increasing magnetic field intensity the (e, C) band shifts to higher energy at a rate proportional to the inverse of the electron effective mass (see Eq. 2.2 and Fig. 2.14), whose value depends on x as discussed in Section 2.6.1; see Fig. 2.15b. At the same time, the energy separation between the transitions related to opposite circular polarization states increases (σ^+ and σ^- transitions are

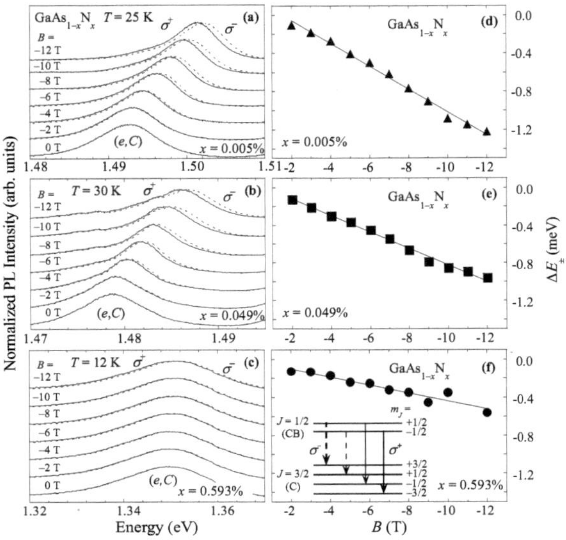

Figure 2.20 (a–c) Polarization-resolved PL spectra in the region of the (e, C) recombination under different magnetic field intensities, recorded for Ga(AsN) samples with $x = 0.005\%$ (a), 0.049% (b), and 0.593% (c). The solid (dashed) lines represent spectra for left (right) circular polarized light and correspond to $\sigma^+(\sigma^-)$ transitions. The negative sign of B indicates an anti-Faraday configuration, where B has the same direction of the emitted light but the opposite way. (d–f) Dependence of the Zeeman splitting $(\Delta E_\pm)$ on B for the same samples shown in (a–c). Solid lines are fits to the data by Eq. 2.3, where g_e^* is the only fitting parameter. We have assumed the value $g_{C3/2} = 0.52$ for the g-factor of the $m_J = \pm 3/2$ carbon levels (see Ref. (Bimberg, 1978)). Inset: Scheme of the levels involved in the (e, C) transition under a magnetic field $B > 0$. CB and C indicate the CB- and the carbon-related level, respectively. J and m_J are the total angular momentum and its projection along the magnetic field direction, respectively. Vertical arrows indicate dipole-allowed transitions in a Faraday configuration, thicker-line arrows indicate transitions actually resolved in the experiment.

associated with left and right circular polarization, respectively). We will refer to this separation as the *Zeeman splitting*, $\Delta E_\pm$.

The energy levels responsible for the (e, C) recombination are the CB minimum, with total angular momentum $J = 1/2$, and the carbon acceptor level, with $J = 3/2$. The four transitions allowed between these levels, in a dipole approximation and for a Faraday configuration, are sketched in the inset of Fig. 2.20f. Transitions may occur between states whose projections of J along the growth direction differ by $\Delta m_J = +1$ or -1, with right (σ^- transition) or left (σ^+ transition) circular polarization of the emitted photons, respectively (Bimberg, 1978). However, only two PL components differing in the polarization state are resolved in our Ga(AsN) samples. In the N-containing samples studied here the linewidth of the (e, C) band is 5–10 meV, that is, much broader than that (≤ 1 meV) reported in N-free, almost intrinsic GaAs (Bimberg, 1978). Moreover, the density of states of the light-hole ($m_J = \pm 1/2$) components of the C level is three times smaller than that of the heavy-hole ($m_J = \pm 3/2$) ones (Bimberg, 1978), thus hampering the spectral resolution of transitions to the $m_J = \pm 1/2$ carbon states. In principle, the lattice mismatch between Ga(AsN) epilayers and the GaAs substrate present in our samples determines a biaxial tensile strain that removes the degeneration in $|m_J|$. The shear potential for the C level is measured to be about half with respect to that of the valence band (Kim et al., 1997). At the highest nitrogen concentration ($x = 0.6\%$); therefore, the strain shifts the light-hole component of the C level only 4 meV above the heavy-hole component (Kim et al., 1997), while the linewidth of the recombination peak is greater than 10 meV. Moreover, at the temperature and power densities of our measurements, carrier population of the heavy-hole component is most likely dominant with respect to the light-hole one, which has a lower density of states. Consequently, only heavy-hole states will be considered in the analysis of Zeeman splitting measurements, namely,

$$\Delta E_{\pm}(B) = E(\sigma^+) - E(\sigma^-) = (-g_e^* + 3g_{C3/2})\mu_B B, \quad (2.3)$$

where μ_B is the Bohr magneton and $g_{C3/2}$ is the gyromagnetic factor of the $m_J = \pm 3/2$ carbon state (we recall that in our case $B < 0$).

Due to the small values of the splitting with respect to the PL linewidth (see Fig. 2.20a–c) for each value of magnetic field intensity $\Delta E_{\pm}$ has been estimated as

$$\Delta E_{\pm} \cong \frac{I_{\sigma^+} - I_{\sigma^-}}{dI_{\sigma^+}/dE}, \quad (2.4)$$

where $I_{\sigma+}(I_{\sigma-})$ is the peak-normalized luminescence intensity corresponding to $\sigma^+(\sigma-)$ transitions at a given magnetic field B,

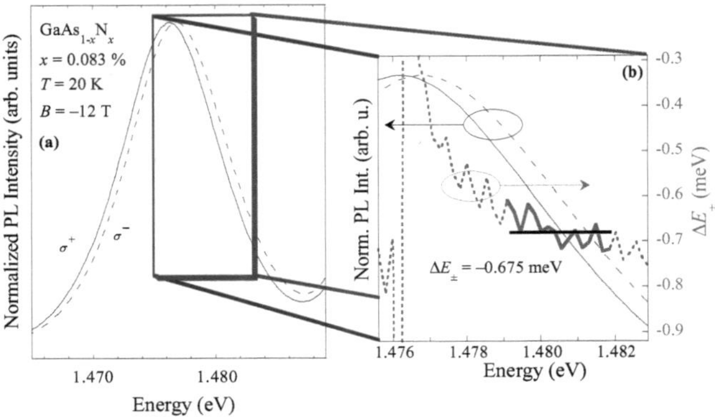

Figure 2.21 Differential method used for the Zeeman splitting determination in Ga(AsN) samples. (a) Polarization-resolved PL spectra at $B = -12$ T for a sample with $x = 0.083\%$. The continuous (dashed) line is the spectrum of the $\sigma^+(\sigma^-)$ polarization component of the emitted light. (b) Zoom of the region of interest highlighted by the rectangular area in (a). The dotted line is the splitting $\Delta E_\pm$, as calculated by Eq. 2.4. An average on the flat zone of $\Delta E_\pm$ (bold line) gives the correct value for the energy splitting of the two components.

and $dI_{\sigma+}/dE$ is the first derivative of the PL signal with respect to the photon energy. This analysis must be limited to the band tails, due to the divergence of $\Delta E_\pm$ at the peak maximum; see Fig. 2.21. To minimize the contribution of N-related transitions, which in some case appear on the low-energy side of the (e, C) band, we focused on the high-energy side of the (e, C) band. The obtained Zeeman splitting values are shown in Fig. 2.20d–f as a function of the applied magnetic field. The Zeeman splitting has a linear dependence on B up to the maximum investigated field intensity ($|B| = 12$ T), with a slope depending strongly on nitrogen concentration. For $|B| < 2$ T, the σ^- and σ^+ spectra are often superimposed.

By taking for $g_{C3/2}$ the value it has in GaAs (0.52 ± 0.04) (Bimberg, 1978), namely, by assuming that the valence band gyromagnetic factor is not affected by N incorporation, g_e^* can be derived by fitting Eq. 2.3 to the Zeeman splitting data. The values of the electron gyromagnetic factor are shown as a function of x in Fig. 2.22a, where virgin and hydrogenated samples are indicated by different symbols. The electron gyromagnetic factor exhibits a sign

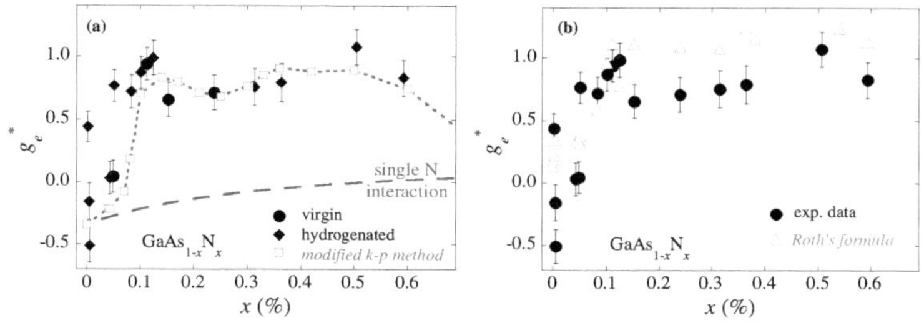

Figure 2.22 (a) Dependence of the electron gyromagnetic factor (g_e^*) on the nitrogen concentration in Ga(AsN). Full dots (diamonds) refer to as-grown (hydrogenated) samples. Open squares connected by a dotted line are the result of a modified $k{\cdot}p$ method, taking into account the interaction of the NC states with the CB minimum. The dashed line is the result of a model including only the interaction with the single-N level. A comparison between the measured compositional dependence of g_e^* (full dots) and that determined by Eq. 2.5 (*Roth's formula,* open triangles) is shown in panel (b).

reversal for $x \geq 0.04\%$ and increases abruptly in a very narrow compositional window, $0.04\% \leq x \leq 0.1\%$. For $x > 0.1\%$, g_e^* has a not well-defined behavior and fluctuates around 0.7 for the highest x values. These results confirm the highly nonmonotonic changes in the Ga(AsN) CB due to the strong hybridization between the CB minimum and the *NC* levels, as previously determined by the compositional dependence of the electron effective mass; see Section 2.6.1. As in the case of the electron effective mass, the N dependence of the electron g-factor in Ga(AsN) is well reproduced by a modified $k{\cdot}p$ method (see the open squares in Fig. 2.22a), which duly takes into account the interaction with the *NC* states. A simple model considering only the effect of the single-N level highly underestimates, instead, the CB modifications induced by nitrogen incorporation; see the dashed line in Fig. 2.22a.

The compositional dependence of the electron gyromagnetic factor shown in Fig. 2.22a closely resembles that of the electron effective mass, previously shown in Fig. 2.15b for the same system. In the framework of a $k{\cdot}p$ model, a direct relationship between g_e^* and m_e^* is expected on the basis of the so-called *Roth's formula* (Roth et al., 1959):

$$\frac{g_e^*(x)}{g_0} - 1 = -\frac{\Delta_0}{3E_g(x) + 2\Delta_0}\left(\frac{m_0}{m_e^*(x)} - 1\right), \qquad (2.5)$$

where g_0 (m_0) is the gyromagnetic factor (effective mass) of electrons in vacuum, $E_g(x)$ is the experimental value of the band-gap energy of Ga(AsN), and Δ_0 is the split-off energy (kept fixed at the GaAs value, 0.341 eV). In Fig. 2.22b the compositional dependence of the experimental g_e^* values is compared with that predicted by Eq. 2.5, where the electron effective mass values employed are those measured in the same samples and shown in Fig. 2.15b. The agreement between the two sets of data is very good, especially if one considers the approximate nature of Eq. 2.5 (obtained by using only the first three bands in a $k \cdot p$ model; see Ref. (Roth et al., 1959)) and the *absence* of any free parameter. This agreement suggests a few considerations. First, since the electron gyromagnetic factor and effective mass are derived from two independent measurements, the experimental consistency provided by the Roth's formula (Fig. 2.22b) further supports the importance of the NC states in determining the electronic properties of Ga(AsN)—as described by the LCINS model (see Refs. (O'Reilly et al., 2009; Masia et al., 2006; Polimeni et al., 2008a; Lindsay and O'Reilly, 2004b) and Chapter 1)—as well as the ability of hydrogen to gradually restore the N-free material properties. Second, the $k \cdot p$ model is based on a perturbative approach, which provides explicit expression for several physical quantities in solids (Kane, 1957). In particular, this model predicts that the electron effective mass and gyromagnetic factor are *smaller* for *smaller* band-gap compounds and alloys. For Ga(AsN) this decrease is counteracted by the diminished value of the optical matrix element describing the coupling between the CB minimum and the valence band maxima, which follows the loss of Γ character of the CB minimum (see Ref. (Polimeni et al., 2008b) and Chapter 1). In addition, the success of the *modified* $k \cdot p$ method to account for the data of m_e^* (Fig. 2.15b), g_e^* (Fig. 2.22a), and their relationship via Roth's formula (Fig. 2.22b) indicates that the Ga(AsN) electronic states can be described in terms of coherent Bloch waves with a well-defined **k** vector, at least at the band extrema, where the Hamiltonian expansion is performed in the $k \cdot p$ method. Therefore, despite the loss of a full translational symmetry deriving from the potential disorder induced by NCs, a description of dilute nitride alloys in terms of conventional semiconductor alloy concepts (such as electron effective mass and g-factor) is still applicable for N concentrations as large as about 1%.

Finally, the findings presented in this section, in particular the substantial indistinguishability between virgin and hydrogenated samples as regard the values of g_e^* and m_e^* for the same values of the effective nitrogen concentrations/emission energies, indicate that hydrogenation represents an effective and extremely powerful tool to finely tune all the electronic and optical properties of dilute nitride alloys.

2.7 Conclusions

In this chapter, we reviewed the effects of hydrogen incorporation in Ga(AsN), (InGa)(AsN), and Ga(PN). The capability of hydrogen to transform the whole physical properties of one material, for example, Ga(AsN), into another, that is, GaAs, represents a unique case in semiconductor physics. Indeed, the formation of N–H complexes into the lattice of dilute nitrides modifies the whole band structure, namely, band-gap energy, CB curvature, and spin properties, in a fully controllable and reversible manner.

The energetic of the N–H complexes responsible for these effects was addressed by thermal annealing and photon-induced dissociation. In the first experiment, we found that the activation energy for thermal dissociation of the N–H complexes follows a Gaussian distribution. This energy is close to those found in the case of very dilute N concentrations (impurity limit), where it was determined that H binds stronger to larger NCs. These results indicate that the giant band-gap red shift induced by N in dilute nitrides should be accounted for by *different* NC configurations rather than by a *single* NC configuration. Laser dissociation experiments showed that the dissociation of N–H complexes in hydrogenated III–N–Vs can be achieved at a laser power corresponding to a local temperature much smaller than that required for thermal dissociation (>250°C). Photon absorption by the N–H complexes mediated by photogenerated carriers provides a mechanism for the band-gap reopening and features interesting applications. By this method, indeed, the band gap of the III–N–Vs can be in-plane-patterned with submicron spatial resolution and high-energy accuracy (see Chapter 8 for others in-plane bang-gap engineering techniques).

Finally, the combined use of hydrostatic pressures and magnetic fields with PL on differently hydrogenated Ga(AsN) samples permitted to trace the compositional evolution of the electron mass and gyromagnetic factor with an unprecedented accuracy. This allowed unveiling of an electronic hierarchy between different NCs as regard the extent of the perturbation these clusters exert on the electronic properties of the GaAs host.

Acknowledgments

We are grateful to A. Miriametro for his skilful and continuous technical support. We thank F. Martelli (at IMM-CNR, Italy), S. Rubini (at TASC-CNR, Italy), and W. Stolz (at Philipps University, Marburg, Germany) for growing the Ga(AsN) samples presented in this chapter. We thank C. W. Tu and H. P. Xin (at the University of California, San Diego, USA) for growing Ga(PN) samples and A. Forchel (at the University of Würzburg, Germany) for providing (InGa)(AsN) samples. We acknowledge the fruitful collaboration on pressure-dependent measurements with P. J. Klar (at Justus Liebig Universität, Giessen, Germany) and on laser-writing and μPL experiments with N. Balakrishnan, O. Makarovsky, and A. Patanè (at the University of Nottingham, UK). We thank E. P. O'Reilly and A. Lindsay (at the Tyndall National Institute, Cork, Ireland) for theoretical modeling and calculations and D. Giubertoni (at Fondazione Kessler, Trento, Italy) for SIMS measurements. We would like also to acknowledge G. Baldassarri Höger von Högersthal, M. Bissiri, M. Felici, F. Masia, and R. Trotta, who gave great contributions at different stages of this research, as well as all the coauthors of this book, with whom we collaborated to unveil the fascinating properties of hydrogenated dilute nitrides. We acknowledge COST Action MP0805 and one of us (GP) the EU under grant agreement no. PIEF-GA-2010-272612 for funding part of this research.

Chapter 3

Effect of H on Defects in Dilute Nitrides

D. Dagnelund, W. M. Chen, and I. A. Buyanova
Department of Physics, Chemistry and Biology, Linköping University, SE-58183 Linköping, Sweden
dagne@ifm.liu.se

In this chapter we review recent experimental findings on the role of H incorporated during postgrowth hydrogenation on defects in GaNP and GaNAs alloys. The main emphasis is on Ga interstitial (Ga_i)-related defects that are known to be among the dominant nonradiative recombination centers in dilute nitrides. From optically detected magnetic resonance (ODMR) studies, hydrogen is shown to cause activation of new Ga_i-related defects in GaNP due to the formation of Ga_i–H complexes. Markedly different H behavior is, however, observed in GaNAs, where H efficiently passivates Ga_i-related defects present in the as-grown material. Possible mechanisms responsible for the observed difference in the H behavior in GaNP and GaNAs are also discussed.

3.1 Introduction

Due to its high chemical reactivity, hydrogen is known to efficiently interact with nearly all types of imperfections and impurities

Hydrogenated Dilute Nitride Semiconductors: Theory, Properties, and Applications
Edited by Gianluca Ciatto

ISBN 978-981-4463-45-4 (Hardcover), 978-981-4463-46-1 (eBook)
www.panstanford.com

present in semiconductors, which most often leads to their passivation (McCluskey and Haller, 1999; Van de Walle and Neugebauer, 2006) and, therefore, a reduction of electrical activity. The passivation results from formation of hydrogen–impurity complexes either by the direct bonding of hydrogen to an impurity atom or by the bonding of hydrogen to a lattice atom in the immediate neighborhood of the impurity. Well-known examples of impurities/defects that can be efficiently passivated by hydrogen include S_{As}, Se_{As}, and Te_{As} donors in GaAs (Pearton et al., 1986), B acceptors (Sah et al., 1983; Pankove et al., 1983) and P donors (Johnson et al., 1986) in Si, the V-O centers in Si and SiGe/Si heterostructures (Chen et al., 1996; Buyanova et al., 1997), Be acceptors in AlGaAs/GaAs quantum wells (QWs) (Buyanova et al., 1996), and silicon dangling bonds (P_b defect) at a Si/SiO_2 interface (Caplan et al., 1979). In fact, the passivation of the P_b defects is absolutely essential for reliable operation of metal oxide semiconductor field-effect transistors—the heart of integrated circuit technology. In some rare cases, H was also found to activate defects and impurities such as the Ga vacancy in GaAs (Capizzi et al., 1993; Amore Bonapasta and Capizzi, 1998; Capizzi et al., 1992) and neutral dopants such as Si and C in ultrapure Ge (Haller et al., 1981).

Hydrogen is also known to greatly affect properties of dilute nitrides such as Ga(In)NAs and GaNP alloys, which are novel materials promising for a variety of applications, as introduced in the previous chapters. Most unexpectedly in these materials, incorporation of H was found to cause dramatic changes in the fundamental band structure by effectively neutralizing all alloy properties caused by the presence of nitrogen and recovering the band-gap energy, electron effective mass, and other parameters of the N-free hosts, as reported in detail in Chapter 2. Furthermore, hydrogen even induces large modification of the structure of dilute nitrides, as will be shown in the following chapters.

Though this unusual H behavior has attracted a great deal of attention over the last decade, studies devoted to effects of hydrogenation on grown-in defects in dilute nitrides emerged only recently (Dagnelund et al., 2011, 2012; Shafi et al., 2011; Dagnelund et al., 2013). These studies were motivated by the fact that defect formation (leading to efficient nonradiative recombination) is known to be very severe in dilute nitrides (Buyanova et al.,

2003a) and, in fact, presents one of the key obstacles for widespread applications of these materials. For example, Wang et al. have recently shown (Wang et al., 2009b) that up to 88% (!) of recombination in Ga(In)NAs alloys suffers nonradiative losses via nonradiative recombination and has identified complexes involving a Ga interstitial (Ga_i) atom as responsible defects. Moreover, growth processes utilized for fabrication of dilute nitrides, such as gas source molecular beam epitaxy (GS-MBE) or metalorganic chemical vapor deposition (MOCVD), usually involve hydrogen (Xin et al., 1999; Kurtz et al., 2001). This means that H is a common contaminant in dilute nitrides and may affect defect formation (Janotti et al., 2003). It is therefore clear that better understanding and control of the grown-in defects and their interaction with hydrogen in dilute nitrides are highly desirable.

In this chapter we will provide an overview of the reported effects of postgrowth H treatment on defects in GaNP and GaNAs alloys. We will show that incorporation of H has a strong effect on intrinsic defects in these materials. In GaNP, it leads to *activation* of new defects, which have a Ga_i atom at their core, as revealed from optically detected magnetic resonance (ODMR) studies (Dagnelund et al., 2011). The observed activation critically depends on the presence of N in the alloy, as it does not occur in GaP with a low level of N doping. Since none of these Ga_i-related defects were previously detected in as-grown GaNP, the activation was attributed to the formation of Ga_i–H complexes. Subsequent thermal annealing of hydrogenated GaNP samples is shown to efficiently *suppress* the Ga_i-related defects. This effect can be tentatively ascribed to dissociation of H–Ga_i complexes. In sharp contrast, in GaNAs hydrogen is found to efficiently *passivate* Ga_i-related defects present in the as-grown material (Dagnelund et al., 2012). This is the expected effect of H acting as a defect passivator. On the other hand, H-induced *activation* of EL2 defects in GaNAs with a $[N] > 0.2\%$ was revealed from deep-level transient spectroscopy (DLTS) studies (Shafi et al., 2011). The summarized results provide a useful insight into the effect of H incorporation on recombination centers in dilute nitrides and will hopefully shed light on control of the defects in these materials by optimizing postgrowth treatments.

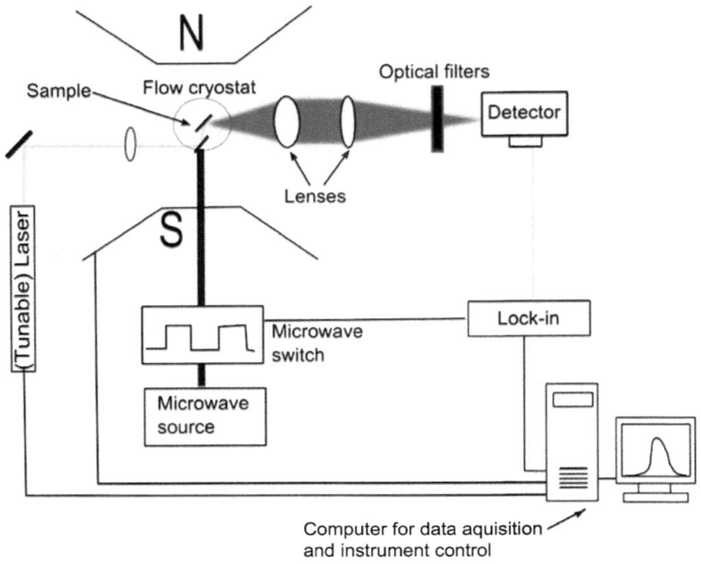

Figure 3.1 Schematic illustration of a typical ODMR setup.

The chapter is organized as follows. In Section 3.2, we describe the samples investigated and the experimental techniques employed for the defect characterization. In Sections 3.3 and 3.4, a detailed description of the effects of H incorporation on defects in GaNP and GaNAs alloys, respectively, is given. Finally, in Section 3.5 we summarize the results and discuss the possible mechanism responsible for the observed difference in the H behavior in GaNP and GaNAs.

3.2 Experiments

3.2.1 *Description of the ODMR Technique*

ODMR is a spectroscopic method in which transitions between spin sublevels of a paramagnetic defect are detected by optical means. A typical ODMR setup is schematically depicted in Fig. 3.1. A traditional way of detecting transitions between spin sublevels is by monitoring a change of the microwave (MW) absorption as a function of an applied magnetic field. The method is then called electron paramagnetic resonance (EPR). The scaling-up of EPR detection from the MW to the optical regime renders great sensitivity to the ODMR technique. As a combination of EPR and photoluminescence (PL), ODMR can provide information on a role of the specific defect, which is detected via magnetic resonance, in the monitored PL emission. In addition, if a sample under study consists of several layers of different semiconductor materials, one can selectively study properties of the these layers by monitoring the corresponding PL emission. These advantages make ODMR suitable for studies of thin films and layered and quantum structures. Both EPR and ODMR are used for obtaining microscopic information on paramagnetic defects, including their chemical origin, identity of neighboring atoms, charge states, symmetry, and localization of the electron wavefunction.

The ODMR method relies on spin selection rules in the monitored recombination process, which links the polarization or the intensity of the emitted light to the population distribution of the spin levels of a paramagnetic defect. Since transitions between these magnetic sublevels are magnetic-dipole-allowed, they can be induced by an externally applied electromagnetic field of the matching frequency (usually in the MW region with magnetic fields commonly available). When the energy splitting between the two sublevels equals the energy of the MW quanta, a spin-resonant transition occurs. The MW-induced transition between the Zeeman sublevels changes the population of these levels and consequently gives rise to a change in the total PL intensity or polarization (Fig. 3.2). This change is defined as an ODMR signal. There are several prerequisites for detecting an ODMR signal. First of all, the studied material should emit light. Secondly, the defect state should be paramagnetic. In addition, an applied MW field should induce a net change in the monitored PL, that is, the spin-split levels should have different populations in the steady state (before the MW is applied). This requirement is also valid for EPR (otherwise there will be no net MW absorption by the sample). Usually, the shorter-lived sublevel has a lower steady-state population. Resonant MW quanta will cause transitions to this level, resulting in a faster decay at the defect. If recombination via the defect is nonradiative, the ODMR signal can be detected as a decrease

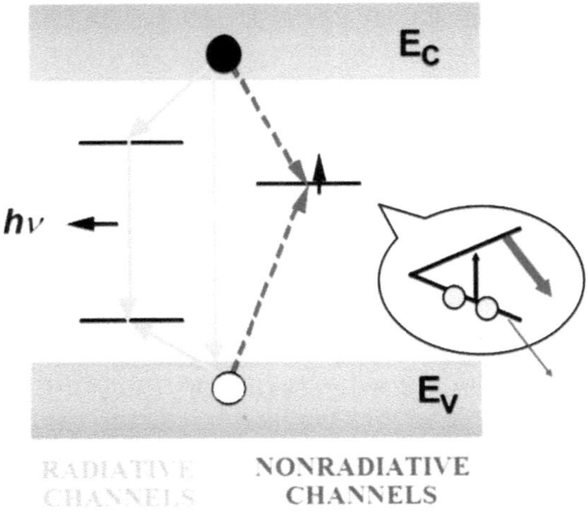

Figure 3.2 An illustration of the origin of a negative ODMR signal. The monitored PL emission (denoted by hν) is decreased by the MW-induced increase of recombination via the defect (the thicker arrow represents faster decay from the Zeeman-split spin sublevel).

of the monitored PL (negative ODMR signal). If, on the other hand, the defect is directly involved in the radiative recombination, the ODMR signal will be positive as the MW-induced transitions accelerate the radiative recombination. ODMR is therefore often used to study defect states that govern carrier recombination. These defects can in fact control key parameters of materials, such as their radiative efficiency and free carrier lifetime. An in-depth review of the ODMR techniques can be found in Ref. (Chen, 2000).

3.2.2 *Ga Interstitial-Related ODMR Signal*

Analysis of ODMR spectra is usually performed with the aid of the spin Hamiltonian that includes electron Zeeman and central hyperfine interaction terms:

$$H = \mu_B \mathbf{B} \cdot \mathbf{g} \cdot \mathbf{S} + \mathbf{S} \cdot \mathbf{A} \cdot \mathbf{I} \tag{3.1}$$

Here μ_B is the Bohr magneton, **S** is the effective electronic spin, **I** is the nuclear spin of the defect, **g** is the electronic **g**-tensor and **A** is the central hyperfine tensor. If the observed ODMR signal is

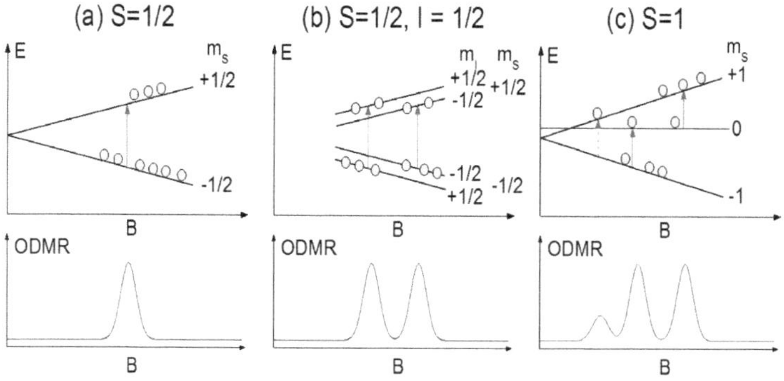

Figure 3.3 (a) Electron Zeeman levels (the upper part of the figure) and expected ODMR spectra (the lower part of the figure) for three cases: (a) $\mathbf{S} = 1/2, \mathbf{I} = 0$; (b) $\mathbf{S} = 1/2, \mathbf{I} = 1/2$; and (c) $\mathbf{S} = 1, \mathbf{I} = 0$. The allowed magnetic dipole transitions ($\Delta m_S = \pm 1$ and $\Delta m_I = 0$) are indicated by the solid arrows. The dashed arrow indicates a spin-forbidden transition (sometimes enabled by mixing of electron wavefunctions).

isotropic, **g** and **A** tensors can be reduced to scalars g and A. The effective spin and the symmetry of the defect can often be deduced from the number of resonances in the ODMR spectra (see Fig. 3.3) and their angular dependences. Chemical identification of a defect is based on measurements of the central and/or ligand hyperfine interactions between the unpaired spin of electron(s) trapped at the defect center and the nuclear spin(s). The hyperfine interaction gives rise to an energy level splitting and, hence, to a splitting of the ODMR lines—see Fig. 3.3b. Since each chemical element has a definite number of isotopes with known nuclear spins, magnetic moments, and natural abundance, observation of the resolved hyperfine structure allows chemical identification of the defect. One excellent example is identification of Ga interstitial-related defects in GaNAs (Thinh et al., 2001). Ga has two isotopes ^{69}Ga and ^{71}Ga with the nuclear spin $\mathbf{I} = 3/2$, natural abundance 60.4% (^{69}Ga) and 39.6% (^{71}Ga), and the ratio of nuclear magnetic moments $\mu(^{71}\text{Ga})/\mu(^{69}\text{Ga}) = 1.27$. The hyperfine interaction between the electron spin $\mathbf{S} = 1/2$ and the nuclear spin $\mathbf{I} = 3/2$ of a Ga atom in the core of the defect gives rise to two overlapping ODMR quadruplets (one for each of the two Ga isotopes) with the relative intensity ratio reflecting their

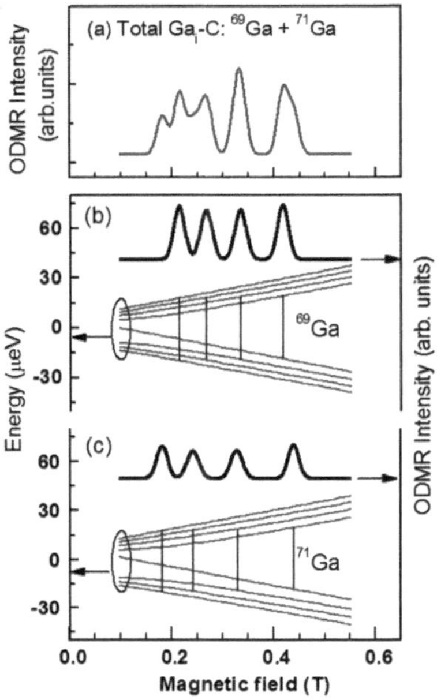

Figure 3.4 (a) Expected ODMR spectra of Ga_i-related defect in the X band. (b and c) Breit–Rabi diagram for $\mathbf{S} = 1/2$ electron, including hyperfine interaction with ^{69}Ga and ^{71}Ga nuclei with $\mathbf{I} = 3/2$. The magnetic dipole transitions are indicated by the solid vertical lines.

natural abundance and their relative hyperfine interaction scaling with the nuclear moment ratio—see Fig. 3.4).

3.2.3 *Samples*

Two sets of GaNP/GaP and GaNAs/GaAs structures grown by GS-MBE are considered in this review. Each set contains several structures with distinctly different N compositions and, therefore, concentrations/types of grown-in defects active in recombination prior to H treatment. The most important growth parameters of the structures are summarized in Table 3.1.

Postgrowth hydrogenation was performed by ion beam irradiation from a Kaufmann source at 300°C, using a low ion energy (100 eV) and a current density of ~10 μA/cm^2. An even lower H

Table 3.1 List of the GaNP and GaNAs samples, with the main growth parameters. The numbers 5 and 20 in the parentheses refer to the thicknesses of the GaNAs and GaAs layers in the multiquantum well (MQW) structure

	GaNP/GaP					GaNAs/GaAs		
Sample	2666	2961	2671	L012	2689	2661	2468	2522
Growth T(°C)	520	520	520	590	520	420	420	420
[N](%)	0.05	0.60	0.80	1.4	2.3	0.7	1.3	1.6
Structure	epi	epi	epi	epi	epi	epi	epi	MQW
Thickness (nm)	250	250	250	100	750	100	110	7 × (5/20)

ion energy of 20 eV was used for one of the $GaN_{0.006}P_{0.994}$ epilayers. The hydrogen doses ranged between 2.7×10^{17} cm^{-2} and 4.5×10^{18} ions/cm^2. To separate the effect of H from that of thermal annealing, a piece of the as-grown $GaN_{0.006}P_{0.994}$ epilayer was annealed at 300°C without H treatment. Moreover, to study the effect of thermal annealing after hydrogenation, a piece of a GaNP epilayer was annealed for 60 minutes at 400°C in an argon ambient.

Secondary ion mass spectrometry (SIMS) was used to confirm H incorporation and to determine distributions of H and N within several representative structures. Results of SIMS measurements for the investigated GaNP/GaP and GaNAs/GaAs structures are summarized in Figs. 3.5 and 3.6, respectively. Depth profiles of N distributions from the as-grown samples confirm the intended sample structures during the growth. For example, uniform distributions of N were found in GaNP (Fig. 3.5) and GaNAs (Fig. 3.6a) epilayers, whereas the N profile for the GaNAs MQW structure (Fig. 3.6b) clearly shows formation of seven periods of GaNAs QWs separated by GaAs barriers. One also notices that the GaNAs epilayer (MQW structure) is capped by a 200 Å thick (500 Å-thick) GaAs layer. The background H concentration in the structures was around $\sim1\times10^{18}$ cm^{-3} close to the buffer layer and gradually increased toward the surface, which is not surprising as H is abundantly present during the GS-MBE growth. All N profiles remained identical in the hydrogenated samples, proving that no out-diffusion of nitrogen occurred during H treatment. On the other hand, the H distributions dramatically changed after the hydrogenation and became fully

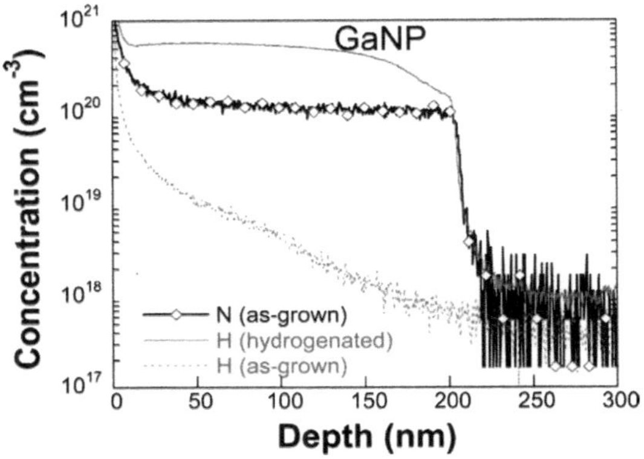

Figure 3.5 Representative SIMS profiles of N and H in the $GaN_{0.008}P_{0.992}$ epilayer before and after H treatment with the dose of $1x10^{18}$ cm^{-3}. Reprinted with permission from Ref. (Dagnelund et al., 2012). Copyright (2012) by the American Institute of Physics.

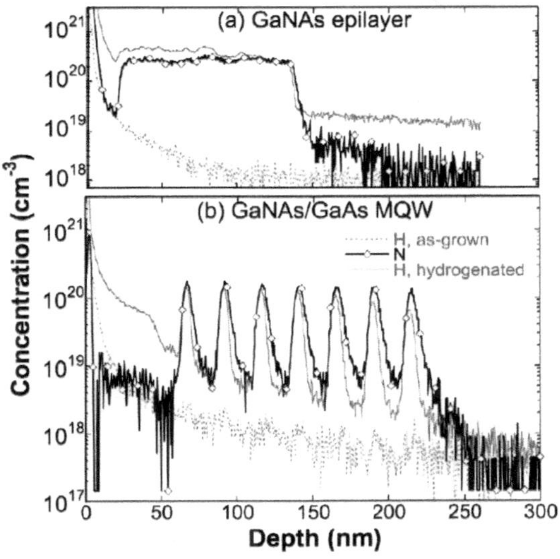

Figure 3.6 Representative SIMS profiles of N and H in the $GaN_{0.013}As_{0.987}$ epilayer (a) and $GaN_{0.016}As_{0.984}$/GaAs MQW (b) before and after H treatment. Reprinted with permission from Ref. (Dagnelund et al., 2012). Copyright (2012) by the American Institute of Physics.

correlated with the N profiles, which is in agreement with the previous studies (Buyanova et al., 2007). Such correlated distributions of H with a certain element/impurity is typical when H passivates this impurity (Pearton, 1994) and, therefore, implied strong bonding of H to N atoms (Berti et al., 2007; Buyanova et al., 2007; Trotta et al., 2009a).

3.3 GaNP Alloys

Figure 3.7 summarizes effects of postgrowth hydrogenation on PL and ODMR spectra of GaNP alloys with different N compositions. The ODMR spectra were monitored by measuring near-band-edge PL emissions due to radiative recombinations at N-related localized states (Buyanova and Chen, 2004) that are shown in Fig. 3.7a. A red shift of these emissions observed with increasing N content reflects a reduction of the band-gap energy of the alloys due to the bowing in the band-gap energy. On the other hand, a blue shift occurring after hydrogenation is due to the H-induced recovery of the energy band gap that can be viewed as a decrease in the effective N concentration due to bonding of H to N (Buyanova et al., 2004). The PL peak position of the hydrogenated GaNP shifts toward its value in the untreated (i.e., not hydrogenated) sample after thermal annealing (see, e.g., sample with [N] = 0.81%). This is attributed to breaking of H–N bonds and subsequent out-diffusion of H. Moreover, the annealing also causes an increase in the PL intensity. All these effects of hydrogenation and annealing on the PL properties in GaNP are discussed in Chapter 2 and, therefore, will not be further discussed here.

To understand effects of hydrogenation and annealing on defects in GaNP, ODMR studies were performed, and the corresponding results are summarized in Fig. 3.7b. Postgrowth hydrogenation was found to cause a dramatic change in intensity of the ODMR spectra in all investigated structures due to appearance of new ODMR signals. Except for the ODMR line denoted as L3 in Fig. 3.7b, all recorded ODMR signals were negative, that is, corresponded to a decrease in intensity of the monitored near-band-edge PL under the spin resonance conditions. This means that the spin-

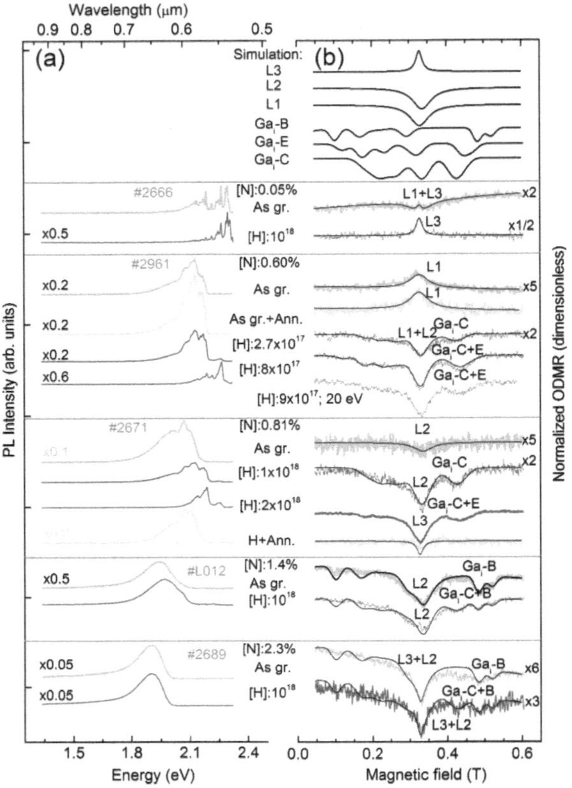

Figure 3.7 (a) Representative PL spectra measured from the studied GaNP epilayers. All spectra are measured with the excitation photon energy exceeding the GaP band gap. (b) Isotropic ODMR spectra measured at the X band (9.2 GHz) by monitoring the PL emissions within the visible spectral range (550–810 nm) shown in Fig. 3.7a. Simulated ODMR spectra are displayed at the top of the figure, whereas the experimentally measured spectra are shown below for the specified N compositions. The simulated ODMR spectra are also overlayed over the experimental spectra (thin black lines). The ODMR intensity is normalized to the PL intensity.

resonance-enhanced recombination via the corresponding defects leads to a decrease in the PL intensity, that is, the defects act as competing recombination centers degrading the optical quality of the alloy (Chen, 2000). The observed changes in the ODMR spectra provided an unambiguous proof that the postgrowth hydrogenation has strong effects on defect properties of the alloy. Before analyzing these effects, however, we would like to provide a brief overview of the detected ODMR signals, which can be attributed to several different paramagnetic centers. They are denoted as Ga_i–B, Ga_i–E, Ga_i–C, and L1–L3 in Fig. 3.7b. The first three signals belong to the defect complexes that have a Ga interstitial atom (Ga_i) at their core, depending on the resolved hyperfine structure (Dagnelund et al., 2011; Wang et al., 2009a,b). The single lines L1–L3, on the other hand, originate from paramagnetic centers with an effective electron spin $\mathbf{S} = 1/2$ and a g-factor close to 2. These centers could not be identified due to a lack of a resolved hyperfine structure. These defects, therefore, will not be further discussed in the chapter.

Spin Hamiltonian parameters of all revealed paramagnetic centers are given in Table 3.2, whereas the corresponding simulated ODMR spectra are shown in Fig. 3.7. For all samples, results of the simulations, including contributions from the specified centers, are also overlaid with the experimentally measured ODMR spectra. The agreement between the simulations and experimental results is rather satisfactory, justifying the assignments of the defects and reliability of the obtained fitting parameters.

As all revealed defects possess a unique set of the spin Hamiltonian parameters, the ODMR technique was employed to individually monitor their formation as a function of N and H content (Dagnelund et al., 2011, 2012, 2013), as will be summarized below.

3.3.1 *Mechanism of Defect Formation*

The most pronounced effect of H incorporation is the hydrogen-induced activation/creation of the Ga_i–C and Ga_i–E defects. This effect is the most apparent in the alloys where no Ga_i defects were detected prior to the hydrogenation (e.g., with [N] $= 0.6\%$ and 0.81%; see Fig. 3.7) but also occurs in alloys with a higher N content (e.g., [N] $= 1.4\%$) where the Ga_i defects were already formed in the as-grown material but in a different configuration, that is, Ga_i–B. Such defect activation by hydrogen is rather unusual and has only been reported in a handful of cases. Interestingly, activation of the Ga_i-related defects is facilitated by N presence, as it is not observed for the alloy with the lowest N content (0.05%).

Table 3.2 Spin Hamiltonian parameters and linewidths obtained from the best fit to the experimental ODMR results. The ratio $A(^{71}Ga)/A(^{69}Ga)$ was chosen as 1.27–1.3, that is, close to that of their nuclear magnetic moments $\mu(^{71}Ga)/\mu(^{69}Ga) = 1.27$

Samples	GaNP/GaP							GaNAs/GaAs		
Defects	L1	L2	L3	$Ga_i - B$	$Ga_i - E*$	$Ga_i - C*$	G1	$Ga_i - A$	$Ga_i - B$	$Ga_i - C$
S	1/2	1/2	1/2	1/2	1/2	1/2	1/2	1/2	1/2	1/2
I	...	...	...	3/2	3/2	3/2	...	3/2	3/2	3/2
g	2.00 ±0.01	1.96 ±0.01	2.01 ±0.01	2.00 ±0.01	2.00 ±0.01	2.00 ±0.01	2.04 ±0.01	2.00 ±0.01	1.99 ±0.01	2.00 ±0.01
$A(^{69}Ga)$ ($x10^{-4}$ cm^{-1})	...	...	...	1150 ±50	830 ±30	620 ±30	...	740 ±40	1250 ±60	620 ±30
Line width (mT)	60 ±6	60 ±6	20 ±2	20 ±2	35 ±4	35 ±4	50 ±5	35 ±4	25 ±3	35 ±4

*In GaNP, Ga_i–C is significantly enhanced and Ga_i–E is only observed after postgrowth hydrogen treatment.

Appearance of the Ga_i-related ODMR signals could be due to a direct *creation* of Ga_i as a result of the H bombardment or H-induced *activation* of the Ga_i defects already present in the GaNP or a combined effect of both. The most straightforward mechanism for defect formation is a kick-out process of a Ga atom in GaNP as a result of a head-on collision with an implanted 100 eV H ion. The maximum energy that can be transferred in such an elastic collision is estimated to be 5.6 eV (Chadderton, 1965). This energy is below the threshold displacement energy of a Ga atom: $\sim$ 8.8 eV (Vavilov et al., 1975). However, considering the large amount of H implanted and possibility of subthreshold defect formation mechanisms (e.g., impurity ionization) (Klinger et al., 1981) or the Baner–James two-step model (Naber and James, 1961), direct introduction of Ga_i defects by H bombardment could still occur. To evaluate the relevance of this mechanism, one of the samples was hydrogenated with H ions having lower kinetic energy (20 eV instead of 100 eV). The ODMR spectrum from this sample ([N] $=$ 0.6%) is shown in Fig. 3.7b and is found to be identical to that observed after hydrogenation using 100 eV H ions, in spite of a five-time decrease in the H ion energy. This indicates that H ion bombardment–induced defect formation is not the main mechanism. Further support for this conclusion is found from ODMR spectra of GaNP with a low N content: even though the same kick-out process is also expected to occur in GaNP independent of the N composition, the Ga_i–C defects were not formed after hydrogenation in GaNP with a low N content, as can be seen from Fig. 3.7. This rules out the bombardment-induced creation of Ga_i in the studied GaNP epilayers.

Therefore, the emergence of Ga_i-related ODMR signals must be due to *activation* of Ga_i defects that were readily present in the epilayers but were inactive in spin-dependent carrier recombination before H treatment. The activation could be accomplished in several ways. It may be related to the H-induced reopening of the alloy band gap, which alters the energy position of the Ga_i-related level with respect to the band edges, for example, from resonance in the conduction band to within the band gap, making it active or more efficient in carrier recombination. However, judging from the PL spectra (Fig. 3.7a), the band-gap energy of the untreated $GaN_{0.006}P_{0.994}$ alloy seems to be similar to that of the $GaN_{0.0081}P_{0.9919}$

after H treatment with a dose of 1×10^{18} cm^2. Yet, the Ga$_i$–C defect is only detected in the latter sample. This enables us to rule out the band-gap reopening alone as the mechanism. Alternatively, H incorporation may affect the Fermi level position in the alloy, changing the charge state of the Ga$_i$ complexes to the spin-active one and, therefore, making possible their detection via ODMR. However, this explanation seems to be somewhat less likely as we have not reliably detected the Ga$_i$–C defect in GaNP alloys with [N] $<$ 2% that were not subjected to hydrogenation. This is in spite of the fact that the studied samples span a wide range in doping and also residual contamination. The second and more probable mechanism is formation of complexes between H atoms supplied by the hydrogenation and a Ga$_i$ atom that was already present in an ODMR-inactive state in the as-grown sample. This might result in a change either in the position of the defect energy level or in its charge state toward the spin-active one, thereby activating the defects in carrier recombination monitored by ODMR. The observed anticorrelation between the ODMR signal intensity and the visible PL intensity signifies the importance of the Ga$_i$-related defects in competing carrier recombination.

We can also conclude that the initial presence of Ga$_i$ should become energetically favorable only in materials containing a large amount of N, as it does not occur in GaP:N with the low N composition of [N] $=$ 0.05%, that is, within the doping limit. However, the question whether the N atom is directly involved as a partner of the complex or barely promotes formation of Ga$_i$ cannot be answered on the basis of the currently available data, unfortunately, due to a lack of the resolved hyperfine structure related to N.

Interestingly, the type of the Ga$_i$ defects activated by H treatment is sensitive to the *ratio* between H and N concentrations, [H]/[N], in the alloys (see Fig. 3.8). For a lower [H]/[N], only one Ga$_i$–C is observed for all N contents up to 2.3%. After further addition of H, however, an additional Ga$_i$-related ODMR signal (Ga$_i$–E) is observed (see Fig. 3.8b). Ga$_i$–E exhibits a larger hyperfine splitting by about 30% as compared to Ga$_i$–C (see Table 3.2). The fact that the appearance and type of the Ga$_i$ defects depend on the [H]/[N] ratio could be explained, for example, if Ga$_i$ is a part of a defect

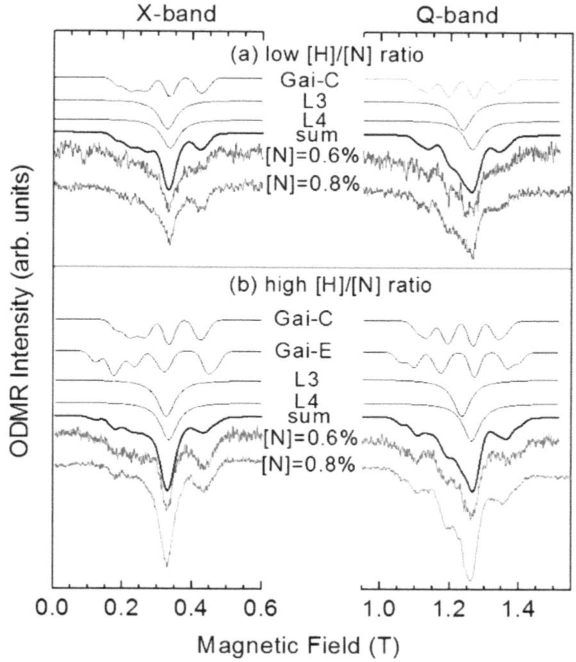

Figure 3.8 Typical X-band and Q-band ODMR spectra, the lower two curves in (a) and (b), measured at 5 K and **B** || [001] by monitoring the PL emissions shown in Fig. 3.7, from the H-treated GaNP epilayers with (a) a low and (b) a high [H]/[N] ratio. The simulated ODMR spectra are displayed by the upper four curves in (a) and the upper five curves in (b), using the spin Hamiltonian parameters given in Table 3.2. Reprinted with permission from Ref. (Dagnelund et al., 2011). Copyright (2011) by the American Institute of Physics.

complex that involves N in a neighboring position. Trapping of H by the N atom could activate the resulting Ga_i defects, such as Ga_i–C at a lower [H]/[N] ratio. Further addition of H should result in bonding of more H atoms at the defects, thereby altering electronic and spin properties of the defects, leading to, for example, the appearance of Ga_i–E. The formation of the aforementioned complexes could occur as a result of either diffusion of H itself or an increased mobility of the Ga_i (or its partners in the complex) owing to an H-induced decrease in the energy barrier for their migration. The former is believed to be more likely, considering the known high mobility of H in semiconductors.

In the case of H–N bonding, the H-induced passivation of N was found to be reversible (Polimeni et al., 2003a) by thermal annealing treatment at moderate temperatures (330°C). This is due to breaking of H–N bonds and subsequent out-diffusion of hydrogen. In an attempt to find out whether the H-induced activation of Ga_i-C defects can also be reversed, one of the hydrogenated samples ([N] = 0.81%) was subjected to postgrowth thermal annealing. It was found that annealing of the H-treated sample caused a dramatic quenching of the Ga_i–C ODMR signal, accompanied by an increase of the PL intensity. This could be tentatively attributed to dissociation of the H atom from the Ga_i–C complex and subsequent out-diffusion of H.

We would like to note that H incorporation also causes a slight reduction of the ODMR signal from the Ga_i–B defect. The observed minor effect in hydrogen passivation of defects that are active in recombination in the as-grown material is, however, not surprising in view of the commonly known ability of H to passivate various deep centers in semiconductors.

3.4 GaNAs Alloys

3.4.1 *ODMR*

To evaluate to which extent the above discussed H-induced effects are common for dilute nitrides, ODMR studies were extended to GaNAs alloys. To facilitate a direct comparison of results for both material systems, GaNAs epilayers were selected with similar N compositions as that in the case of GaNP. The results of the performed ODMR measurements are summarized in Fig. 3.9, whereas Fig. 3.10 presents typical spectra of PL emissions used for detection of the ODMR signals. The near-band edge emission in GaNAs is due to recombination of excitons trapped by potential fluctuations of the conduction band edge (Buyanova et al., 1999), whereas a broad PL band in the NIR spectral range is related to defects of unknown origin. Similar to GaNP alloys, H incorporation was found to cause a blue shift of the near-band-edge emission

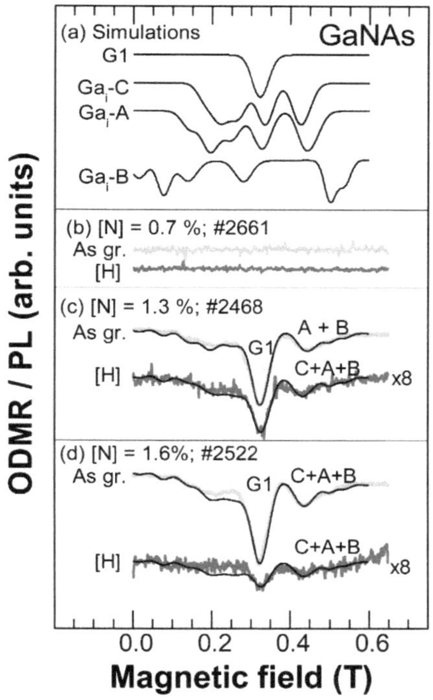

Figure 3.9 ODMR spectra from GaNAs structures detected at 5 K by monitoring the near-band-edge emissions. The ODMR signals are isotropic and are normalized to the PL intensity. Simulated ODMR signals are displayed in (a), whereas the experimentally measured spectra are shown in (b–d) for the specified N compositions. The simulated ODMR spectra are also shown in (b–d) by black, thin curves. The spin Hamiltonian parameters used in the simulated spectra are given in Table 3.2. Reprinted with permission from Ref. (Dagnelund et al., 2012). Copyright (2012) by the American Institute of Physics.

due to combined effects of the H-induced band-gap reopening and suppression of alloy fluctuations (Polimeni et al., 2001a; Buyanova et al., 2003b).

In addition, a strong reduction in the intensity of the defect-related PL band was observed, likely due to the hydrogen-induced passivation of the corresponding radiative centers. The ODMR spectra from the as-grown samples were found to be very similar when recorded via either the near-band-edge or the defect-

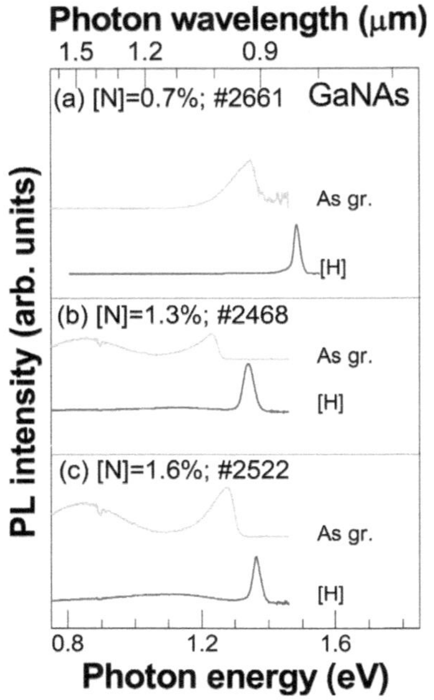

Figure 3.10 Typical PL spectra measured at 5 K before and after postgrowth hydrogenation from (a) a GaNAs epilayer with [N] = 0.7%, (b) a GaNAs epilayer with [N] = 1.3%, and (c) GaNAs/GaAs MQW with [N] = 1.6%. The spectra are calibrated by the instrument response. A blue shift of the near-band-edge PL, observed for the GaNAs MQW structure with the highest N content of 1.6% as compared, for example, with the epilayer sample with [N] = 1.3%, is due to the quantum confinement effect. Reprinted with permission from Ref. (Dagnelund et al., 2012). Copyright (2012) by the American Institute of Physics.

related emissions. They contain a number of lines related to several ODMR signals. As in the case of GaNP, the revealed ODMR signals have a negative sign, which evidences that the involved defects act as efficient recombination channels competing with the radiative recombination. All ODMR signals were suppressed in the hydrogenated structures.

Spin Hamiltonian parameters of the defects were obtained on the basis of an analysis of the measured ODMR spectra by using the spin Hamiltonian given in Eq. 3.1 and are summarized in Table 3.2. The spectra were found to contain a single-line signal (denoted as G1 in Fig. 3.9) from a paramagnetic center (**S** = 1/2) of unknown origin, as well as multiline signals from several Ga_i-related defects. The latter were identified on the basis of the resolved hyperfine structure and are denoted as Ga_i–A, Ga_i–B, and Ga_i–C (Wang et al., 2009b; Vorona et al., 2006). We underline that the same names were given for the Ga_i-related interstitial complexes in GaNP and GaNAs alloys due to the similarity in their hyperfine interaction strengths. The exact local surrounding of the defects, however, may differ between these materials.

From Fig. 3.9 one can conclude that the effects of hydrogen incorporation in GaNAs are distinctly different from that observed in the GaNP alloys. First of all, we notice that the main effect of hydrogenation here is dramatic quenching of the Ga_i-related ODMR signals, which implies efficient passivation of the Ga_i-related defects by H. This behavior is more "traditional," as hydrogen is known to effectively passivate various point defects in semiconductor materials. On the other hand, defect activation, if any, could only be observed for the GaNAs alloy with [N] = 1.3%—see Fig. 3.9, where a relative contribution of Ga_i–C in the measured ODMR spectrum increases in the hydrogenated material. However, since this increase is accompanied by an overall strong quenching of the ODMR intensity, it is hard to reliably conclude whether Ga_i–C was indeed activated by H or it just becomes more visible due to the stronger quenching of the Gai–A, Gai–B signals after H treatment. To resolve this issue a separate study of as-grown samples has been conducted at a higher MW frequency. The results (displayed in Fig. 3.11) show that Ga_i–C and Ga_i–A have a nearly equal ODMR signal intensity prior to hydrogenation. Thus, no activation of Ga interstitials in GaNAs is observed. In the case when Ga_i–C was readily formed in the as-grown sample its intensity is found to be reduced by hydrogenation (Fig. 3.9d).

3.4.2 *DLTS*

DLTS is a powerful technique that is conventionally used to gain information about the electrical signatures of defects such as activation energy (E_a), capture cross section (σ_n), and the defect

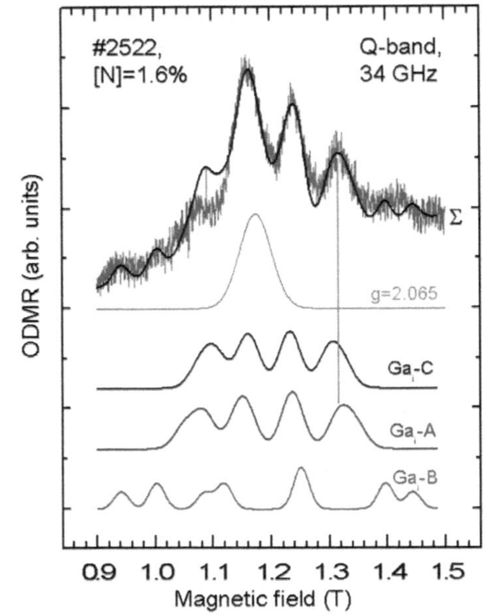

Figure 3.11 ODMR spectra (top) obtained from $GaN_{0.016}As_{0.984}$ at 5 K and 34 GHz, together with the simulated ODMR lines (bottom).

concentration (N_t). In a recent DLTS study, Shafi et al. characterized the effect of postgrowth hydrogenation on deep levels in GaNAs alloys with the N content up to 0.8%. A summary of the observed defect/impurity levels along with their concentration and activation energy is presented in Table 3.3. Most levels coincide with those found in untreated GaAs. Some levels are activated by hydrogen incorporation (indicated by underlined concentration values in Table 3.3), while others are passivated (indicated by "P" in Table 3.3). Below we present a brief summary of results discussed in Ref. (Shafi et al., 2011).

The deep trap with $E_a = 0.80 \pm 0.04$ eV was observed in both GaAs and GaNAs epilayers and was attributed to the defect center EL2, which is a double donor with a near midgap energy level that involves an As antisite (As_{Ga}). The trap concentration was found to decrease with increasing N content but was enhanced by postgrowth hydrogenation for the sample with $[N] \geq 0.4\%$.

Table 3.3 Trap density and activation energy (E_a) as obtained from standard DLTS and Laplace DLTS data (Ref. (Shafi et al., 2011)) from as-grown and hydrogenated GaNAs epilayers. Level density "P" indicates that the defect level was passivated below the detection limit by postgrowth hydrogenation. Bold, underlined densities indicate a substantial *activation* of traps upon hydrogenation

Level (model)	E_a (eV)	Level density, N_t (10^{14} cm^{-3})						
		[N]=0%	0.2%		0.4%		0.8%	
		As gr.	As gr.	H	As gr.	H	As gr.	H
EL2 (As_{Ga})	0.80±0.04	29	7.6	3.5 P	5.1	136		**6.4**
EL3 (O_{As})	0.52±0.04		2.1	630	3.5	P		
EL5	0.37±0.09				2.4	P	35	9.1 P
EL6 (NN_x)	0.34±0.07		22	P			1.9	P
EN2, M3	0.59±0.05						5.8	**26**
A3(NN_i)	0.28±0.02		1.7	P				-
EN3 (NN_x)	0.44±0.09		1.5	P			7.0	P

The trap with the energy level at ~0.52 eV below the conduction band edge was observed in $GaN_{0.002}As_{0.998}$, and its concentration increased with N content. Shafi et al. attributed this level to the EL3 trap previously observed in GaAs and ascribed to a substitutional oxygen atom on an arsenic site (O_{As}). The effect of hydrogenation on this trap is not clear: its concentration increased by 2 orders of magnitude in the [N] = 0.2% sample, while it disappeared upon hydrogenation in the sample with [N] = 0.4%.

The levels EL6, A3, and EN3 were tentatively associated by the authors with multi-N complexes responsible for the PL lines, which are observed in very diluted GaNAs and can be quenched by hydrogenation. These levels were completely passivated upon hydrogenation.

The concentration of the EL5 trap goes to zero or sizably decreases upon hydrogenation.

Finally, the concentration of the EN2 trap increases upon hydrogenation. This trap has similarities with the metastable hydrogen-related M3/M4 defects in GaAs.

3.5 Conclusions

In this chapter we have analyzed the effects of postgrowth H treatment on defects in both GaNP and GaNAs. The results of ODMR studies presented in Sections 3.3 and 3.4 clearly show that H has markedly different effects on the Ga_i-related defects in GaNP and GaNAs alloys. H incorporation in GaNP is found to lead to *activation* of new Ga_i-related defects due to the formation of Ga_i–H complexes, whereas in GaNAs hydrogen is found to efficiently *passivate* Ga_i-related defects present in the as-grown material. Though the exact physical mechanism for this difference is currently not fully understood and requires future theoretical studies, we may speculate that a possible reason could be a difference in the band-gap energies of these materials. It is possible that in both materials H interacts similarly with Ga_i complexes, but the energy-level positions of the formed defects relative to the band edges are significantly different. For example, because of the smaller band gap of GaNAs as compared with that of GaNP, the defect energy level of the formed complex in GaNAs may lie close to either the valence or the conduction band edge (or even outside the band gap). This would strongly reduce the role of the corresponding center in carrier recombination, making it undetectable by the ODMR technique. Further efforts are needed to shed light on these questions.

DLTS results show that H incorporation in GaNAs efficiently quenches several defects (EL5, EL6, A3, and EN3). Some of these defects were attributed to multi-N complexes that can be efficiently passivated due to bonding between nitrogen and hydrogen. Other defects, such as EL2 and EL3, exhibit more puzzling behavior as the defects can either be activated or passivated by H, depending on the N content. This anomalous behavior is not observed in the ODMR studies. It is currently not understood and requires further investigation.

Chapter 4

Theory of Hydrogen in Dilute Nitrides

Aldo Amore Bonapasta and Francesco Filippone
CNR-ISM, via Salaria Km 29.300, C.P. 10, 00015 Monterotondo Stazione (RM), Italy
aldo.amore@ism.cnr.it

At variance with the mechanisms responsible for the passivation of typical shallow dopants in GaAs, where H bonding with the impurity or to its neighbors permits a recovery of the "natural" chemical valence of the impurity itself, in the case of the N isoelectronic impurity in GaAs, H induces no changes in the N chemical valence. It breaks, instead, two Ga–N bonds to form complexes sharing a common, quite stable NH_2 core characterized by a local C_{2v}-like symmetry, thus strongly perturbing the interactions of N with its four Ga neighbors. A theoretical picture is presented here, where the N effects on the GaAs electronic properties, as well as their passivation upon hydrogenation, are closely related to the local N–Ga interactions and to the perturbations induced by H bonding on the N environment. The formation of multi-H complexes with N, predicted by theory, is also discussed together with the support given to their existence by recent experimental findings.

Hydrogenated Dilute Nitride Semiconductors: Theory, Properties, and Applications
Edited by Gianluca Ciatto

ISBN 978-981-4463-45-4 (Hardcover), 978-981-4463-46-1 (eBook)
www.panstanford.com

4.1 Introduction

Many experimental and theoretical studies have been dedicated to the investigation of the properties of atomic hydrogen introduced in semiconductors. Such a great effort is motivated by two peculiar features of hydrogen. First, atomic H presents intrinsic, fascinating properties: it is the smallest atom in nature, and it is also characterized by a highly reactive character and a basically amphoteric behavior, as shown by the existence, in different chemical environments, of both the H^+ and H^- ionic species, in addition to the neutral H^0 form (Cotton and Wilkinson, 1988). Second, hydrogen has significant effects on semiconductor properties:

(i) It generally behaves as an interstitial and amphoteric impurity that can compensate both acceptor and donor shallow impurities as well as deep defects.
(ii) It diffuses in the crystal lattice and binds to impurities or to their neighboring atoms by strongly perturbing their chemical environment. This leads to the formation of H-impurity complexes in the semiconductor matrix as well as to a full electronic passivation of the impurity itself. In other words, H wipes out from the forbidden energy gap the impurity induced electronic levels (Pankove and Johnson, 1991).
(iii) Theoretical studies have suggested the existence of an H electronic level in the band structure of a semiconductor, which represents a sort of universal reference level, that is, it has a constant position with respect to the vacuum level, independent of the host semiconductor (Van de Walle and Neugebauer, 2003; Kilic and Zunger, 2002). This implies that, in addition to its passivation effects, H can induce a level resonant in the conduction band (CB) in materials having a suitably small energy gap (e.g., InN), thus realizing an *n*-type doping of the host semiconductor (Van de Walle et al., 2010).

In the last two decades, many studies have been also dedicated to the investigation of the properties of nitrogen and its complexes with hydrogen in III–V–N dilute alloys, like in $GaAs_yN_{1-y}$, GaP_yN_{1-y}, and $In_xGa_{1-x}As_yN_{1-y}$, where the group V element is partially replaced by N (Henini, 2005; Trotta et al., 2012b). The interest in these materials stems from several surprising phenomena induced by N atoms in III–V–N dilute alloys as well as by the parallel, surprising effects produced by the introduction of atomic hydrogen. As already shown in details in Chapter 1, in III–V–N dilute alloys, small percentages of N (1–4%) induce indeed a striking reduction of the energy gap (E_g) of the host III–V semiconductor, much stronger than a linear dependence. Moreover, the electron effective mass at the CB edge attains abnormally high values. Finally, a tensile strain, that is, a reduction of the lattice constant of the host semiconductor, is also induced by N. This last seems the only N effect compatible with the virtual crystal approximation (VCA). On the hydrogen side, largely addressed in Chapter 2, it was discovered that H incorporation strongly modifies the band gap of III–V–N alloys; as an example, in $GaAs_yN_{1-y}$, it leads to the recovery of the E_g value of the N-free host. Such a remarkable effect on the band gap is accompanied by a H-induced variation of the electron effective mass, which clearly indicates that the entire band structure of the semiconductor is profoundly affected by H. Furthermore, even the structural characteristics of H-irradiated $GaAs_yN_{1-y}$ undergo major modifications. A reversal of strain from tensile to compressive and an effective re-establishment of local ordering, as found by X-ray diffraction, X-ray absorption fine structure measurements (see Chapters 6 and 7), and optical emission spectra, were indeed observed in hydrogenated GaAsN samples (Ciatto et al., 2009; Berti et al., 2007; Polimeni et al., 2003b). The N neutralization (or passivation) by H is rather surprising because, in principle, the simple models generally used to explain the H passivation of shallow impurities seem to exclude the occurrence of H effects on isoelectronic impurities, like N. In this regard, further theoretical investigations have clarified the origin of N passivation by H in $GaAs_yN_{1-y}$ by predicting also some novel effects, like the formation of multi-H complexes resulting from the N–H interaction (Amore Bonapasta et al., 2007). The existence of such complexes has been recently supported by very accurate infrared (IR) spectroscopy investigations (see Chapter 5) (Wen et al., 2012).

Here, we try to put the basis for a sound theoretical framework, which should clarify the relationships between the general properties of H in semiconductors, the properties of N–H complexes in $GaAs_yN_{1-y}$ alloys, and the H effects on the properties of such semiconductors.

In this regard, after a description of the theoretical methods given in Section 4.2, we reconsider some general properties of H in III–V semiconductors in Section 4.3, while in Section 4.4 we briefly review experimental and theoretical results on the properties of N and N–H complexes. In Sections 4.5 and 4.6 we discuss some features of N–H complexes and the mechanisms of N passivation, respectively. Finally, we draw conclusions in the last section.

4.2 Theoretical Methods

In general, the results reported here, like total energies or atomic geometries, have been achieved by using density functional theory–local density approximation (DFT-LDA) methods (D. Marx and J. Hutter, 2000). The chemical bonding in the N–H complexes has been investigated by analyzing atomic distances and atomic displacements, as well as the distribution of the total (valence) electronic charge density. For more details, see below.

The formation energy of N–nH complexes, useful in determining the stability of a given complex configuration, has been estimated by analogy with the case of defects in a semiconductor (Amore Bonapasta and Giannozzi, 2000). Thus, in a supercell approach, the formation energy per H atom $\Omega[\text{N–}n\text{H}]^q$ of a N–nH complex involving nH atoms and a charge q can be written as

$$\Omega[\text{N–}n\text{H}]^q = \frac{1}{n}\left[E[\text{N–}n\text{H}]^q - E[\text{N}] - n\mu_\text{H} + q(E_\text{F} + E_\text{V})\right]$$

where $E[\text{N–}n\text{H}]^q$ is the total energy of the simulation supercell (e.g., a 64-atom GaAs supercell) containing one N atom and n H atoms; μ_H is the chemical potential of hydrogen, which is generally assumed equal to half of the energy of the H_2 molecule in the vacuum; and q is the positive (negative) charge on the complex, namely, the number of electrons transferred from (to) the complex to (from) an electron reservoir with a chemical potential (or Fermi level) E_F referred to the top of the valence band (VB), E_V (i.e., $E_\text{F} = 0$ corresponds to a Fermi level at the top of the VB).

The above expression of the formation energy allows the comparison of the energies of different charge states of a defect at different values of the Fermi energy and, therefore, to identify the charge state lowest in energy, (i.e., the stable form of a defect) at a given value of the Fermi energy. That expression permits also to estimate the transition energy of a defect electronic state, that is, the Fermi energy value $\epsilon^{n/n'}$ for which the occupation number of the defect state in the energy gap changes from n to n'. Such a value is given indeed by the Fermi energy position where the formation energies of the n and n' charge states of the defect are equal (Van de Walle et al., 2001). The $\epsilon^{n/n'}$ level is related to the position of an electronic state induced by a defect in the energy gap, as observed experimentally, and may be used to determine whether and how the defect behaves as acceptor or donor (Mattioli et al., 2010). In the $GaAs_{0.97}N_{0.03}$ alloy, (i.e., the alloy simulated by 64-atom supercells containing one N atom), substitutional N atoms and N–H complexes have been considered as isolated defects and the corresponding formation energies and $\epsilon^{n/n'}$ values have been calculated.

The $\epsilon^{n/n'}$ values should be compared with the GaAs energy gap, which is generally underestimated in LDA. In some cases, $\epsilon^{n/n'}$ values have been compared with the experimental band gap. Here, we report results where the energy gap has been estimated by calculating the $\epsilon^{0/-}$ value in the N-free compound material. Moreover, this value has been checked by calculating the $\epsilon^{+/0}$ value for the Si shallow donor (i.e., a substitutional Si atom at a Ga site, Si_{Ga}). This shallow donor induces a level close to the bottom of the GaAs CB: the corresponding $\epsilon^{+/0}$ value provides therefore a reference to locate the N and N–H levels in the energy gap and allows the comparison of consistent results obtained within a same approach. In the present case, transition energies play a significant role. They can be used indeed to determine the position of the electronic levels induced by N and N–H complexes in the energy gap and, therefore, to follow the evolution of the band gap of a host material induced by the incorporation of N and the subsequent formation of N–H complexes.

In theoretical simulations reported hereafter, total energies have been calculated by using periodic boundary conditions and supercell simulations, separable ab initio pseudopotentials

(Gonze et al., 1991), plane wave basis sets, the special-points technique for **k**-space integration, and the exchange–correlation functional of Ceperley–Alder as implemented in the Quantum-Espresso suite of programs (Giannozzi et al., 2009). Ultrasoft pseudopotentials have been used in the case of nitrogen (Vanderbilt, 1990). Geometry optimizations have been performed by fully relaxing the positions of all the atoms of a supercell by minimizing the atomic forces. The results concerning the geometries, the formation energies, and the transition energies of the N–H complexes have been achieved by using 64-atom supercells, the (4,4,4) **k**-point Monkhorst–Pack mesh, and cutoffs of 22 Ry.

Vibrational frequency values have been calculated by performing a full vibrational analysis via Lanczos diagonalization of the Hessian matrix, (Filippone and Parrinello, 2001) exploiting the variational formalism of density functional perturbation theory (Putrino et al., 2000). This procedure has required further geometry optimizations of the N–H complexes. In detail, these calculations have been performed in LDA by using the exchange–correlation functional through Padé approximation (Goedecker et al., 1996), as implemented in the CPMD code (CPMD, 2012). Separable ab initio relativistic pseudopotentials of the Goedecker–Hutter–Hartwigsen type have been used for Ga and As (Hartwigsen et al., 1998), a Von Barth–Car type pseudopotential has been used for H, and a norm conserving the Martins–Troullier pseudopotential has been used in the case of N (Troullier and Martins, 1991). In this case, the results have been achieved by using 64-atom supercells, energy, and frequency values at the Γ point and cutoff of 50 Ry in plane wave basis sets. No appreciable differences have been found between the geometries of the N–H complexes calculated by the two different computer codes used.

Given the limits of LDA in the evaluation of semiconductor energy gaps mentioned above, some notes on recent developments of DFT regarding in particular the use of hybrid functionals are worth. In fact, in theoretical investigations of the H effects on the properties of III–V–N alloys, as it will be clarified in the following sections, hybrid functionals can give useful indications on the H passivation of the N effects.

DFT grounds on the Hohenberg–Kohn theorems (Hohenberg and Kohn, 1964), which legitimate the use of electronic density as the basic variable for describing atomic, molecular, and solid systems. The energy can be written as a functional of density itself, but the exact form of the kinetic ($T[\rho]$) and electron–electron ($V_{\text{ee}}[\rho]$) interaction terms is unknown. To circumvent this problem it is possible to adopt the Kohn–Sham approach; the N electron density is decomposed into N orbitals describing noninteracting electrons. In this way we can compute accurately the kinetic energy term, $T_{\text{s}}[\rho]$, which misses electron correlation, and the Coulomb term, $J[\rho]$, which misses electronic exchange.

What is left out of these two functionals can be put into the exchange and correlation functional, defined as follows:

$$E_{\text{xc}}[\rho] = T[\rho] - T_{\text{s}}[\rho] + V_{\text{ee}}[\rho] - J[\rho].$$

Again, we do not know its exact form. Nevertheless, it can be calculated accurately for a homogeneous electron gas, giving rise to the widely famous LDA, where the functional is dependent only on the local value of the density, $E_{\text{xc}}^{\text{LDA}}[\rho]$. This approximation works rather well in describing several features of atomic, molecular, and solid systems, such as bond geometries, and may be improved by adding a dependence on the gradient of the density itself, $E_{\text{xc}}^{\text{GGA}}[\rho, \nabla\rho]$ (the so called generalized gradient approximation [GGA]).

Between the major failures in LDA/GGA-DFT we have, for instance, underestimation of chemical reaction barriers and energy band gaps or overestimation of binding energies in charge transfer complexes. They owe to a common root, the delocalization error typical of approximate functionals, due to the dominant Coulomb term (Cohen et al., 2008). Since the functional is formally exact, we stress that the errors come only from our inability to describe its correct form.

A useful description of the delocalization error can be obtained if we consider the behavior of the energy functional with respect to fractional charges. Electrons have a discrete nature; therefore the exact energy functional follows the number of electrons as a series of straight lines. The corresponding LDA curve, is, instead, convex between integers (see Fig. 4.1). Since the derivative of the energy

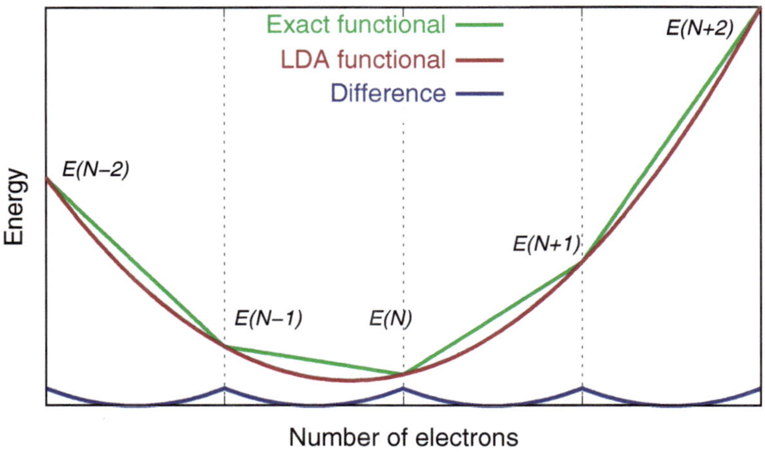

Figure 4.1 Schematic behavior of E vs. N in the DFT exact formulation (green) and LDA formulation (red), with (blue) their difference.

with respect to the occupation numbers gives electron affinity or ionization potentials, it is easy to draw the connections with LDA failures. One possibility to correct such misbehavior is the addition of a quadratic term that forces a discrete occupation of the orbitals; this procedure is known as the Hubbard U correction, (Anisimov et al., 1997; Cococcioni and de Gironcoli, 2005; Kulik et al., 2006) and is particularly suitable for describing electronic and magnetic properties of transition metal impurities (e.g., Fe or Mn) in GaN or GaAs host matrices (Amore Bonapasta et al., 2005; Filippone et al., 2011).

A different route to correct such errors is represented by the nowadays widely used hybrid functionals. In such functionals a part of Hartree–Fock (HF) exchange is added to the LDA/GGA exchange-correlation functional

$$E_{\mathrm{xc}}^{\mathrm{hyb}}[\rho] = C_{\mathrm{c}}^{\mathrm{LDA}} E_{c}^{\mathrm{LDA}}[\rho] + C_{\mathrm{x}}^{\mathrm{LDA}} E_{\mathrm{x}}^{\mathrm{LDA}}[\rho] + C_{\mathrm{x}}^{\mathrm{HF}} E_{\mathrm{x}}^{\mathrm{HF}}[\rho]$$

In this case, since the HF behavior of E *vs.* N is concave, the application of HF exchange results in error cancellation against LDA. There are several recipes for hybrid functionals, depending on the LDA/GGA functional used as a basis and on the coefficients used in the linear combination of components. One of the most used is the

Heyd–Scuseria–Ernzerhof (HSE) functional (see Ref. (Janesko et al., 2009) and references therein), particularly suited for solid-state simulations, since it splits the exchange operator into short-range and long-range parts. Hybrid functionals have been successfully used in the study of III–V semiconductors (Virkkala et al., 2012; Janotti et al., 2012) and are the basis of some of the results presented hereafter. The most important handicap coming with hybrid functionals is the rather heavy computational cost (especially for computer codes based on plane waves and reciprocal space integrations), given the need to integrate the exchange operator in real space for every **k**-point in the mesh; in solid-state calculations this part can become a very important bottleneck for the whole simulation.

4.3 General Properties of H in III–V Semiconductors

As mentioned above, the general chemical properties of atomic hydrogen indicate that it can exist in the H^+, H^0, and H^- forms. Then, preliminarily, we can try to speculate about the forms we expect to observe for an interstitial atomic H introduced in a III–V lattice. As an example, let us consider the GaAs crystal lattice. From a theoretical point of view, GaAs is described as a perfect zinc-blende lattice (see Fig. 4.2) where Ga and As atoms form polarized, covalent chemical bonds. With this chemical bonding, the electronic

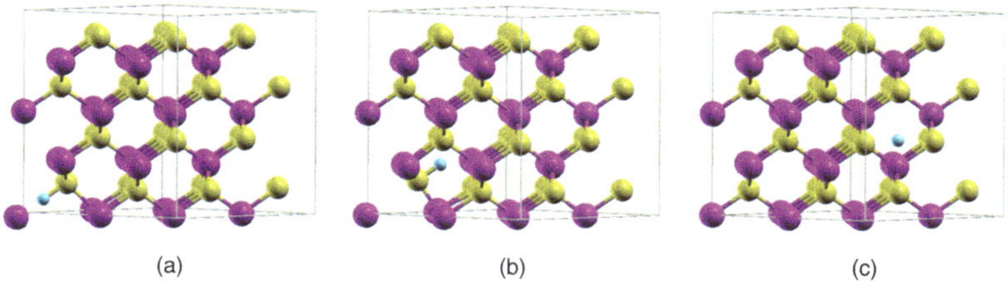

Figure 4.2 Relevant interstitial sites for a H atom in the GaAs lattice. The BC (bond-centered), AB–As (antibonding [AB] site on the As side), and Td(Ga) (tetrahedral site in a cage with nearest-neighboring Ga atoms) sites are shown in (a), (b), and (c), respectively. Ga, As, and H atoms are represented by magenta, yellow, and light blue spheres, respectively.

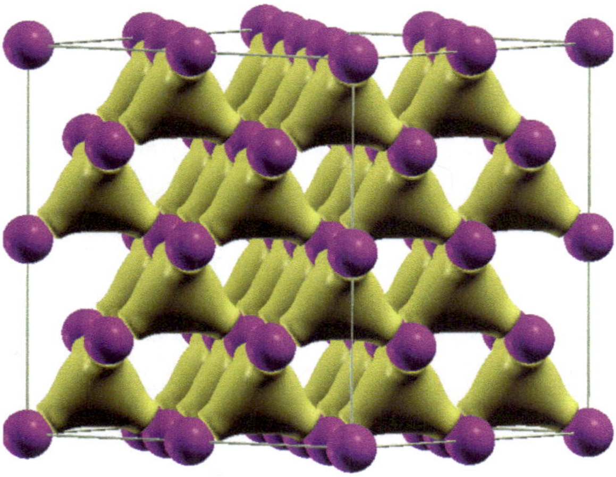

Figure 4.3 Isosurface of the valence charge density in GaAs. Ga atoms are represented by magenta spheres. An accumulation of the electronic charge around the As atoms hides these atoms, which are located at the centers of the isosurface tetrahedra. The isosurface is sampled at a charge density value of 0.04 $e/(au)^3$.

charge accumulates along and around the Ga–As bond axes, with a progressive increase of its density on going from the Ga to the As site (Amore Bonapasta and Pavesi, 1996). As shown in Fig. 4.3, high-charge-density regions, along the Ga–As bonds, coexist with low-density regions, corresponding to the cages surrounding tetrahedral interstitial sites, like the Td(Ga) site, which has Ga atoms as nearest-neighboring atoms (see Fig. 4.2), or the Td(As) sites.

Thus, the following simple considerations can be attempted: i) A highly reactive species like H^0 will hardly exist as an isolated interstitial species in the GaAs lattice. ii) The H^+ ion could reach a stable form when located close to the high-density regions along the Ga–As bonds. iii) The H^- ion, where H has a He-like electronic configuration, could be stable when located at a Td(Ga), that is, at the center of a low-density region where it is surrounded by slightly positive Ga neighbors. Despite its simplicity, the above qualitative picture well describes the essence of the properties of H in GaAs and, in general, of H in IV and III–V semiconductors, as indicated by pioneering DFT investigations performed, among the others, by Van

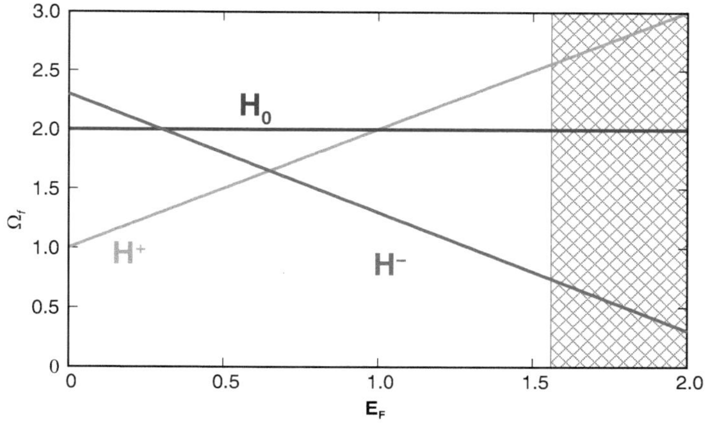

Figure 4.4 Formation energy (Ω_f) of different charge states of H in GaAs as a function of the Fermi energy (E_F). The zero of the Fermi energy is the top of the VB. All values are given in eV.

de Walle (Van de Walle et al., 1989; Pankove and Johnson, 1991) and Pavesi and Giannozzi (Pavesi and Giannozzi, 1992). In GaAs, in fact, first, these theoretical studies indicated that H^+ has a stable location at the bond-centered (BC) site (Fig. 4.2a,[a]) H^- has a stable location at the Td(Ga) site (Fig. 4.2c), and H^0 has a stable location at the As AB site (AB-As), (Fig. 4.2b), that is, a site aligned to the Ga–As bond axis, out of the bond and close to the As atom. Second, formation energies calculated as a function of the Fermi energy E_F clarified the conditions permitting the existence of each of such three H forms (see Fig. 4.4): H^+ forms when the E_F is lower than the midgap (as in the case of *p*-type doped GaAs), H^- forms for higher values of E_F (*n*-type doped GaAs), while H^0 never forms.

Figure 4.4 also permits us to estimate the $\epsilon^{+/-}$ transition level, which is given by the E_F value corresponding to the crossing point between the H^+ and H^- formation energy curves, at about 0.6 eV from the top of the VB to be compared with a GaAs E_g estimated equal to 1.56 eV. These results indicate that H can compensate

[a]in GaAs the BC site is displaced toward the As atom. This site is referred to as a BC site because it was first introduced for H in silicon, where it is at the center of a Si–Si bond.

shallow or deep acceptors (A) having $\epsilon^{0/-}$ levels lower in energy than the $\epsilon^{+/-}$ of H, giving rise to H^+ and A^- pairs. Similarly, H can compensate shallow or deep donors (D) having $\epsilon^{+/0}$ levels higher in energy than the H $\epsilon^{+/-}$ level, giving rise to H^- and D^+ pairs. In a slightly different perspective, it can be considered that a neutral H located at the AB–As site induces a singly occupied electronic level in the energy gap, as shown by calculated electronic eigenvalues (not reported here). This level has a donor character since, at the electron loss, H forms a H^+ ion located at the BC site, where it can be efficiently screened by the electronic cloud of the Ga–As bond. A measure of the strength of this donor character is given by the position of the hydrogen $\epsilon^{+/0}$ transition level in Fig. 4.4. Similar considerations can be done for the acceptor character of a neutral H atom at the Td(Ga) site. Thus, H is an amphoteric impurity that presents a very peculiar feature: its acceptor/donor character depends on its low- or high-charge-density interstitial location. In a perfectly intrinsic GaAs, H atoms can form H^+ and H^- pairs together with H_2 molecules (located at Td interstitial sites).

The value of the hydrogen $\epsilon^{+/-}$ transition level, together with the qualitative picture we have just given, fully agrees with the H universal level proposed in Ref. (Van de Walle and Neugebauer, 2003), where the $\epsilon^{+/-}$ level is independent of the host material. Such a universal location is directly related, indeed, to the relative energies of the H^+ and H^- species embedded in the high- and low-charge-density regions, respectively, of a semiconductor. These regions show almost the same features through different covalent semiconductors (including IV, III–V, and II–VI semiconductors), thus justifying a H behavior almost independent of the atomic species forming the host crystal lattice.

Coming back to the H passivation effects, in the presence of a shallow impurity an isolated H atom can assume an ionic form to compensate the impurity, thus leading to the formation of H^+-A^- or H^--D^+ pairs. In turn, H ions can easily diffuse in a crystal lattice, as indicated also by a pioneering theoretical work for H^+ in crystalline silicon (Buda et al., 1989). Diffusing H ions undergo the Coulomb attraction exerted by the compensated impurity, bind to the impurity, and passivate its effects. Several experimental studies have supported such a theoretical picture, confirming the

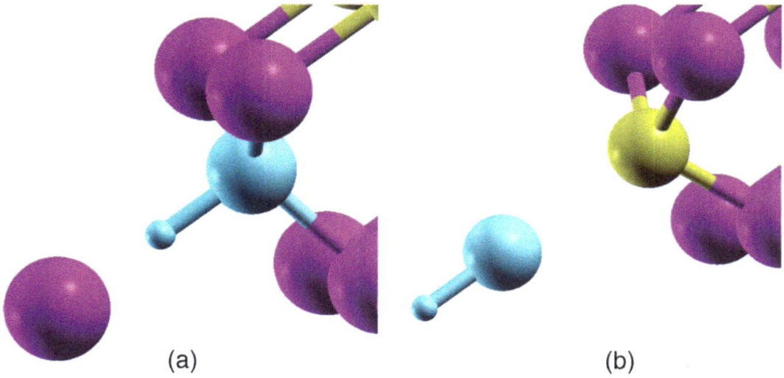

Figure 4.5 Atomic configurations of some H–Si complexes in GaAs. (a) A Si_{As}–H_{BC}(–Ga) complex, formed by H with a Si acceptor and (b) an (As)–Si_{Ga}–H_{AB} complex, formed by H with a Si donor. Ga, As, and Si atoms are represented by magenta, yellow, and light blue spheres, respectively. The H atom is represented by a small, light blue sphere.

passivation by hydrogen of shallow impurities and the formation of H–impurity complexes (Pankove and Johnson, 1991). In particular, IR spectroscopy measurements and theoretical calculations have given information on the features of the H bonding as well as on the configurations of H–impurity complexes. The properties of complexes formed by H with shallow impurities can be exemplified in the cases of Si and C dopants in GaAs.

A substitutional Si in the anionic site of GaAs (Si_{As}) behaves as a shallow acceptor. H compensates the Si_{As}, diffuses as a H^{+} ion, and binds to the ionized acceptor by forming a "on-line" Si_{As}–H_{BC}(–Ga) neutral complex (see Fig. 4.5a). In this complex, H forms a strong bond with the Si atom and weakly interacts with the Ga atom. This induces a strong perturbation of the Si properties, which can be schematized in terms of a H atom saturating a silicon dangling bond (DB). Such an atomic arrangement implies that the Si and Ga atoms reach their "natural" valence of four and three, respectively, thus accounting for a full neutralization of the Si acceptor behavior, since the acceptor level is wiped out from the energy gap (Pavesi and Giannozzi, 1991). It can be also noted that H at a BC site corresponds to the stable configuration for an ionized donor H.

Similarly, an (As)–Si_{Ga}–H_{AB} on-line complex (see Fig. 4.5b) forms in the case of a Si_{Ga} donor by involving a H acceptor. In this complex, the As atom has a weak interaction with the Si one; thus it seems once more possible to apply the "natural" valence criterion: the As reaches its valence of five and the Si its valence of four, thus accounting for the passivation of the Si donor. The H located at the AB site points toward a low-density region by recalling the configuration of an ionized acceptor H at the Td(Ga) site.

The above pictorial models effectively explain the H passivation of different shallow dopants in III–V semiconductors (Pankove and Johnson, 1991), although they do not fully describe the features of H–dopant interactions in these materials. For instance, in the case of the C shallow dopant in GaAs, theoretical results indicate both similarities and differences with respect to the Si dopant (Amore Bonapasta, 1991). When we have a C_{As} shallow acceptor, H forms a C_{As}–H_{BC}(–Ga) complex similar to that formed with the Si acceptor. Once more, H passivates C_{As}, and this effect can be explained again by a simple model where a C_{As} DB pointing toward a Ga neighbor is *saturated* by a H atom. This leads to the structure of the C_{As}–H_{BC}(–Ga) complex shown in Fig. 4.6a, where a strong C–H bond is formed by a H located close to a BC site and the C and Ga chemical valences (four and three, respectively) are fully

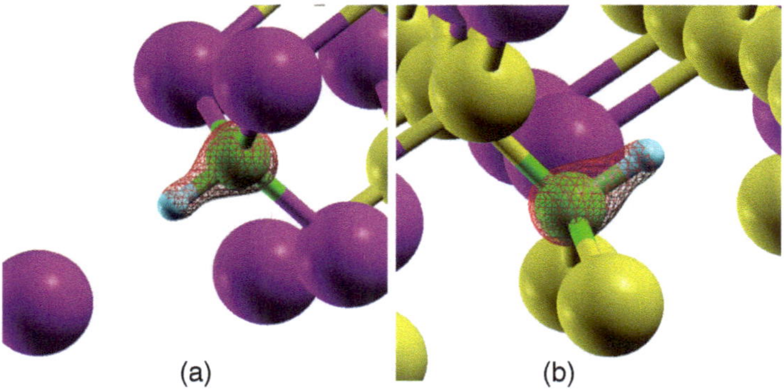

Figure 4.6 Atomic configurations and isosurfaces of valence charge density for some H–C complexes in GaAs. (a) A C_{As}–H_{BC}(–Ga) complex, formed by H with a C acceptor and (b) a C_{Ga}–H_{BC}–(As) complex, formed by H with a C donor. Ga, As, C, and H atoms are represented by magenta, yellow, green, and light blue spheres, respectively. The isosurfaces are sampled at a charge density value of 0.2 $e/(au)^3$.

satisfied. On the contrary, in the case of the C_{Ga} shallow donor, theory indicates the formation of a C_{Ga}–H_{BC}(–As) complex (see Fig. 4.6b) at variance with the Si donor case. Different interactions of H with Si and C come from different chemical properties of the two dopants. First, C is more electronegatve than Si, so C–H bonds are significantly stronger than Si–H bonds. Second, the C atom is smaller in size than Si; therefore, it is possible to insert a H atom at the BC site of a C_{Ga}–As bond and to form a strong C–H bond together with a negligible H–As interaction, thus satisfying the "natural" valence criterion outlined above. Figures 4.6a and 4.6b show a significant accumulation of electronic charge around the C–H pair in both the C_{Ga}–H and C_{As}–H complexes, respectively, which clearly indicates that a strong C–H bonding dominates the properties of both C–H complexes in GaAs. These features of the C–H interaction will be recalled when discussing the interaction of H with the N impurity in GaAsN alloys.

4.4 N Impurity and H–N Complexes in Dilute III–V Semiconductors

In this section we mostly discuss the GaAsN alloy, seen as a representative of the other III–V–N alloys, and briefly review the experimental and theoretical results about N and N–H complexes.

As anticipated in Chapter 1, the remarkable band-gap bowing experimentally observed in III–V–N dilute alloys like InGaAsN has been first described by using the "band anticrossing" (BAC) model (Shan et al., 1999a). We remind here that the main assumption behind the BAC model originates from the strong interaction between a N-induced localized resonant state above the CB edge of the host material and the CB itself, which pushes the CB edge lower and causes a band-gap reduction. Such a picture can be complicated by the presence of random nitrogen clusters, which can form a spectrum of N-related states, as shown in large-supercell calculations (Kent et al., 2002; Lindsay and O'Reilly, 2004b).

The effect of nitrogen on III–V nitrides has been also successfully described by tight-binding (TB) calculations (Chapter 1) (O'Reilly et al., 2002; O'Reilly and Lindsay, 2010).

A different explanation of the band-gap bowing induced by N has been provided by the empirical pseudopotential method (Kent et al., 2002; Kent and Zunger, 2001b). In contrast to the BAC model, the results of empirical pseudopotential calculations indicate that a modification of the bottom of the CB is due to nitrogen-induced mixing of the CB valley states.

Recently, the N effects on the CB edge of the host GaAs have been also investigated by using HSE hybrid functionals in a supercell approach (Virkkala et al., 2012). Hybrid functionals give band gaps near the experimental ones. In the case of N, it is expected therefore that this theoretical approach improves the estimate of the defect level positions with respect to the band edges given by LDA or GGA calculations. As a matter of fact, HSE results show, for a GaAsN alloy, the presence of a N-induced level resonant in the CB and a reduction of the GaAs energy gap in agreement with the experimental findings.

The structure and properties of N–H complexes in hydrogenated GaAsN have been the subject of a wealth of papers that we briefly review here. Such studies led to a somewhat puzzling picture and raised questions about the proposed theoretical models (Henini, 2005; Trotta et al., 2012b). One of the first models proposed by theory was the so-called N–H_2^*, where two H atoms are bonded to N and to a neighboring Ga atom, respectively, aligned to the N–Ga bond axis (Kim and Chang, 2002; Janotti et al., 2002b; Amore Bonapasta et al., 2002; Janotti et al., 2002a; Amore Bonapasta et al., 2003, 2004; Dixon et al., 1998). Such a complex was challenged by IR absorption measurements (described in detail in Chapter 5) supporting a dihydrogen complex with both the H atoms bonded to a single N atom (Jiang et al., 2004). Then, a so-called C_{2v} complex (i.e., an NH_2 complex with a configuration that locally reminds a C_{2v} symmetry) was also proposed by theoretical simulations (Amore Bonapasta et al., 2003). An improvement of such a complex concerned a small but remarkable difference in its local geometry, which assumed the canted or asymmetric form shown in Fig. 4.7d.[a] The canted C_{2v} complex was believed to account for the IR results (Du et al., 2005; Fowler et al., 2005), for X-ray absorption spectroscopy (XAS)

[a]Note that different groups have used a different nomenclature for this canted form: the asymmetric C_{2v} complex is described by a C_{1h} symmetry in Chapter 5.

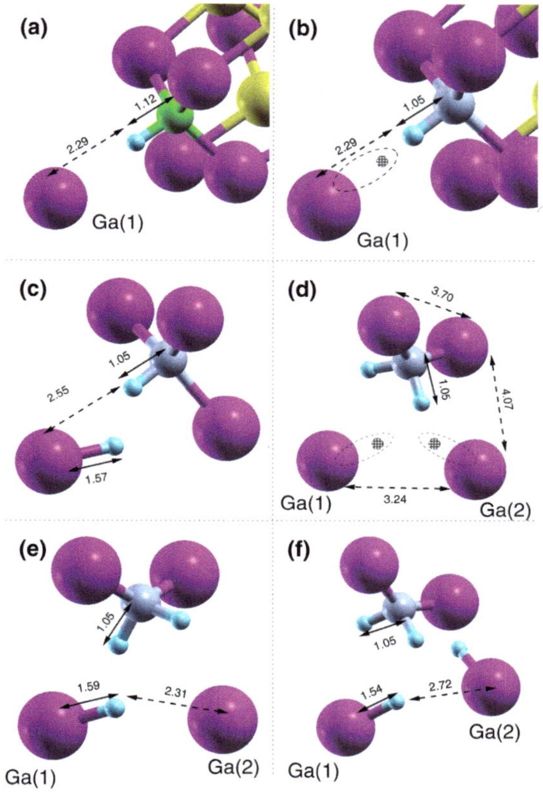

Figure 4.7 Atomic configurations of some N-H and C–H complexes. (a) C_{As}–H_{BC}–Ga(1) and (b) [N–H_{BC}] with a sketched DB on Ga(1), (c) [N–H_{BC}]–H_{Ga}, (d) asymmetric C_{2v} with sketched DBs on Ga(1) and Ga(2), (e) C_{2v}–H_{Ga}, and (f) C_{2v}–$2H_{Ga}$. BC indicates a bond-centered site of H. Ga, As, C, N, and H atoms are represented by magenta, yellow, green, light, and small light blue spheres, respectively. Atomic distances are given in Å.

results (Ciatto et al., 2005a), as well as for the recovery of both the GaAs energy gap (Amore Bonapasta et al., 2003) and lattice parameter (Polimeni et al., 2003b) (the results of different structural characterizations will be detailed in Chapters 5 and 6). However, a high-resolution X-ray diffraction (HRXRD) and photoluminescence (PL) spectroscopy study (described in Chapter 7) questioned again the theoretical models by showing that the lattice overshooting observed in the hydrogenated GaAsN is *related* to a combined effect of N and H, but *cannot be attributed* to the C_{2v} complex (Bisognin et al., 2006; Trotta et al., 2012b). In fact, this study demonstrated that moderate annealing is able to gradually reduce the lattice parameter overshooting and recover the GaAs unit cell without removing the C_{2v} complexes (full details on this study in Chapter 7). In conclusion, some other complex, different from C_{2v}, should be responsible for the observed overshooting.

At this point, on the grounds of first-principle DFT results, a novel multi-H model for hydrogen–nitrogen complexes was proposed, comprising one or two satellite H atoms next to the canted C_{2v} complex (Amore Bonapasta et al., 2007). In the following Sections 4.5 and 4.6, we will present the main features of this new model and explain in detail how it can fully account for all of the mentioned experiments. Many of the reported results are taken from reference (Amore Bonapasta et al., 2007). Some novel results regarding the H passivation of N will be also presented.

4.5 Energetics, Charge States, Formation Mechanisms, and Vibrational Properties of H–N Complexes

The essence of the multi-H model can be illustrated with a brief and schematic comparison of the local properties of the *isoelectronic* N_{As} with those of the C_{As} shallow acceptor in GaAs.

As discussed in Section 4.3, the C_{As} passivation can be explained by considering a H atom *saturating* a C DB pointed toward a Ga neighbor. This schematic picture of the H bonding is illustrated in Fig. 4.8-I, where a H^+ ion first induces the opening of the C–Ga bond formed by a C_{As}^{-1}-ionized acceptor and the formation of a C dangling bond carrying an electron pair and, then, saturates such a DB. This leads to the structure of the neutral C_{As}–$H_{BC} \cdots Ga(1)$ complex shown in Fig. 4.7a, where the "natural" valence criterion is fully satisfied (atomic distances and charge density distribution calculated for the C_{As}–$H_{BC} \cdots Ga(1)$ complex are reported in Figs. 4.7a and 4.9a, respectively, for an easier comparison with the corresponding quantities in N–H complexes).

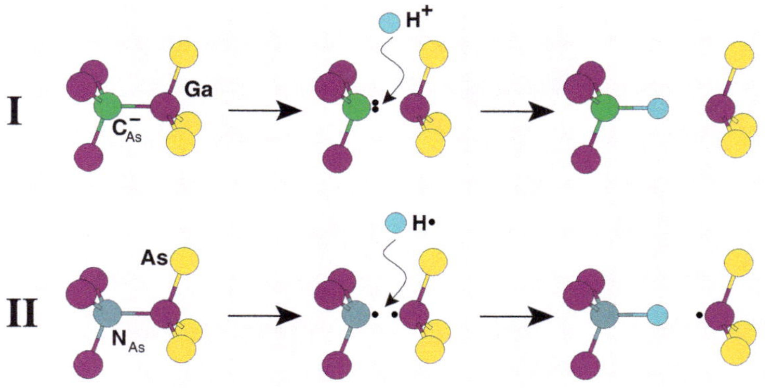

Figure 4.8 A simplified scheme for hydrogenation of impurities in GaAs. Ga, As, N, C, and H atoms are represented by magenta, yellow, light blue, green, and cyan spheres, respectively. In scheme I, a C_{As}–Ga bond opens at the arrival of H^+ ion; a fully occupied DB forms on the ionized C_{As}; such a DB is saturated by the H atom, and the "natural" valences of C_{As} and Ga atoms are satisfied. In scheme II, a similar mechanism is described for the N_{As} isoelectronic impurity; neither H nor N_{As} is ionized, and, therefore, the N_{As}–Ga bond opening leaves an unsaturated DB behind.

A thoroughly different situation occurs when a H atom binds to the N isoelectronic impurity, forming the N_{As}–$H_{BC} \cdots$ Ga(1) complex shown in Fig. 4.7b. In this case, in fact, H forms a strong H–N bond, akin to the just described H–C bond (compare distances and charge densities in Figs. 4.7a and 4.9a with the corresponding quantities in Figs. 4.7b and 4.9b). However, as schematized in Fig. 4.8-II, for an isoelectronic impurity like the N atom, the breaking of the N–Ga bond induced by a H atom leads to two DBs belonging to the N and the Ga atoms, respectively, carrying one electron each. Thus, for every N–H bond formed, an unsaturated DB on the Ga neighbor is left, in the resulting neutral complex, as sketched in Fig. 4.7b and 4.8-II. Similarly, two DBs are induced on the Ga atoms of the two broken Ga–N bonds in the canted C_{2v} complex (see Fig. 4.7d). Evidence of the existence of two unsaturated Ga DBs in the C_{2v} complex is given by the different geometry estimated for the C_{2v}^0 and C_{2v}^{+2} complexes. In the former case, the Ga(1)–Ga(2) distance in Fig. 4.7d, 3.24 Å, is significantly shorter than the host Ga–Ga

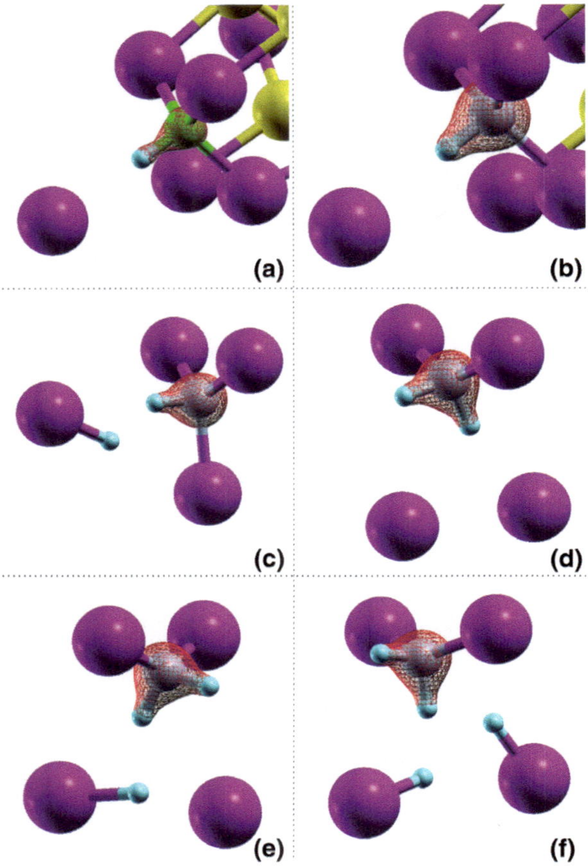

Figure 4.9 Isosurfaces of valence charge density of the N–H and C–H complexes shown in Fig. 4.7. (a) C_{As}–H_{BC}–Ga(1), (b) [N–H_{BC}], (c) [N–H_{BC}]–H_{Ga}, (d) asymmetric C_{2v}, (e) C_{2v}–H_{Ga}, and (f) C_{2v}–2H_{Ga}. BC indicates a bond-centered site of H. Ga, As, C, N, and H atoms are represented by magenta, yellow, green, light, and small, light blue spheres, respectively. The isosurfaces are sampled at a charge density value of 0.2 $e/(au)^3$.

distances (4.07 Å in Fig. 4.7d); this is an indication that Ga(1) and Ga(2) atoms interact to form a weak bond involving their DB electrons. On the contrary, in the C_{2v}^{+2} complex (not shown in the figure), the same Ga(1)–Ga(2) distance grows to 4.02 Å, since the two DB orbitals relative to the Ga atoms involved in the complex

Table 4.1 Complex formation energies per H atom, Ω_f, estimated at the values of the Fermi energy, E_F, 0 (i.e., the top of the VB), $E_g/2$, and E_g (1.56 eV). Ω_f of neutral complexes are given in the last row

	Complexes			
E_F	N–H^+_{BC}	C^{+1}_{2v}	C_{2v}–H^+_{Ga}	C_{2v}–$2H^+_{Ga}$
0	−0.40	0.01	−0.13	0.12
$E_g/2$	0.38	0.40	0.13	0.31
E_g	1.16	0.80	0.39	0.51
	0.27	0.31	0.12	0.04

have been discharged of their (formerly) unpaired electrons, and the interaction with each other is now hindered (Du et al., 2005; Amore Bonapasta et al., 2007). The effects of H bond breakage and formation have relevant consequences, since the H-induced DBs can be saturated by further H atoms, leading to multi-H complexes. Complexes carrying one and two satellite H atoms (see Figs. 4.7e and 4.7f) will be referred to as C_{2v}–H_{Ga} and C_{2v}–$2H_{Ga}$, respectively.

The importance of N–H complexes in the description of structural and electronic effects in hydrogenated GaAsN alloys is strictly related to their formation energies and stabilities. Referring to complexes shown in Fig. 4.7, formation energies, Ω_f, are reported in Table 4.1 and bonding and dissociation energies are reported in Table 4.2. These quantities have to be considered all together to determine, as a function of the electron chemical potential, the right charge state of a N–H complex. This, in turn, is needed in the description of a formation or dissociation process of a complex.

Table 4.1 reports formation energies Ω_f per H atom calculated for some positively charged N–H complexes estimated for a Fermi energy, E_F, ranging from 0 (i.e., top of the VB), to $E_g/2$, and E_g (present estimate 1.56 eV). Since Ω_f values of the neutral complexes, given in Table 4.1's last row, are smaller than the values corresponding to the positively charged complexes when $E_F = E_g/2$, in intrinsic GaAsN the complexes have to be considered in the neutral charge state.

Table 4.2 For a complex formation (e.g., $N + H^{+}_{BC} \rightarrow N\text{–}H^{+}_{BC}$), the corresponding energy difference $\Delta E = E[N] + E[H^{+}_{BC}] - E[N\text{–}H^{+}_{BC}]$ is given in the upper part of the Table. For a complex dissociation (e.g., $N\text{–}H_{BC} \rightarrow N + H_{AB}$), the corresponding $\Delta E = E[N\text{–}H_{BC}] - E[N] - E[H_{AB}]$ is given in the lower part. Positive (negative) values indicate exothermic (endothermic) reactions

Complex	Reacting species (products)	ΔE (eV)
$N\text{–}H^{+}_{BC}$	$N+H^{+}_{BC}$	1.05
C^{+1}_{2v}	$N\text{–}H_{BC}+H^{+}_{BC}$	0.90
$[N\text{–}H_{BC}]\text{–}H^{+}_{Ga}$	$N\text{–}H_{BC}+H^{+}_{BC}$	0.60
$C_{2v}\text{–}H^{+}_{Ga}$	$C_{2v}+H^{+}_{BC}$	1.55
$C_{2v}\text{–}H^{+}_{Ga}$	$[N\text{–}H_{BC}]\text{–}H_{Ga}+H^{+}_{BC}$	1.00
$C_{2v}\text{–}2H^{+}_{Ga}$	$C_{2v}\text{–}H_{Ga}+H^{+}_{BC}$	0.61
$[N\text{–}H_{BC}]$	$N+H_{AB}$	−1.58
$[N\text{–}H_{BC}]\text{–}H_{Ga}$	$N\text{–}H_{BC}+H_{AB}$	−1.61
C_{2v}	$N\text{–}H_{BC}+H_{AB}$	−1.61
$C_{2v}\text{–}H_{Ga}$	$C_{2v}+H_{AB}$	−1.13
$C_{2v}\text{–}2H_{Ga}$	$C_{2v}\text{–}H_{Ga}+H_{AB}$	−1.50
$C_{2v}\text{–}2H_{Ga}$	$C_{2v}+H_2$	−1.35

With this result and those in Table 4.2 we can discuss possible mechanisms of complex formation. First, let us assume that atomic H is introduced in an intrinsic GaAsN sample. Atomic H should form H^{+}-H^{-} pairs, and the mobile H^{+} ions should be attracted by the N atoms because the partial negative charge they carry is larger than that of the As atoms. Then, $N\text{–}H^{+}_{BC}$ complexes should form, favored by the corresponding exothermic reaction (see Table 4.2). But from Table 4.1 we know that $N\text{–}H_{BC}$ complexes forming in a positive charge state become neutral when $E_F = E_g/2$, as in the intrinsic materials considered here. Even neutral $N\text{–}H_{BC}$ complexes are quite stable, as shown by their dissociation energy, $\Delta E = 1.58$ eV (see Table 4.2). Then, H^{+} ions trapped at the N sites evolve into neutral $N\text{–}H^{0}_{BC}$ complexes, thus implying that further H^{+} ions need to form to compensate the H^{-} and satisfy the charge neutrality condition. These H^{+} ions, in turn, may react with $N\text{–}H^{0}_{BC}$ complexes to form C^{+1}_{2v} complexes, which are also stable (see Table 4.2) and change their charge state to neutral in intrinsic materials. The above mechanism can be applied to all of the considered N–H complexes: for example, C_{2v} complexes initially formed in the charge state +1 become neutral (for $E_F = E_g/2$) before reacting with a further H^+ ion and so on. In principle, H^+ formation terminates when N–H complexes reach the highest number of H atoms, that is, four atoms, and H^- species are thoroughly compensated.

In general, the mentioned experimental measurements (Trotta et al., 2012b) are performed in slightly *p*-doped GaAsN samples containing $\approx 10^{16}$ cm^{-3} acceptors (A) against a 3% ($\approx 10^{21}$ cm^{-3}) of N atoms (Polimeni, 2007). In these materials, a slightly different mechanism can be hypothesized. At the beginning of hydrogenation, a Fermi energy $E_F < E_g/2$ favors the acceptor compensation by H and the formation of H^+ ions far from the negative acceptors A^-. This implies that E_F is raised to $E_g/2$. Diffusing H^+ ions hardly reach the A^- acceptors, because they are *trapped* preferably by the predominant N atoms and form N–H^+_{BC} complexes. Once more, these complexes become neutral in the presence of compensated acceptors, thus requiring the formation of further H^+ ions to satisfy the charge neutrality condition. As described just above, these ions can form C^{+1}_{2v} complexes and develop therefore into multi-H complexes. H^+ formation stops when N–H complexes achieve the maximum number of H atoms and further H^+ ions reach the acceptors and passivate them by forming H–A complexes. At the end of this process, *p*-doped GaAsN samples become intrinsic and further introduced H atoms form H^+–H^- pairs and hydrogen molecules. In agreement with the above mechanisms, the ΔE values in the upper part of Table 4.2 involve a H^+ ion located at its stable BC site in GaAs (Pavesi and Giannozzi, 1992) as a species participating to the complex formation. On the other hand, reaction products coming from the dissociation of neutral N–H complexes include neutral H atoms in their stable AB site in GaAs (H_{AB}) (Pavesi and Giannozzi, 1992).

The ΔE values estimated for the complex formation suggest that diffusing H^+ ions successively form complexes (see Table 4.2): N–H^+_{BC}, C^{+1}_{2v}, [N–H_{BC}]–H^+_{Ga}, C_{2v}–H^+_{Ga}, and C_{2v}–2H^+_{Ga}. The C_{2v}–2H_{Ga} complex coming from the full saturation of the H-induced DBs in the C_{2v} complex is proposed therefore as the predominant neutral complex at the end of the hydrogenation procedure.

N–H complexes having three or more H atoms bonded to the same N atom (not included in Table 4.2) are metastable or unstable.

Dissociation energies (ΔE values in the lower part of Table 4.2) indicate that multi-H complexes involving one or two satellite H atoms, like C_{2v}–$2H_{Ga}$, are quite stable ($\Delta E < 0$), and their stability is comparable to that of the single–H complex N–H_{BC} of Fig. 4.7b. However, given the smaller absolute value of their dissociation energies, they should be the first to be removed upon annealing.

Finally, the vibrational frequencies, calculated for the two stretching and scissoring N–H modes in the three C_{2v}-derived complexes (Fig. 4.7d–f), are all in a quite good agreement with the experimental IR lines reported in Table 4.4. The small differences between the *three* vibrational frequencies of these three complexes suggest that these C_{2v}-derived complexes, when alternatively present in hydrogenated GaAsN samples, may give hardly distinguishable IR signals. Notwithstanding, a recent study, where accurate multiple-internal-reflection IR measurements, HRXRD, and PL measurements have been performed in hydrogenated and deuterated samples of GaAsN diluted alloys and combined with theoretical calculations (Wen et al., 2012), has related the changes from compressive to tensile strain induced by H in GaAsN to the formation of N–H complexes containing more than two H atoms. In detail, the results of that study (better addressed in Chapter 5) suggest that several multi-H structures involving a C_{2v}-like core and having different thermal stabilities can be formed with relative concentrations depending on the hydrogenation temperature. These results partially question previous experimental results fixing to three the number of H atoms involved in the formation of N–H complexes (Trotta et al., 2012b; Berti et al., 2007). They give also strong support to the multi-H theoretical model by suggesting that the number of satellite H atoms (one or two) around a C_{2v} core can be affected by the conditions of the hydrogenation procedure.

4.6 Mechanism of N Passivation

The H-induced electronic passivation of N may be studied by means of transition energy levels, as explained in Section 4.2. First, the

Table 4.3 Transition energy $\epsilon^{n/n'}$ values (in eV), frequency values (in cm^{-1}), and $\Delta a/a$ values (%) estimated for the investigated N–H complexes. Stretching and scissoring frequencies of the H modes are indicated by ν and δ, respectively. H-1 and H-2 indicate H atoms bonded to N; *neg* indicates a negative value

Complex	$\epsilon^{+/0}$	$\epsilon^{0/-}$	ν H-1	ν H-2	δ	$\Delta a/a$
C_{2v}	0.60	1.50	3143	2846	1444	0.00
C_{2v}–H_{Ga}	0.75	1.36	3189	2805	1435	0.42
C_{2v}–$2H_{Ga}$	*neg*	1.69	3167	2884	1439	0.39
[N–H_{BC}]–H_{Ga}	*neg*	1.47	–	2853	–	−0.02
N–H^0_{BC}	0.67	0.95	–	2805	–	−0.33
N–H^+_{BC}			3213	–	–	−1.17
Exp.[(a)]			3195	2967	1447	0.48

(a) See Ref. (Jiang et al., 2004).

$\epsilon^{0/-}$ transition states calculated for GaAs and $GaAs_{0.97}N_{0.03}$, 1.56 eV and 0.84 eV, respectively, are assumed here as estimates of the corresponding energy gaps. Then, observing the $\epsilon^{0/-}$ values calculated for the C_{2v}, C_{2v}–H_{Ga}, and C_{2v}–$2H_{Ga}$ complexes, shown in Table 4.4, we see that in these complexes the N effects on the electronic structure are passivated, recovering the GaAs energy gap. On the contrary, the [N–H_{BC}] complexes have no passivating effects. The case of the C_{2v}–H_{Ga} complex deserves a brief comment. In our previous work (Amore Bonapasta et al., 2007), an N passivation induced by such a complex was considered uncertain. Further careful calculations have dissipated any doubt: i) an $\epsilon^{0/-}$ value slightly larger than the previous one has been estimated (see Table 4.4), and ii) most important, it has been clarified that this value is smaller than the estimated GaAs energy gap only for a limit of the used approach. The $\epsilon^{0/-}$ states imply indeed the evaluation of the energy of a negatively charged system. At variance with the C_{2v} and C_{2v}–$2H_{Ga}$ complexes, in the C_{2v}–H_{Ga} complex, the added electron is accommodated by an electronic level related to the unsaturated Ga DB. This leads to an $\epsilon^{0/-}$ value, 1.36 eV, slightly smaller than the values estimated for the two former complexes as well as than the GaAs energy gap. Notwithstanding, this value is much higher

Table 4.4 Transition energy $\epsilon^{n/n'}$ values (in eV), frequency values (in cm^{-1}), and $\Delta a/a$ values (%) estimated for the investigated N–H complexes. Stretching and scissoring frequencies of the H modes are indicated by ν and δ, respectively. H-1 and H-2 indicate H atoms bonded to N; *neg* indicates a negative value

Complex	$\epsilon^{+/0}$	$\epsilon^{0/-}$	ν H-1	ν H-2	δ	$\Delta a/a$
C_{2v}	0.60	1.50	3143	2846	1444	0.00
C_{2v}–H_{Ga}	0.75	1.36	3189	2805	1435	0.42
C_{2v}–$2H_{Ga}$	*neg*	1.69	3167	2884	1439	0.39
[N–H_{BC}]–H_{Ga}	*neg*	1.47	-	2853	-	−0.02
N–H_{BC}^{0}	0.67	0.95	-	2805	-	−0.33
N–H_{BC}^{+}			3213	-	-	−1.17
Exp.[a]			3195	2967	1447	0.48

(a) See Ref. (Jiang et al., 2004).

than the estimated GaAsN energy gap, thus indicating a passivating capability also of the C_{2v}–H_{Ga} complex.

A recovery of the GaAs energy gap caused by the formation of each of the C_{2v}-derived complexes has been also confirmed by preliminary (still unpublished) results achieved by calculating the electronic energy levels of supercells simulating GaAs, the $GaAs_{0.97}N_{0.03}$ alloy, and the formation of the C_{2v} complexes, as well as of the N–H_{BC} complex with the HSE hybrid functional. These calculations find a GaAs energy gap of 1.25 eV and its lowering at 0.67 eV in the GaAsN alloy similarly to the results reported in a recent HSE investigation of the same alloy (Virkkala et al., 2012). Our results also show a recovery of the GaAs energy gap at the value of 1.30 eV induced by each of the three C_{2v} complexes. No band-gap recovery has been found, instead, when simulating the N–H_{BC} complex.

The structure and chemical bonding of C_{2v}-like complexes have great importance in the discussion of the neutralization mechanisms of N electronic effects in GaAsN alloys, which lead to band-gap and lattice constant recovery and, possibly, to lattice overshooting. In this regard, keeping in mind the results shown in Figs. 4.7 (in particular (d–f), and 4.9, we stress the following features of the

C_{2v}-like complexes: i) The three C_{2v}-derived complexes include a same C_{2v} core characterized by quite short N–H bond distances (Fig. 4.7) and by an electronic charge accumulation around the N–H pairs (Fig. 4.9). Both features indicate the formation of strong bonds, similar to those formed by C_{As} in GaAs (see Figs. 4.7a and 4.9a). Such similarities originate from the common properties of the C and N atoms, which have both a small size, large electronegativity (note the charge accumulation on the C and N atoms in Fig. 4.9), and the same affinity to form strong bonds with H in molecular systems. ii) The structure and chemical bonding of the C_{2v} core imply a strong perturbation of the isolated N environment. In fact, when the N impurity is "naked," its four Ga neighbors displace from their position in the GaAs lattice toward the N atom, reaching a Ga–N distance, 2.10 Å, significantly shorter than the lattice Ga–As bonds, 2.50 Å. When two Ga–N bonds are broken due to the formation of a C_{2v} core, the N atom displaces toward its two still bonded Ga neighbors, displacing the N–H_2 complex as a whole block and reaching Ga–N bond lengths of 2.06 Å, closer to the optimal value given by the Ga and N covalent radii, 1.96 Å.[a]

The above considerations suggest that N passivation is closely related to the strong perturbation of the N environment (change of local structure and chemical bonding) produced by the formation of the C_{2v} core. As a check, a calculation of the electronic structure (i.e., a simple wavefunction optimization) has been performed with the atomic arrangement fixed at the C_{2v} complex configuration, while removing the two H atoms in the complex. The GaAsN electronic structure, in this particular arrangement, shows quite similar to the case of the C_{2v} complex, that is, N is passivated and the energy gap is restored. Therefore, most of the N effects can be ascribed to its local interactions with the Ga neighbors, while their passivation seems mainly related to the breaking of two N–Ga bonds and to the consequent relax of the two remaining N–Ga bonds.

All together, the above results indicate a key role of the C_{2v} core in the passivation of the N electronic effects, which is realized indeed

[a]Ga has a covalent radius of 1.26 Å, while N has a covalent radius of 0.70 Å (Phillips, 1973).

by each of the three complexes sharing this atomic arrangement around the N atom.

The other important observable after hydrogenation is the neutralization of the effects that N insertion in GaAs induces on the lattice. Table 4.4 reports the $\Delta a/a$ values calculated for the investigated N–H complexes, where a is the lattice constant calculated for bulk GaAs and Δa is the difference ($a^{\perp}$-a), $a^{\perp}$ being the lattice constant along the [001] direction. A $\Delta a/a$ value of 0.48% is the lattice overshooting measured in hydrogenated $GaAs_{1-y}N_y$ samples with $y = 3.20\%$, close to the value $y = 3.12\%$ characterizing our GaAsN supercells (Bisognin et al., 2004, 2006). The values in Table 4.4 show that while the C_{2v} complex fully recovers the GaAs lattice constant, only the C_{2v}–H_{Ga} and C_{2v}–$2H_{Ga}$ complexes induce an appreciable lattice overshooting, comparable to the experiment. A value of -1.08% estimated for the lattice contraction induced by N in $GaAs_{0.97}N_{0.03}$ is also in very good agreement with an experimental value of -1.18% (Bisognin et al., 2004, 2006). H_2 molecules do not induce any overshooting.

All the above results indicate that the C_{2v} complex and its satellite H children play a significant role in the description of the properties of hydrogenated GaAsN alloys. The complexes with satellite H atoms should predominate in hydrogenated GaAsN, neutralize the N electronic effects, thanks to the common C_{2v} core, and cause lattice overshooting. Mild annealing treatments (described in detail in Chapter 7) can induce their dissociation leading to C_{2v} complexes and H_2 molecules. The C_{2v} complexes recover the GaAs lattice constant and passivate the electronic N effects as well. Annealing at higher temperatures removes the C_{2v} complexes and restores the N effects on the GaAs lattice constant and band structure.

The above theoretical picture fully accounts, therefore, for the H passivation effects and the evolution of the electronic and structural properties with the temperature observed in hydrogenated GaAsN and agrees with XAS results compatible with the formation of C_{2v}-derived complexes (Ciatto et al., 2009). Moreover, a difference of 0.22 eV between the dissociation energies of the C_{2v}–$2H_{Ga}$ and C_{2v} complexes (which can be viewed as a lower bound of the activation energy for the complex dissociation) favorably compares with the difference of 0.2–0.3 eV estimated between the corresponding, experimental activation energies (Bisognin et al., 2006).

4.7 Conclusions

The mechanisms responsible of the H passivation of N effects on the GaAs properties have been clarified by performing a comparison with the corresponding mechanisms leading to the passivation of typical shallow dopants in the same material.

The H neutralization of the effects of shallow dopants on the electronic structure of GaAs can be schematized in terms of breaking of a dopant–neighbor bond and H saturation of the resulting dopant dangling bond. This leads indeed to the recovery of the "natural" chemical valence of both the impurity and its neighbor.

On the contrary, when considering a single H atom interacting with the N isoelectronic impurity, the Ga–N bond opening and the formation of the H–N bond lead to an unsaturated DB on the Ga neighbor. Similarly, the formation of an NH_2 complex leads to two unsaturated DBs on the Ga neighbors and, then, to the formation of multi-H complexes involving one or two H satellites.

In all of the multi-H complexes, two H atoms break two Ga–N bonds to form complexes sharing a common, quite stable NH_2 core showing a local C_{2v} symmetry. This strongly perturbs the N local interactions with its four Ga neighbors. In the present theoretical picture, the effects of N on the host GaAs properties originate from such N–Ga local interactions, while such effects are neutralized mainly by the strong perturbation induced by H bonds on the N environment. Finally, it is shown that the formation of multi-H complexes with N, predicted by theory, can be supported by recent experimental findings.

Chapter 5

Microscopic Structure of N–H_n Complexes in Dilute Nitride Semiconductors Revealed by Their Vibrational Properties

Michael Stavola and W. Beall Fowler
Department of Physics and Sherman Fairchild Laboratory, Lehigh University, Bethlehem, Pennsylvania 18015, USA
mjsa@Lehigh.edu

The addition of hydrogen to dilute III–N–V alloys causes a surprising shift of the semiconductor band gap. Vibrational spectroscopy and complementary theory reveal the structures and microscopic properties of the nitrogen- and hydrogen-containing defects that give rise to these remarkable effects caused by hydrogen.

5.1 Introduction

Several of the dilute III–N–V alloys have attracted much attention because of a novel reduction of the band gap caused by the addition of N (see Chapter 1), which can be exploited in technological applications such as long-wavelength lasers in the

Hydrogenated Dilute Nitride Semiconductors: Theory, Properties, and Applications
Edited by Gianluca Ciatto

ISBN 978-981-4463-45-4 (Hardcover), 978-981-4463-46-1 (eBook)
www.panstanford.com

telecommunications range and multijunction solar cells (Henini, 2005; Buyanova and Chen, 2004). The further addition of hydrogen to the dilute $GaAs_{1-y}N_y$ and $GaP_{1-y}N_y$ alloys, unlike the case of other semiconductors, allows tuning the band gap of the alloys (Chapter 2), which has led to novel strategies for the fabrication of band-gap engineered structures with nanometer spatial resolution (Felici et al., 2006a; Trotta et al., 2008, 2011, 2012b).

In GaAs and GaP, both nitrogen and hydrogen are lighter than the host atoms and give rise to local vibrational modes (LVMs). Vibrational spectroscopy (Stavola, 1998; McCluskey, 2000) becomes, therefore, an ideal probe of the microscopic structures of nitrogen- and hydrogen-containing defects in these III–V compounds. The primary goal of our studies of the vibrational properties of hydrogenated III–N–V alloys has been to determine the microscopic structures and transformations of the nitrogen- and hydrogen-containing defects that give rise to the surprising band-gap shifts caused by hydrogen.

Commercially available Fourier transform infrared (IR) spectrometers allow IR spectra to be measured with high resolution and a high signal-to-noise ratio over a broad spectral range. Our experimental studies have been made with a Bomem DA3 spectrometer equipped with a KBr beamsplitter and InSb and HgCdTe detectors (77 K) for measurements in the mid-IR region. A Si bolometer (4.2 K) was used for measurements in the spectral range of 400–1000 cm^{-1}. The addition of uniaxial stress and polarized light helps to reveal further structure-sensitive information (Stavola, 1998). In this case, a custom, stainless-steel, push-rod apparatus driven by a pneumatic cylinder was used to apply stresses of known magnitude to oriented samples.

In other experiments, IR measurements were made in a multiple-internal-reflection (MIR) geometry at 4.2 K to increase the signal-to-noise ratio of our spectra.

5.2 Vibrational Properties of the H–N–H Complex

A microscopic explanation for the band-gap shifts caused by nitrogen and hydrogen in dilute III–N–V alloys attracted the

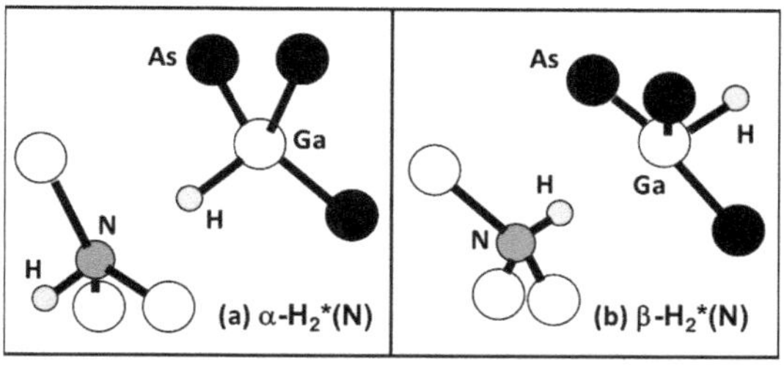

Figure 5.1 The (a) α-H_2^*(N) and (b) β-H_2^*(N) complexes in GaAsN or GaPN. From Ref. (Jiang et al., 2004).

immediate attention of theory. Calculations performed independently by several groups predicted that a defect known as H_2^* would be stabilized by the presence of nitrogen in III–N–V alloys and that this defect would eliminate the reduction of the band gap caused by the presence of nitrogen (Dixon et al., 1998; Kim and Chang, 2002; Janotti et al., 2002a,b; Amore Bonapasta et al., 2002). The H_2^* structure, described in Chapter 4, has one H atom at a bond-centered site and a second H atom at an antibonding site, both aligned along the same <111> axis. Two H_2^*(N) structures with C_{3v} symmetry are possible (Fig. 5.1), α-H_2^*(N) and β-H_2^*(N). Each of these defect structures contains one N–H bond and one Ga–H bond.

IR absorption experiments (Jiang et al., 2004),[a] were performed to test the H_2^*(N) structures predicted by theory. However, this data and its interpretation took us in a new direction—to a canted H–N–H structure with C_{1h} symmetry.

5.2.1 *Effect of H Irradiation on Nitrogen Modes*

A vibrational mode for a N atom substituting for As in $GaAs_{1-y}N_y$ has been observed by Raman and IR spectroscopies (Prokofyeva et al., 1998; Alt et al., 2000, 2001). IR spectra for GaAs samples implanted with nitrogen are shown in Fig. 5.2. The implantation of ^{14}N gives

[a] In certain samples with their Fermi level near the midgap, Clerjaud et al. observed N–H vibrational lines not seen in our later experiments (Hahn et al., 1994; Clerjaud et al., 1996). These results have subsequently been interpreted as arising from a trigonal defect consisting of substitutional N bonded to a single H (Dixon et al., 1998; Kim and Chang, 2002).

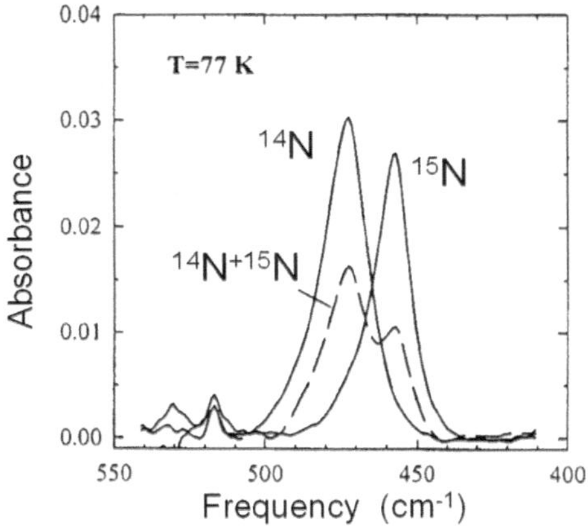

Figure 5.2 IR spectra (77 K) of a GaAs sample implanted with ^{14}N (2×10^{16} cm^{-2}), a GaAs sample implanted with ^{15}N (2×10^{16} cm^{-2}), and a GaAs sample coimplanted with ^{14}N (10^{16} cm^{-2}) and ^{15}N (10^{16} cm^{-2}). An unimplanted GaAs sample was used for reference. From Ref. (Alt et al., 2000).

rise to a vibrational band at 473 cm^{-1}, and the implantation of ^{15}N gives rise to a vibrational band at 458 cm^{-1}, confirming the role of nitrogen in the center. The coimplantation of ^{14}N and ^{15}N gives these same vibrational bands without any evidence for the vibrational coupling of these modes, indicating that the center contains a single nitrogen atom. Similarly, the presence of nitrogen in $GaP_{1-y}N_y$ gives rise to a vibrational band at 496 cm^{-1} (Fig. 5.3) that lies among other GaP phonon-related absorption (Buyanova et al., 2004; Kleekajai et al., 2008). The intensity of the 496 cm^{-1} band increases as the nitrogen content of the dilute alloy increases (Kleekajai et al., 2008).

What is the effect of hydrogen (or deuterium) irradiation on the nitrogen vibrational mode that was discovered for III–N–V alloys without hydrogen? Figure 5.4a shows the nitrogen LVM at 473 cm^{-1} for an as-grown $GaAs_{0.992}N_{0.008}$ sample. Upon hydrogen irradiation, the nitrogen mode is nearly eliminated (Jiang et al., 2004). (Hydrogen irradiation was performed with a Kaufman ion source with the samples held at 300°C to promote H indiffusion and with an ion current density of $\sim$ 10 μA/cm^2.) Difference spectra are shown in Fig. 5.4b for hydrogen- and deuterium-irradiated samples

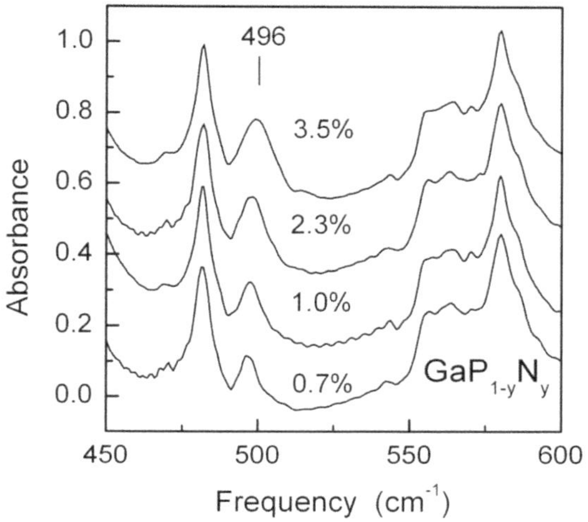

Figure 5.3 IR spectra (4.2 K) measured for $GaP_{1-y}N_y$ layers with different N concentrations.

with the as-grown $GaAs_{1-y}N_y$ sample used for reference (Jiang et al., 2004). In this case, the nitrogen LVM that is eliminated by H (or D) irradiation is seen as a negative peak. Similar results have been observed for $GaP_{1-y}N_y$ samples treated with hydrogen or deuterium (Kleekajai et al., 2008).

Figure 5.5 shows the annealing behavior of the N modes for $GaP_{0.993}N_{0.007}$ and $GaAs_{0.992}N_{0.008}$ samples that had been irradiated with deuterium (Kleekajai et al., 2008). Remembering that these are difference spectra with an as-grown sample used for reference, the reduction in the intensity of the nitrogen mode absorption caused by deuterium irradiation is seen to be eliminated by an anneal at 400°C for both the $GaP_{1-y}N_y$ and $GaAs_{1-y}N_y$ samples. The thermal stabilities of the hydrogenated nitrogen centers have been found to be correlated with the band-gap shift caused by H irradiation and probed by photoluminescence measurements (Jiang et al., 2004; Kleekajai et al., 2008). The annealing behavior of the nitrogen LVM is an excellent benchmark for other vibrational lines containing N and H in the dilute III–N–V alloys.

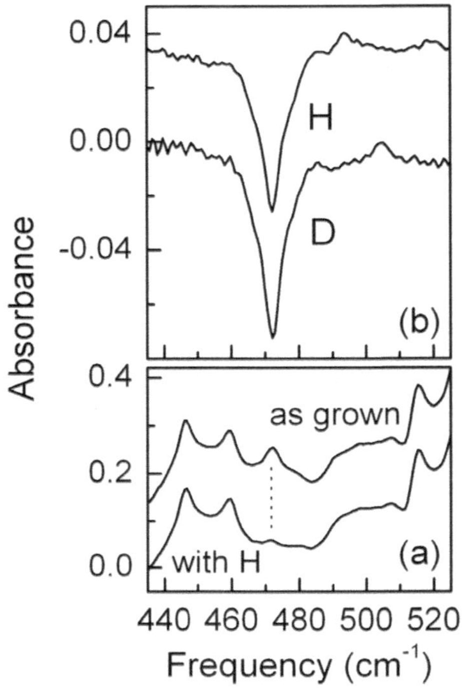

Figure 5.4 IR spectra (4.2 K) showing the N mode for $GaAs_{0.992}N_{0.008}$ samples. Spectra for an as-grown sample and a hydrogenated sample are shown in (a). Difference spectra for hydrogenated and deuterated samples are compared in (b). From Ref. (Jiang et al., 2004).

5.2.2 *Two Weakly Coupled N–H Modes*

The introduction of H into $GaP_{1-y}N_y$ produces two H-stretching LVMs with frequencies (4.2 K) of 2955 cm^{-1} and 3202 cm^{-1} (Fig. 5.6) (Kleekajai et al., 2008).

The substitution of D for H produces the corresponding D-stretching modes, confirming the role of H in the LVMs. Similar results were found for $GaAs_{1-y}N_y$ (Jiang et al., 2004). Tables 5.1 and 5.2 contain lists of LVM frequencies for $GaP_{1-y}N_y$ and $GaAs_{1-y}N_y$ samples that had been irradiated by hydrogen or deuterium. The ratio of H-stretching to D-stretching mode frequencies is found to be $r = \omega_H/\omega_D = 1.34$, a value that is typical of hydrogen

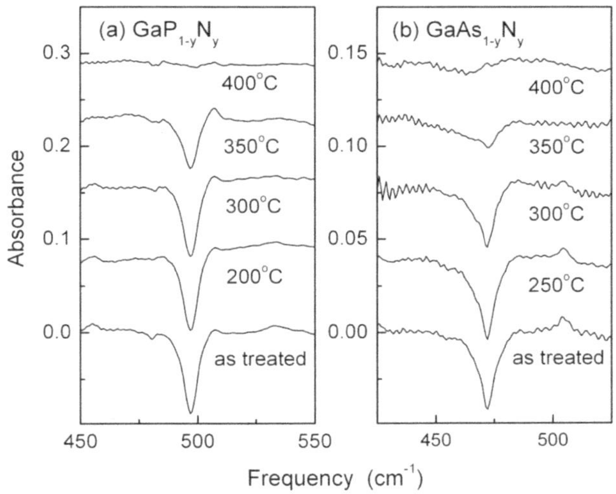

Figure 5.5 Absorbance spectra (4.2 K) for (a) a deuterated $GaP_{0.993}N_{0.007}$ sample and (b) a deuterated $GaAs_{0.992}N_{0.008}$ sample that were subsequently annealed (30 min) at the temperatures shown. Samples that did not contain D were used for reference, so the N mode whose intensity is reduced by deuteration is seen as a downward peak in the spectrum. From Ref. (Kleekajai et al., 2008).

bonded to a light-mass element. This frequency ratio and the large values of the mode frequencies caused us to assign the two vibrational lines at 2955 cm^{-1} and 3202 cm^{-1} in $GaP_{1-y}N_y$ to N–H stretching vibrations (Kleekajai et al., 2008) and similarly for the corresponding vibrational lines seen for $GaAs_{1-y}N_y$ (Jiang et al., 2004).

Annealing data for a $GaP_{0.993}N_{0.007}$ sample that had been irradiated with deuterium are shown in Fig. 5.7 (Kleekajai et al., 2008). The N–D modes at 2211 cm^{-1} and 2381 cm^{-1} (that correspond to the N–H modes at 2955 cm^{-1} and 3202 cm^{-1}) are found to be unstable for an anneal at 400°C, similar to the behavior of the nitrogen mode at 496 cm^{-1} that reappears when H (or D) is annealed away (Fig. 5.5). The similarity of these annealing results is consistent with assigning the N–H (N–D)-stretching modes at 2955 (2211) cm^{-1} and 3202 (2381) cm^{-1} to the defect or

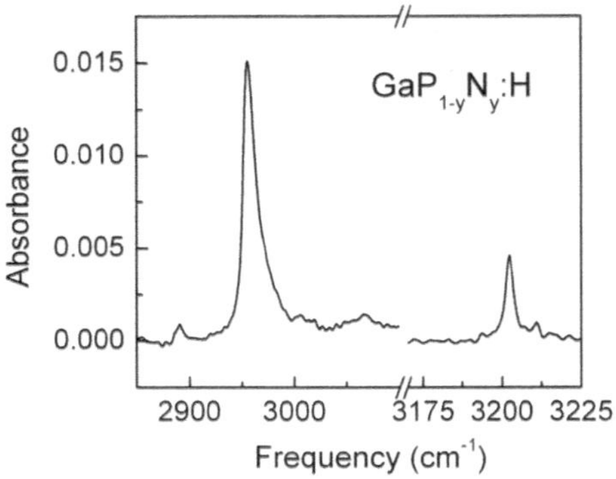

Figure 5.6 IR spectrum (4.2 K) showing the H-stretching modes for a hydrogenated $GaP_{0.993}N_{0.007}$ sample. From Ref. (Kleekajai et al., 2008).

Table 5.1 Experimental vibrational frequencies (in units cm^{-1}) for the H and D modes seen in hydrogenated and deuterated $GaP_{0.993}N_{0.007}$. The frequency ratio $r = \omega_H/\omega_D$ is also given. Weak lines are indicated by a (w). From Ref. (Kleekajai et al., 2008)

Assignments	ω(H)	ω(D)	r
Stretch	3202	2381	1.345
Stretch	2955	2211	1.336
2nd harmonic	2891(w)	2150(w)	1.345
in plane wag	1458	1082	1.348
out of plane wag	1069	-	-

defects responsible for the band-gap shifts produced by hydrogen irradiation. The N–H and N–D modes in $GaAs_{1-y}N_y$ also show similar annealing behavior (Jiang et al., 2004).

The vibrational spectra of hydrogenated $GaP_{1-y}N_y$ and $GaAs_{1-y}N_y$ reveal two N–H IR lines and no Ga–H lines. These results alone do not determine whether or not the two $H_2^*(N)$ defects predicted by theory (Dixon et al., 1998; Kim and Chang, 2002; Janotti et al., 2002a,b; Amore Bonapasta et al., 2002) are responsible for the observed IR spectrum. Without further information, it is possible

Table 5.2 Experimental vibrational frequencies (in units cm^{-1}) for the H and D modes seen in hydrogenated and deuterated $GaAs_{0.9913}N_{0.0087}$. The frequency ratio $r = \omega_H/\omega_D$ is also given. Weak lines are indicated by a (w). From Ref. (Kleekajai et al., 2008)

Assignments	ω(H)	ω(D)	r
Stretch	3195	2376	1.345
Stretch	2967	2217	1.338
2nd harmonic	2868(w)	2137(w)	1.342
in plane wag	1447	1076	1.345
out of plane wag	1068	798	1.338
out of plane wag	1057	–	–

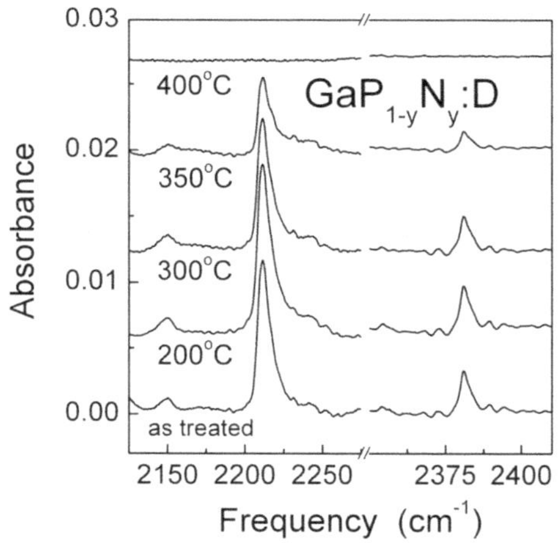

Figure 5.7 IR spectra (4.2 K) showing the D-stretching modes for a deuterated $GaP_{0.993}N_{0.007}$ sample that was subsequently annealed (30 min) at the temperatures shown. From Ref. (Kleekajai et al., 2008).

that the two N–H modes arise from the N–H bonds of α-H_2^*(N) and β-H_2^*(N). The absence of Ga–H IR lines might be explained by the possibilities that these lines are weak or perhaps are too broad to be distinguished. These considerations apply also to the corresponding defects containing deuterium.

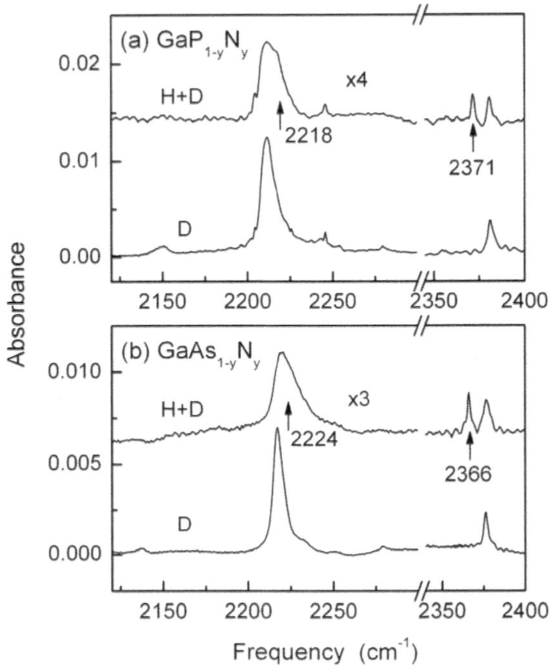

Figure 5.8 Absorbance spectra (4.2 K) showing the D-stretching modes for (a) $GaP_{0.993}N_{0.007}$ and (b) $GaAs_{0.9913}N_{0.0087}$ samples. In both panels (a) and (b) the lower spectrum was measured for a deuterated sample and the upper spectrum was measured for a sample treated with both H and D. From Ref. (Kleekajai et al., 2008).

Spectroscopic results that more strongly constrain the possible structure(s) of the nitrogen- and hydrogen-containing defects in the III–N–V alloys are obtained from samples that contain both H and D (Jiang et al., 2004; Kleekajai et al., 2008). Figure 5.8a compares spectra for a $GaP_{0.993}N_{0.007}$ sample irradiated with D alone with a sample irradiated with both H and D. The $GaP_{0.993}N_{0.007}$ sample containing H and D shows the two N–D lines at 2381 cm^{-1} and 2211 cm^{-1} characteristic of $GaP_{0.993}N_{0.007}$ treated with D alone. However, Fig. 5.8a shows two additional N–D lines at 2218 cm^{-1} and 2371 cm^{-1}. Figure 5.8b shows similar data for a $GaAs_{0.9913}N_{0.0087}$ sample treated with both H and D.

The IR spectra shown in Fig. 5.8 are characteristic of a *single* defect with two inequivalent N–D-stretching modes that are weakly coupled (Jiang et al., 2004; Kleekajai et al., 2008). Some defects in a $GaP_{1-y}N_y$ sample treated with H and D will contain D alone, giving the N–D lines at 2381 cm^{-1} and 2211 cm^{-1}. Other defects will contain both H and D. In this case, there are N–H and N–D modes that become dynamically decoupled to produce new IR lines, one N–D line for the defect with N–H(1) and N–D(2) bonds and a second N–D line for the defect with N–D(1) and N–H(2) bonds (samples treated in H and D will also have a set of corresponding decoupled N–H lines).

The vibrational coupling that is observed by experiment for the two inequivalent N–D bonds in deuterium-irradiated III–N–V alloys requires that both be associated with the same defect (Jiang et al., 2004), and rules out the possibility that they arise from the two different $D_2^*(N)$ structures predicted by theory. Other nitrogen- and deuterium-containing defects must be considered. Furthermore, the anticorrelation of the annealing behavior of the weakly coupled N–H- (or N–D)-stretching modes with the annealing behavior of the 496 cm^{-1} mode assigned to substitutional N indicates that the NH_n (or ND_n) center contains a single N atom.

5.2.3 *A Canted H–N–H Center and Its Vibrational Properties: Theory*

Different versions of density functional theory have been used by several groups to investigate the N–H_n system. Most of our calculations have utilized the CRYSTAL06 code, depending on Gaussian basis sets (Dovesi et al., 2006). We chose a hybrid DFT approach with a gradient-corrected approximation to the exchange-correlation functional (Becke's B3LYP potential (Becke, 1988) with 20% exact exchange and Lee–Yang–Parr correlation (Lee et al., 1988)). The calculations were carried out in a periodic supercell with 32 or 54 host atoms. Further details may be found in Refs. (Kleekajai et al., 2008) and (Fowler et al., 2005).

An early study by Bonapasta et al. suggested an H–N–H configuration to exist in a +2 charge state under certain circumstances, in both GaAs and GaP (Amore Bonapasta et al., 2003). In this case, two

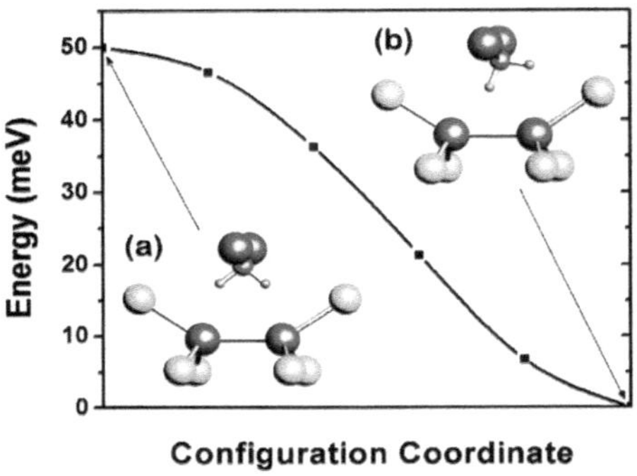

Figure 5.9 The calculated canting pathway for the H–N–H center in $GaAs_{1-y}N_y$ indicating that the uncanted structure is unstable. The insets show the structures of the (a) uncanted and (b) canted H–N–H complexes. From Ref. (Du et al., 2005).

adjacent N–Ga bonds were broken and an H^+ inserted at each bond-centered position. Both N–H bonds would be equivalent in this C_{2v}-symmetry defect, and the calculations of Bonapasta et al. indicated that this symmetric configuration would persist as electrons were trapped at the defect.

Experiments, however, demanded that the two N–H bonds be inequivalent (Jiang et al., 2004). How could this happen? Independent calculations for $GaAs_{1-y}N_y$ by our group (Fowler et al., 2005) and by Du et al. (Du et al., 2005) yielded the surprising result that spontaneous breaking of the C_{2v} symmetry of the H–N–H defect would occur and that the H–N–H complex would tilt, or cant, about the N and within the original H–N–H plane, to a C_{1h} symmetry,[a] yielding the required inequivalency. Figure 5.9 from Du et al. illustrates this relaxation and the accompanying reduction in energy (Du et al., 2005) (the magnitude of the relaxation energy will be discussed in Section 5.3.2). Further evidence for the H–N–H structure came from X-ray absorption near-edge structure (XANES) studies by Ciatto et al. (Ciatto et al., 2005a). These experiments and their interpretation ruled out H_2^* structures and were shown to be

[a]Note that different groups have used a different nomenclature for the structure with C_{1h} symmetry. The C_{1h} complex is called "asymmetric C_{2v}" in Chapters 4 and 6.

consistent with an H–N–H structure, either symmetric or canted (see Chapter 6).

Du et al. found that the neutral H–N–H complex is less stable than either H_2^* defect (Du et al., 2005), a result that we as well as Ciatto et al. also found (Fowler et al., 2005; Ciatto et al., 2005a). However, their calculations also indicated that in the +2 charge state, H–N–H is more stable, again consistent with our results. Du et al. explained the existence of the less stable neutral H–N–H structure by suggesting that the defect would first form as H–N–H in the +2 charge state and that after trapping two electrons the barrier for relaxation into H_2^* would be too large to overcome so that the canted metastable H–N–H structure would persist (Du et al., 2005).

Vibrational frequencies were calculated by both groups (Kleekajai et al., 2008; Fowler et al., 2005; Du et al., 2005). There are two high-frequency stretch modes, two in-plane bend modes (scissors and rocking), and two out-of-plane bend modes. By geometry, the coupling of the in-plane bends is expected to be larger than that of the out-of-plane bends, so if four bend modes are observed, the highest and lowest should correspond to in-plane modes and the other two to out-of-plane modes. This is what is found theoretically (Kleekajai et al., 2008).

But not all four bend modes are seen experimentally (Fig. 5.10), so their identification is nontrivial. However, GaP has also been studied by our group, both experimentally and theoretically, with both H and D isotopes, and the correspondence between the two hosts is remarkably similar. The observation of bend modes in the two hosts for different hydrogen isotopes has allowed us to make compelling arguments as to their identification. Details of the arguments as well as of the vibrational calculations are given in Ref. (Kleekajai et al., 2008). Experimental frequencies and their assignments are given in Tables 5.1 and 5.2 (Kleekajai et al., 2008).

5.3 Piezospectroscopy of the H–N–H Center

The information about a defect's microscopic properties that can be gleaned from vibrational spectra is greatly enhanced by the

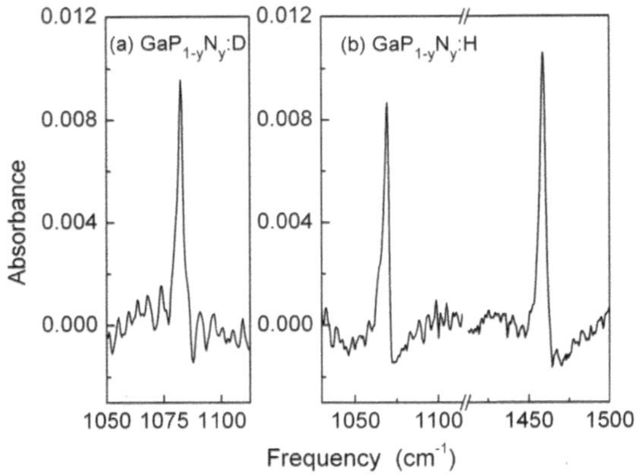

Figure 5.10 Absorbance spectra (4.2 K) showing (a) the D wagging mode for a deuterated $GaP_{0.993}N_{0.007}$ sample and (b) the H wagging modes for a hydrogenated $GaP_{0.993}N_{0.007}$ sample. From Ref. (Kleekajai et al., 2008).

application of uniaxial stress and the use of polarized light (Stavola, 1998). A defect with symmetry lower than its host will have several crystallographically equivalent orientations. The application of uniaxial stress can lower this orientational degeneracy, causing vibrational lines to split into components that can be selectively excited by polarized light. For stresses applied along a few directions of high symmetry, the number of stress-split components and their relative intensities can reveal the symmetry of the defect and the direction of its transition moment (Kaplyanskii, 1963). Furthermore, if the temperature is sufficiently high, the defect can be aligned among its possible orientations that have been made inequivalent by stress. This alignment and its annealing kinetics can be probed with polarized absorption measurements to reveal symmetry properties and the barrier(s) to microscopic defect motion.

5.3.1 *Symmetry of the H–N–H Center*

$GaAs_{1-y}N_y$ epitaxial layers with a reduced nitrogen content (0.15%), and consequently narrower N–D vibrational absorption lines than are shown in Fig. 5.8 above, were grown on 2 mm thick (001) GaAs substrates for uniaxial stress experiments (Wen et al., 2010).

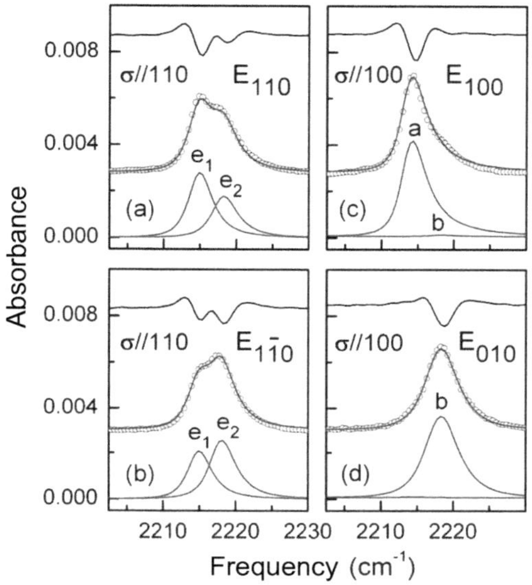

Figure 5.11 Effect of stress applied along the [110] and [100] directions for the 2217 cm^{-1} line of the D–N–D defect in $GaAs_{0.9985}N_{0.0015}$. Spectra (shown by open circles) were measured near 4 K with a resolution of 1 cm^{-1}. The result of a fit to the individual split components is shown by the solid line superposed on the measured IR data. The upper trace in each panel shows the second derivative of the measured spectrum. The incident light was polarized with the directions of the electric vector that are shown. The magnitudes of the stresses along the [110] and [100] directions were 230 MPa and 250 MPa, respectively. From Ref. (Wen et al., 2010).

Samples were prepared with their long axes along either the [100] or the [110] direction for the application of stress. These samples were studied with light incident along the [001] direction and with polarization either parallel or perpendicular to the stress axis.

Spectra (4.2 K) presented in Fig. 5.11 show splittings of the 2217 cm^{-1} N–D-stretching line for both the [110] and [001] stress directions (Wen et al., 2010). Other IR lines of the D–N–D center, for example the 2376 cm^{-1} line, are less sensitive to stress and do not show any resolvable splitting. For the [110] stress direction, the 2217 cm^{-1} line is split into two components, e_1 and e_2. For the [100] stress direction, the 2217 cm^{-1} line is split into two components, a and b.

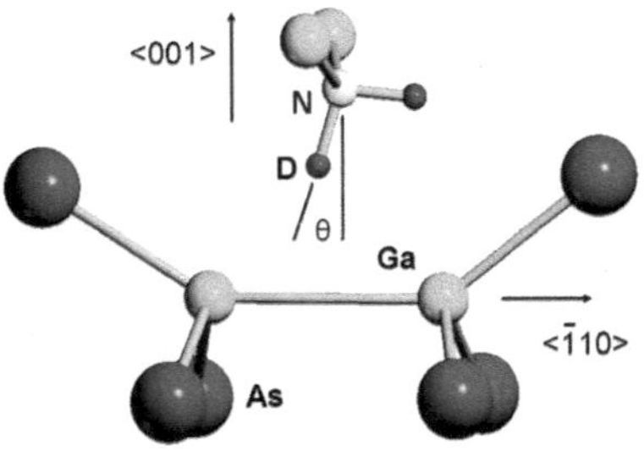

Figure 5.12 The relaxed D–N–D defect in $GaAs_{1-y}N_y$ (figure drawn using MOLDRAW, Refs. (Ugliengo, 2006; Ugliengo et al., 1993)). The angle, θ, of the N–D bond that gives rise to the 2217 cm^{-1} line is measured with respect to the <001> axis that is perpendicular to the <110> primary symmetry axis of the C_{1h} center. From Ref. (Wen et al., 2010).

The symmetry of the D–N–D center and the direction of the transition moment of the N–D-stretching mode at 2217 cm^{-1} can be determined from the number of stress-split components and their relative intensities. The stress-splitting patterns for centers with different symmetries are tabulated in the classic work of Kaplyanskii (Kaplyanskii, 1963). The behavior under stress for the case of C_{1h} symmetry has been expanded upon by Davies et al. (Davies et al., 1987). A comparison of the data in Fig. 5.11 with these results shows that the 2217 cm^{-1} line is due to a defect with C_{1h} symmetry, in agreement with the canted D–N–D structure predicted by theory. (A model for the D–N–D center is shown in Fig. 5.12. This C_{1h} defect has its <110> primary symmetry axis normal to the plane of the figure and has a {110} reflection plane.)

The shift rates for the stress-split components seen in Fig. 5.11 and their relative intensities are presented in Table 5.3. Wen et al. determined the parameters $A_1 = -6.4 \pm 2$ cm^{-1}/GPa, $A_2 = 9.6 \pm 2$ cm^{-1}/GPa, and $A_4 = 6.9 \pm 2$ cm^{-1}/GPa (Wen et al., 2010). A_3 could not be determined from the stress data because the components g and f have intensities too small to be seen for the sample orientations that were available to us. The sign of A_4 can be positive or negative.

A C_{1h} center can have its transition moment along the <110> primary axis or along any direction in the {110} reflection plane.

Table 5.3 Uniaxial stress perturbations for a C_{1h} center for the [100] and [110] stress directions and a [001] viewing axis. The columns on the right give the theoretical results for the intensities of the stress-split absorption lines for specific angles of the transition moment direction that lies in the plane that is normal to the main <110> symmetry axis of a C_{1h} center. Results are given for $\theta = 0°$ (left column) and $\theta = 7°$ (right column), where θ is the angle between the transition moment direction and the <001> axis perpendicular to the <110> symmetry axis of the center. See Ref. (Wen et al., 2010)

Stress	Component	Shift rate	E//100 : E//010	
[100]	a	A_1	4:0	3.9:0.03
	b	A_2	0:4	0.06:4.0
			E//110 : E//1$\bar{1}$0	
[110]	e_1	$(A_1+A_2)/2\text{-}A_4$	2:2	2.3 : 1.6
	e_2	$(A_1+A_2)/2+A_4$	2:2	1.6 : 2.3
	f	A_2 - A_3	0:0	0.03:0.03
	g	$A_2 + A_3$	0:0	0.03:0.03

The relative intensities of the stress-split components for the C_{1h} case were determined by Davies et al. as a function of the angle θ that specifies the direction of the transition moment in the reflection plane (Davies et al., 1987) (the angle θ is measured with respect to the <001> direction that is perpendicular to the <110> primary axis of the defect). The intensity ratios for the e_1 and e_2 components seen for stress applied along the [110] direction were determined by fits of the line shapes shown in Fig. 5.11 (Wen et al., 2010). For E//110, the ratio is $I_{||}(e_1):I_{||}(e_2) = 1.45:1$. For E//1$\bar{1}$0, the ratio is $I_{\perp}(e_1):I_{\perp}(e_2) = 1:1.36$. From these results, the angle of the transition moment was determined to be $\theta = 7° \pm 4°$.

The intensities of the a and b components seen for stress applied along the [100] direction are also consistent with an N–D bond angle with a small value for θ.

The inequivalent N–D modes of the D–N–D center are only weakly coupled, so an independent oscillator model is appropriate.

In this case, the transition moment direction gives the angle of the N–D bond whose vibrational mode is being probed. Theory finds that the 2217 cm^{-1} mode is due to the N–D bond shown in Fig. 5.12 that is oriented near the <001> direction (Kleekajai et al., 2008; Fowler et al., 2005).

Calculations using CRYSTAL06 predict the angle θ to be $18^\circ \pm 6^\circ$ (Fowler et al., 2005). The corresponding value found by Du et al. appears (from their Fig. 3) to be approximately 22° (Du et al., 2005). Thus the theoretical values agree with each other, but both are larger than found experimentally.

5.3.2 *Canting Distortion of the H–N–H Center*

The relative intensities of the components e_1 and e_2 seen for [110] stress depend on the temperature of the sample when the stress is applied, because the D–N–D complex can become aligned by stresses applied at elevated temperatures (Wen et al., 2010). For the data shown in Fig. 5.11, the sample was cooled to 4.2 K in the absence of stress. Stress was then applied at low temperature where the center cannot reorient. However, if [110] stress is applied at elevated temperatures, the relative intensities of the e_1 and e_2 components can be changed. And it turns out that the energy of the barrier to reorientation for the D–N–D complex is equal to the energy lowering produced by the canting relaxation of the defect.

Spectra presented in Fig. 5.13 show how the intensities of the e_1 and e_2 components are changed by stress applied at elevated temperatures (Wen et al., 2010). In these experiments, the sample was cooled to 4.2 K in the absence of stress, and a [110] stress was then applied. The sample was then annealed under stress at the temperatures shown for 30 minutes and cooled to 4.2 K for an IR absorption measurement to be made with polarized light. For annealing temperatures at 25 K and below, the relative intensities of the e_1 and e_2 components were unchanged by the stress anneal. After a 30-minute anneal at 35 K, the stress-induced changes in the intensities of the e_1 and e_2 components had reached equilibrium. Figure 5.14 shows the total intensities of the e_1 and e_2 components (determined by fits of the measured line shapes and summed over polarizations) as a function of the annealing temperature with the

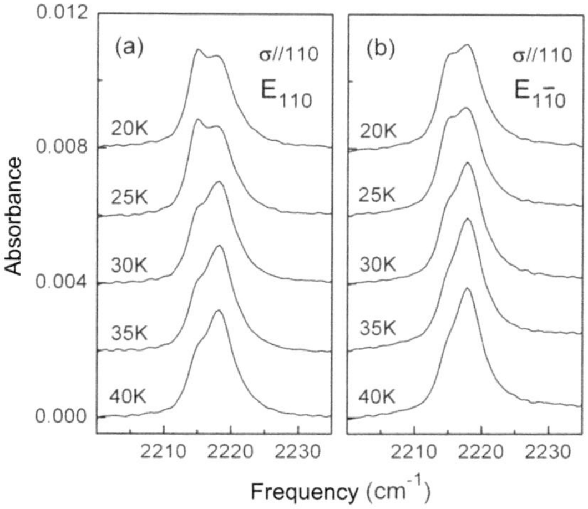

Figure 5.13 Effect of [110] stress (230 MPa) on the relative intensities of the components of the 2217 cm^{-1} line of the D–N–D center in $GaAs_{0.9985}N_{0.0015}$. Stress was applied initially at 4 K. The sample was then annealed for 30 minutes with the stress maintained at the temperatures that are shown. The IR spectra (4 K) shown were measured with polarized light following each anneal. From Ref. (Wen et al., 2010).

stress applied. This data show that the total intensity, $I_{tot}(e_1) + I_{tot}(e_2)$, remains unchanged during these experiments.

Similar experiments for the a and b components produced by [100] stress did not produce any observable alignment, even for stresses applied at temperatures as high as room temperature (Wen et al., 2010). For the C_{1h} defect shown in Fig. 5.12, there are 12 possible center orientations. For the [110] stress direction, four of these orientations give the f and g components that could not be seen for the sample orientations available for our experiments. The remaining eight orientations make up the e_1 and e_2 families with four defect orientations each. The <110> principal axes for the e_1 and e_2 families all make the same angle with respect to the [110] stress direction. What distinguishes the e_1 family of defects from the e_2 family is the sign of the cant angle θ.

Under [110] stress, the center orientations labeled e_1 have a larger ground-state energy than the orientations labeled e_2. If

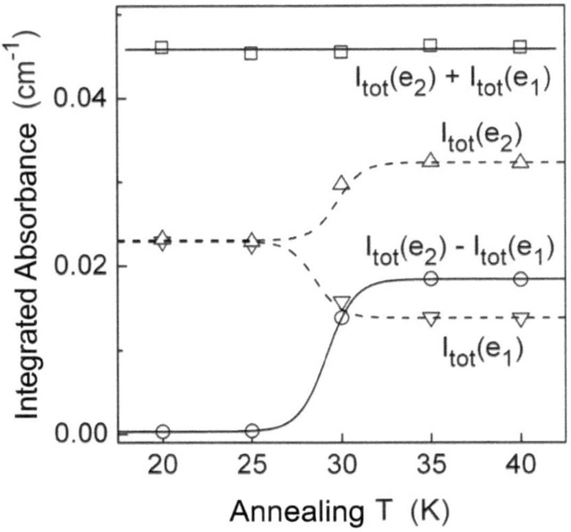

Figure 5.14 Isochronal annealing (30 min, under [110] stress) results from Fig. 5.13, for $I_{tot}(e_1)$, $I_{tot}(e_2)$, $I_{tot}(e_2)+I_{tot}(e_1)$ and $I_{tot}(e_2)-I_{tot}(e_1)$, as a function of the annealing temperature.

the temperature is sufficiently high for the e_1 and e_2 centers to reorient and sufficient time is allowed for thermal equilibrium to be established, the e_1 and e_2 components will become populated according to their Boltzmann factors. The results plotted in Fig. 5.14 show that a stress alignment of $[I_{tot}(e_2)\text{-}I_{tot}(e_1)]/[I_{tot}(e_2)+I_{tot}(e_1)] = 0.4$ can be achieved for a [110] stress of 230 MPa applied for 30 minutes at 40 K (Wen et al., 2010). The data in Figs. 5.13 and 5.14 show that 78% of the total possible alignment is achieved in 30 minutes at 30 K. This result permits the reorientation rate, $k_r = 0.77\times10^{-3}\ s^{-1}$, to be estimated for the temperature 30 K (Wen et al., 2010). This rate is presumed to be due to hopping-over of the barrier between orientations (Fig. 5.15) and is given by

$$k_r = \nu_0 \exp(-E_A/k_B T) \tag{5.1}$$

If the attempt frequency is taken to be a typical phonon frequency, $\nu_0 = 10^{-13}\ s^{-1}$, an energy barrier of $E_A = 96$ meV is obtained from Eq. 5.1 (Wen et al., 2010).

Figure 5.15 shows that the reorientation barrier between the e_1 and e_2 configurations is the energy lowering due to the canting

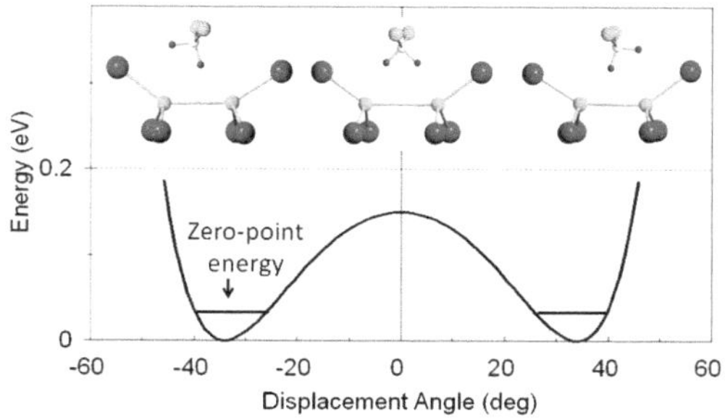

Figure 5.15 Double-well potential for the H-N-H center in $GaAs_{1-y}N_y$, depending on theoretical results from CRYSTAL06. The displacement angle of 34° is consistent with the theoretical canting angle of 21° (Wen et al., 2010) (see Fig. 5.12). Figure drawn using MOLDRAW, Refs. (Ugliengo, 2006; Ugliengo et al., 1993).

relaxation of the D–N–D center. Calculations using CRYSTAL06 predict the difference between the barrier height and the zero-point energy for rocking motion of the relaxed canted state to be 120 meV, as shown in Fig. 5.15 (Wen et al., 2010). This is somewhat larger than the experimental value but is in better agreement with experiment than the 50 meV barrier height obtained by Du et al. (Du et al., 2005) (from which the zero-point energy of ~ 27 meV would be subtracted to compare with the experiment).

5.3.3 *Response to a Challenge to the D–N–D Structure: Sensitivity of the N–D Modes to Stress*

Alt et al. found that the 2217 cm^{-1} line assigned to the D–N–D center in $GaAs_{1-y}N_y$ appeared to be absent from IR spectra measured for some of their deuterated samples where the other lines assigned to the D–N–D center were present (Alt et al., 2009). This observation caused Alt et al. to conclude that the 2376 cm^{-1} and 2217 cm^{-1} lines are due to two different N–D centers. We favor an alternative explanation, depending on the high sensitivity of the 2217 cm^{-1} line to stress, for the apparent absence of the 2217 cm^{-1} line from some of the spectra measured by Alt et al.

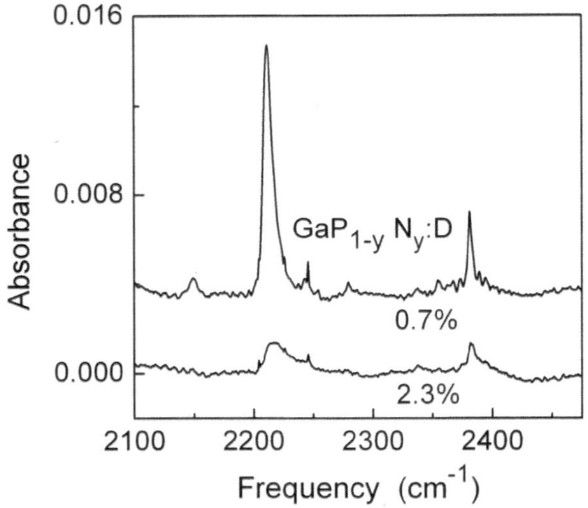

Figure 5.16 IR absorbance spectra (4.2 K, resolution 1 cm^{-1}) showing the N–D-stretching modes for 800 nm thick epitaxial layers of (upper) $GaP_{0.993}N_{0.007}$ and (lower) $GaP_{0.977}N_{0.023}$ grown on GaP substrates. The samples for the upper and lower spectra had been deuterated at a sample temperature of 300°C with doses of 1.0×10^{19} and 1.7×10^{19} impinging ions/cm^2, respectively.

Our experiments for the IR lines of the D–N–D center under stress showed that the frequency of the IR line at 2217 cm^{-1} is substantially more sensitive to applied stress than any of the other IR lines of the center (Wen et al., 2010). For example, the 2217 cm^{-1} line showed an observable stress splitting, while the sharper 2376 cm^{-1} line studied in the same samples showed no stress-induced splitting at all. This sensitivity of one of the N–D modes to strain can also be seen in Fig. 5.16 for $GaP_{1-y}N_y$:D samples with different N contents. In Fig. 5.16, the sample with $y = 0.7\%$ shows strong, sharp IR lines at 2211 cm^{-1} and 2381 cm^{-1} (in this case, the 2211 cm^{-1} and 2381 cm^{-1} lines correspond to the same N–D modes as the 2217 cm^{-1} and 2376 cm^{-1} lines in $GaAs_{1-y}N_y$:D, respectively). For the sample with greater N content, $y = 2.3\%$, the effect of random strain that accompanies the increased N content is much greater for the 2211 cm^{-1} line; the 2211 cm^{-1} line is broadened by a factor of near 4, while the 2381 cm^{-1} line is broadened by only a factor of 1.3. The ratios of the areas of the 2211 cm^{-1} and 2381 cm^{-1} lines are estimated to be 9:1 for both the $y = 0.7\%$ and $y = 2.3\%$ samples, indicating that the changes in the linewidths do not change the relative intensities of the lines. This unpublished data was measured as part of a study of H–N–H centers in $GaP_{1-y}N_y$ performed by Kleekajai et al. (Kleekajai et al., 2008).

The greater sensitivity of the 2211 cm-1 line to stress, and therefore to random strains, can explain the results of Alt et al. (2009). Figure 4 of their paper shows two spectra, the first where both the 2376 cm^{-1} and 2217 cm^{-1} lines are present and the second where only the 2376 cm^{-1} line is seen and the 2217 cm^{-1} line appears to be absent Alt et al. (2009). However, for the spectrum where the 2217 cm^{-1} line appears to be absent, the 2376 cm^{-1} line is much broader, by a factor near 3 (estimated from the published spectrum Alt et al. (2009)), than is the case for the other spectrum shown in the same figure, presumably due to greater random strain in the sample. We estimate that a random strain sufficient to broaden the 2376 cm^{-1} line by a factor of 3 would broaden the 2217 cm^{-1} line, with its greater sensitivity to stress, by a factor of approximately 8. This line broadening would cause the 2217 cm^{-1} line to apparently disappear, not because it is actually absent, but rather because it becomes impossible to distinguish the 2217 cm^{-1} line from other broad unrelated IR features that are present in the spectra of Alt et al.

We strongly favor the assignment of the 2217 cm^{-1} and 2376 cm^{-1} lines seen in $GaAs_{1-y}N_y$:D to a pair of inequivalent, N–D-stretching modes of a D–N–D center (and similarly for the 2211 cm^{-1} and 2381 cm^{-1} lines of $GaP_{1-y}N_y$). The absence of the 2217 cm^{-1} line in a few of the spectra measured by Alt et al. (Alt et al., 2009) can be explained consistently by the sensitivity of this line to stress, as was measured by Wen et al. (Wen et al., 2010), and the consequent increased sensitivity of its linewidth to random strains. The D–N–D structure is, therefore, an appropriate starting point for the consideration of defect structures that contain more than two D atoms that will be considered in the following section.

5.4 Microscopic Explanation of Compressive Strain

The formation of an H–N–H complex has been found by vibrational spectroscopy (Jiang et al., 2004; Kleekajai et al., 2008), XANES (Ciatto et al., 2005a) and theory (Fowler et al., 2005; Du et al., 2005; Ciatto et al., 2005a) to explain the band-gap shift of dilute III–N–V alloys caused by hydrogen irradiation. Vibrational spectroscopy, performed in conjunction with uniaxial stress, has provided an elegant strategy to reveal detailed microscopic information about the deuterated C_{1h} D–N–D complex and to test theory (Wen et al., 2010). However, there are further puzzles.

The addition of N to GaAs to form a dilute $GaAs_{1-y}N_y$ alloy creates a tensile strain within the GaAs lattice. When H or D is added at 300°C, the tensile strain is eliminated along with the band-gap shift caused by N (Polimeni et al., 2003b). With the further addition of H or D, the strain within the lattice becomes compressive (Bisognin et al., 2006; Berti et al., 2007; Bisognin et al., 2007, 2008)! (Furthermore, some experiments suggested that N–H_n complexes with $n > 2$ can be formed (Buyanova et al., 2007).) If H or D is added at reduced temperature, the compressive strain becomes even more pronounced (Wen et al., 2012). The temperature regime near 220°C is especially interesting because sharp interfaces can be defined by the introduction of H or D to make possible the fabrication of band-gap engineered structures with nanometer-scale resolution (Trotta et al., 2008) (see Chapter 8).

High-resolution X-ray diffraction (HRXD) studies of the annealing behavior of the compressive strain within the III–N–V lattice, performed in conjunction with photoluminescence studies of the band-gap shift, showed that the compressive strain could be eliminated by a long anneal at 250°C without affecting the band gap (Berti et al., 2007). An additional annealing treatment at 328°C eliminated the D-induced band-gap shift. These results, discussed in detail in Chapter 7, showed that different defect processes with different thermal stabilities cause the compressive strain and the band-gap shift.

A suite of experimental tools was used to investigate the deuterium content of $GaAs_{1-y}N_y$:D, showing a compressive strain (Berti et al., 2007). These results suggested the formation of an N–D_n

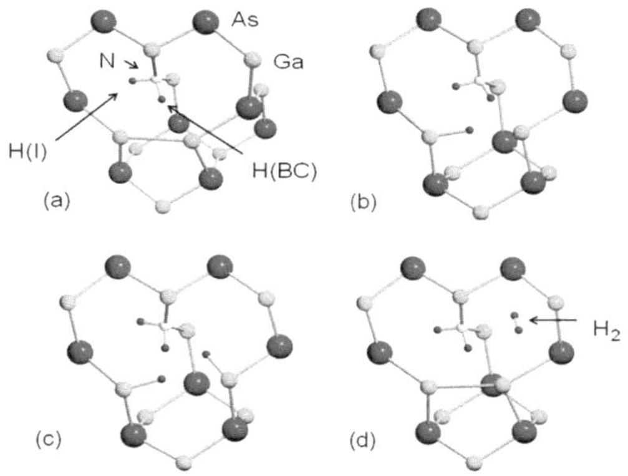

Figure 5.17 Computed structures for GaAs:N:H_n. (a) H–N–H, (b) H–N–H plus one Ga–H, (c) H–N–H plus two Ga–H, and (d) H–N–H plus interstitial H_2 in an adjacent (xy) cell. From Ref. (Wen et al., 2012). Figure drawn using MOLDRAW (Ugliengo, 2006; Ugliengo et al., 1993).

complex with three D atoms. Theory (see Chapter 4) proposed a D–N–D structure with additional D atoms bonded to nearby Ga to form D–N–D–(GaD) and D–N–D–$(GaD)_2$ structures (Amore Bonapasta et al., 2007) (see structures of Fig. 5.17b,c). XANES results were found to be consistent with D–N–D-like structures with additional D atoms bonded nearby (Ciatto et al., 2009).

Vibrational spectroscopy and complementary theory may then be used to probe the microscopic structures of N- and D-containing defects in GaAsN:D that lead to compressive strain within the lattice.

5.4.1 *IR Spectroscopy of $GaAs_{1-y}N_y$:D Showing Compressive Strain*

A set of $GaAs_{0.987}N_{0.013}$ samples was deuterated at different temperatures between 200°C and 280°C for study by HRXRD and IR absorption experiments (Wen et al., 2012). The X-ray diffraction rocking curves in Fig. 5.18a show that samples deuterated with doses between 3.0 and 5.0×10^{18} ions/cm^2 have compressive strains with greater magnitudes for samples deuterated at lower temperatures (HRXRD will be addressed in Chapter 7). IR absorption spectra for these samples are shown in Fig. 5.18b.

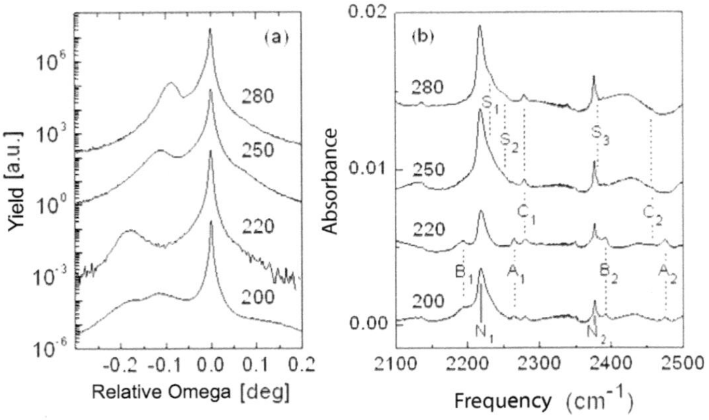

Figure 5.18 (a) (004) X-ray diffraction rocking curves for 230 nm thick samples of $GaAs_{0.987}N_{0.013}$ after deuterium irradiation with doses between 3.0 and 5.0×10^{18} impinging ions/cm^2 at the temperatures shown. (b) IR absorbance spectra (4.2 K, resolution of 1 cm^{-1}) for the N–D-stretching modes of the $GaAs_{0.987}N_{0.013}$ samples whose X-ray diffraction rocking curves are shown in panel (a). From Ref. (Wen et al., 2012).

Table 5.4 Frequencies, in cm^{-1}, of additional IR lines observed for $GaAs_{1-y}N_y$ epitaxial layers deuterated at temperatures between 200°C and 280 °C (Wen et al., 2012)

B_1	S_1	S_2	A_1	C_1	S_3	B_2	C_2	A_2
2190	2224	2251	2264	2280	2379	2389	2459	2476

The IR lines labeled N_1 and N_2 are due to the bare D–N–D complex in $GaAs_{1-y}N_y$:D. A number of additional lines labeled A, B, C, and S are also seen whose frequencies are listed in Table 5.4.

To probe the introduction and annealing behaviors of these new IR lines with a greater signal-to-noise ratio, a $GaAs_{0.9913}N_{0.0087}$ epitaxial layer was grown on a thick substrate to make IR measurements in an MIR geometry possible (Fig. 5.19a). This sample was deuterated with a dose of 6×10^{18} ions/cm^2, giving the bottom IR spectrum shown in Fig. 5.19b (Wen et al., 2012). The lines N_1 and N_2 at 2217 cm^{-1} and 2376 cm^{-1} due to the bare D–N–D center are introduced along with the new lines A, B, and C. Increasing

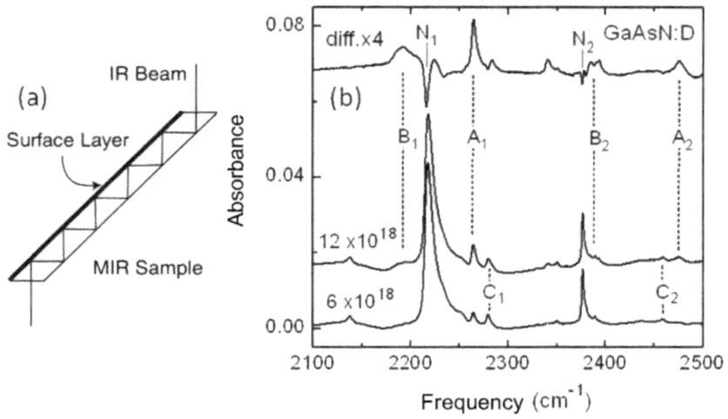

Figure 5.19 (a) The MIR geometry used to increase the signal-to-noise ratio of IR absorption measurements. (b) IR absorbance spectra measured in an MIR geometry for a $GaAs_{0.9913}N_{0.0087}$:D layer. The two lowest spectra are for the same sample after it was deuterated at 220°C with the different doses that are indicated. The upper trace shows the difference of the spectra measured for the different D doses. From Ref. (Wen et al., 2012).

the D dose further to 12×10^{18} ions/cm^2 gives rise to the center spectrum in the panel. The difference of the spectra obtained for the different deuterium doses is shown at the top. These spectra show that adding D strengthens the A, B, and C lines, while weakening the lines N_1 and N_2. The formation of new N–D_n centers at the expense of the bare D–N–D center is suggested by these results.

Spectra are shown in Fig. 5.20a for a $GaAs_{0.9913}N_{0.0087}$ sample deuterated at 220°C with a total D dose of 12×10^{18} ions/cm^2 (Wen et al., 2012). A set of 30-minute anneals was performed, with the temperature increasing in 50°C steps, to probe the thermal stabilities of the IR lines introduced by deuteration. Difference spectra in Fig. 5.20b show the changes that occur between successive annealing steps. There is little change in the N–D spectra until an annealing temperature of 200°C. The results in Fig. 5.20b show that the lines A_1 and A_2 begin to be annealed away at 200°C while the lines N_1 and N_2 increase in strength. At 250°C, the annealing of the A_1 and A_2 lines is completed, the B_1 and B_2 lines are also annealed away, and the C_1 and C_2 lines begin to be annealed away. As the A, B, and C lines are decreased in intensity at 250°C, the lines N_1 and N_2 arising from the bare D–N–D center further gain intensity. In the anneal at 300°C, the lines C_1 and C_2 are annealed

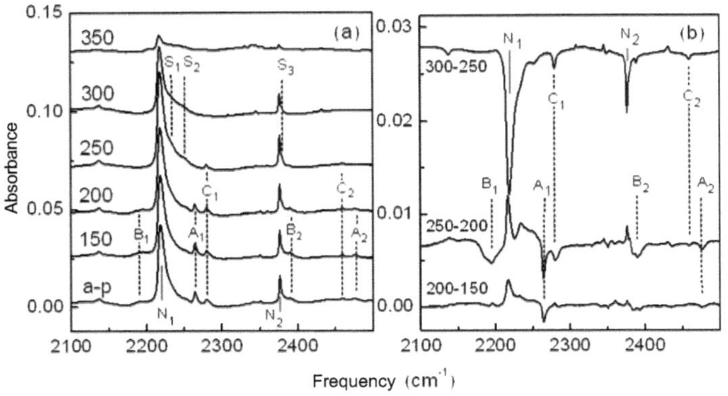

Figure 5.20 (a) IR absorbance spectra (4.2 K) measured in an MIR geometry for a $GaAs_{0.9913}N_{0.0087}$:D layer that had been deuterated at 220°C with a dose of 1.2×10^{19} impinging ions/cm^2. This sample was annealed sequentially at the temperatures indicated (°C) for 30 minutes. (a-p is for the as-prepared sample.) (b) Difference spectra for selected sequential annealing steps shown in panel (a). These difference spectra reveal the changes in spectral-line intensity that occur between the particular annealing steps that are indicated. Ref. (Wen et al., 2012).

away, and the lines N_1 and N_2 of the bare D–N–D defect also begin to disappear.

These results show that the new IR lines, seen for samples with compressive strain, are introduced (Fig. 5.19) or annealed away (Fig. 5.20) in pairs (Wen et al., 2012). Furthermore, the changes seen for the new A, B, and C lines upon annealing are anticorrelated with the changes seen for the N_1 and N_2 lines of the bare D–N–D center. These spectroscopic results suggest that the addition of sufficient deuterium to produce a compressive strain creates new D–N–D centers that are perturbed by the additional D atoms bonded nearby. These new centers give rise to two, new, N–D lines each, shifted by a few tens of wavenumbers from the N–D lines of the bare D–N–D center.

We also note shoulders to the N_1 and N_2 lines labeled S_1, S_2, and S_3. Wen et al. suggested an assignment of these shoulders to additional perturbed D–N–D centers but with a smaller perturbation than produces the A, B, and C lines with their greater frequency shifts from the IR lines of their bare D–N–D parents (Wen et al., 2012).

The IR data from the annealing experiments (Wen et al., 2012) show that a single $N-D_3$ center is not solely responsible for the compressive strain seen within $GaAs_{1-y}N_y$ deuterated at reduced temperature (220°C), as was suggested previously (Berti et al., 2007). Instead, these results suggest that several perturbed D–N–D centers are formed with additional D bonded near a D–N–D core (Wen et al., 2012). Each new perturbed D–N–D center gives rise to a pair of N–D lines shifted from its bare D–N–D parent.

These annealing results are also consistent with HRXRD studies that found a distribution of binding energies for $N-D_n$ centers that give a compressive strain in deuterated $GaAs_{1-y}N_y$ (Bisognin et al., 2007), suggesting that more than one type of center is formed. Theoretical analysis may be used to gain insight into what the microscopic structures of these perturbed D–N–D centers might be.

5.4.2 *$N-H_n$ Centers with $n>2$ Built from an H–N–H Core Structure*

Bonapasta et al. carried out calculations on GaAs:N with two, three, and four H atoms, using density functional theory (Amore Bonapasta et al., 2007) (see Chapter 4 for details). Wen et al. subsequently used CRYSTAL06 to replicate and extend these calculations, predicting structures, relative energies of different metastable configurations, and vibrational frequencies for these plus others, including two (H–N–H and $H-N-H(GaH)_2$) that contain a neighboring interstitial H_2 molecule in two adjacent locations, labeled (xy) and (z), respectively (Wen et al., 2012) (z is the <001> direction that is perpendicular to the <110> primary symmetry axis of the H–N–H core of the defect). Figure 5.17 shows that each system contains a canted H–N–H core. These predicted structures agree well with those of Fig. 1 of Ref. (Amore Bonapasta et al., 2007), despite the fact that the two calculations involved independent approaches. It seems clear that the canted H–N–H core is a robust constituent of all of the considered defects, in good agreement with the XANES results, which will be presented in the following chapter.

Table 5.5 Relative energies of different metastable configurations of GaAs:N:H$_n$, computed by CRYSTAL06 (Wen et al., 2012). NH_3 is H–N–H–(Ga–H) and NH_4 is H–N–H–(Ga–H)$_2$. The labels in parentheses refer to the labels of structures shown in Fig. 5.17

Lower	Higher	Energy difference (eV)
NH_4(c)	NH_3 + 0.5xH_2 distant (b)	1.22
NH_4(c)	NH_2 + H_2(x,y) (d)	1.94
NH_4(c)	NH_2 + H_2 distant	2.07
NH_4(c)	NH_2 + H_2(z)	2.21

Relative stabilities for various configurations computed by Wen et al. (Wen et al., 2012) with CRYSTAL06 are shown in Table 5.5. They found that the four-H defect (Fig. 5.17c) is more stable than the three-H defect (Fig. 5.17b), which, in turn, is more stable than the H–N–H defect with interstitial H_2 (e.g., Fig. 5.17d). Furthermore, the H–N–H defect with an H_2 in an adjacent (xy) cell is slightly more stable than H–N–H with a distant H_2. However, H–N–H with H_2 in a different adjacent (z) cell (structure not shown in Fig. 5.17) is slightly less stable than H–N–H with a distant H_2. But because the relative energies for defects that involve H_2 in different cells are only of order 0.1 eV, their relative stabilities are not definitely established.

It is somewhat difficult to assign specific vibrational modes to a particular structure, because their frequencies are determined by the weakly coupled N–D modes of the D–N–D unit at the core of each structure. Nonetheless, the calculated signs and magnitudes of the frequency shifts of the N–D modes that occur when additional D atoms are added can provide clues that suggest specific assignments (Wen et al., 2012).

Harmonic vibrational frequencies computed by Wen et al. were scaled by factors obtained from anharmonicity calculations using CRYSTAL06 under identical conditions for an NH_2 molecule (Wen et al., 2012). The resulting frequencies for the H–N–H (or D–N–D) defects, 3110 (2291) cm^{-1} and 2795 (2052) cm^{-1}, respectively, then served as the bases for comparison as more complex defects were considered. Calculated frequency shifts for selected D–N–D defects

Table 5.6 Computed frequency shifts, in cm^{-1}, of the D–N–D modes with additional D atoms (Wen et al., 2012). Configurations are shown in Fig. 5.17: for example, (b)-(a) gives the shifts of the higher and lower modes for D–N–D with and without an additional D attached to a Ga. Results from Ref. (Amore Bonapasta et al., 2007) are estimated by isotope-shifting, D for H, the values given for H defects in Table III of that paper

	Present calculations		Ref. (Amore Bonapasta et al., 2007)		Experiment		
	high	low	high	low		high	low
(b)-(a)	25	−33	34	−30	B's	13	−27
(c)-(a)	44	119	18	28	A's	100	47
					C's	83	63
(d)-(a)	31	9	-	-	S_1,S_3	3	7
					S_2,S_3	3	34

are given in Table 5.6, along with values estimated from the H–N–H frequencies given in Table III of Ref. (Amore Bonapasta et al., 2007).

Corresponding experimental frequency shifts from Table 5.4 are also shown. Wen et al. argued that these results suggest that the lines B_1 and B_2 arise from the three-D defect and that either A_1 and A_2 or C_1 and C_2 involve the four-D defect (Wen et al., 2012). They suggested, furthermore, that S_1, S_2, and S_3 involve D–N–D with an adjacent D_2 molecule and that either the A pair or the C pair may involve the four-D defect with an adjacent D_2 molecule.

5.5 Conclusion

The vibrational motions of nitrogen and hydrogen have revealed, step by step, the microscopic properties of the NH_n centers responsible for the remarkable band-gap shifts of the dilute nitrides caused by hydrogen. The observation of two inequivalent, weakly coupled N–H modes (Jiang et al., 2004) provided the essential experimental clue that with corresponding theory (Fowler et al., 2005; Amore Bonapasta et al., 2003; Du et al., 2005; Ciatto et al., 2005a) supported a canted H–N–H structure with C_{1h} symmetry as a cause of the hydrogen-induced band-gap shifts.

Polarized IR absorption measurements, made in conjunction with uniaxial stress, provided an elegant test of the predictions of theory. The C_{1h} symmetry of the H–N–H center was confirmed and the angle of the N–H bond and the energy lowering caused by the canting relaxation of the defect were determined (Wen et al., 2012).

The addition of hydrogen to $GaAs_{1-y}N_y$ samples causes the stress within the lattice to change from tensile to compressive as NH_n centers with $n > 2$ are formed (Bisognin et al., 2006; Berti et al., 2007). IR measurements made in an MIR geometry have a sufficiently high signal-to-noise ratio to study the formation and microscopic structures of these defects that contain more than two H atoms. Experiment and complementary theory find that the H–N–H structure is robust, acting as a core for defects with additional hydrogen atoms bonded nearby (Wen et al., 2012; Amore Bonapasta et al., 2007).

Vibrational spectroscopy and its interpretation by theory, when combined with the other structure-sensitive probes surveyed in this volume, reveal a compelling, atomic-scale model of the H–N–H centers in dilute nitride semiconductors that shift the band gap.

Acknowledgments

We are especially grateful to Mario Capizzi and Antonio Polimeni for introducing us to dilute nitrides and for carrying out this enjoyable and productive collaboration with us on H in these fascinating materials. We are grateful to Fan Jiang, Suppawan Kleekajai, and Lanlin Wen, who performed the experimental work described here as part of their PhD studies, and to Kevin Martin and Kathryn Washer, who collaborated on parts of the theoretical work. We acknowledge the contributions of our many collaborators who grew special samples that made our experiments possible (S. Rubini and F. Martelli) and made complementary experimental measurements (R. Trotta, G. Bisognin, and M. Berti) that helped provide a context for our IR measurements. Work performed at Lehigh University was supported by NSF Grant No. 1160756.

Chapter 6

Structure of Nitrogen–Hydrogen Complexes from X-Ray and Synchrotron Radiation Techniques

Gianluca Ciatto[a] and Federico Boscherini[b]
[a]*Synchrotron SOLEIL, L'Orme des Merisiers, Saint-Aubin, BP 48, F-91192 Gif sur Yvette CEDEX, France*
[b]*Department of Physics and Astronomy, University of Bologna, viale C. Berti Pichat 6/2, 40127 Bologna, Italy*
gianluca.ciatto@synchrotron-soleil.fr

X-ray techniques played a key role in the understanding of the modifications of the structure of dilute nitrides induced at different length scales by hydrogenation. X-ray diffraction provided the scientific community with the first evidence that the modifications of the electronic properties discussed in Chapter 2 go in parallel with a remarkable variation of the alloy structure. Advanced synchrotron radiation X-ray spectroscopy techniques coupled with ab initio simulations based on density functional theory calculations provided firm evidence that the effects of hydrogenation of dilute nitrides are related to the formation of N–H complexes of C_{2v} symmetry, in agreement with infrared (IR) absorption, and that virtually any single N atom incorporated in the alloy is involved in the formation of these complexes. X-ray spectroscopies constitute

Hydrogenated Dilute Nitride Semiconductors: Theory, Properties, and Applications
Edited by Gianluca Ciatto

ISBN 978-981-4463-45-4 (Hardcover), 978-981-4463-46-1 (eBook)
www.panstanford.com

an innovative approach in the analysis of the three-dimensional structure of defects in semiconductors, unique in its capacity to inherently probe and weight all the atomic geometries of the N local environment and highly complementary to other structural tools discussed elsewhere in this book. Finally, these techniques provide a "local view" of the modifications of the electronic structure consequent to hydrogenation, specific to the dilute species at the origin of the effects, and complementary to the "average" electronic properties addressed by the more standard probes used in Chapter 2.

6.1 Introduction

This chapter illustrates the role played by X-ray techniques, in particular synchrotron radiation–based ones, in the analysis of the structure of nitrogen–hydrogen complexes in dilute nitrides. Section 6.2 introduces the X-ray diffraction (XRD) experiments that gave the first evidence of a modification of the structure driven by hydrogenation. Section 6.3 provides a theoretical and experimental framework for understanding the main results presented in this chapter, which were obtained by X-ray spectroscopy; the section explains the physics at the basis of X-ray absorption spectroscopy, with specific reference to the application to the subject of impurities in semiconductors, diffraction anomalous fine structure (DAFS), and X-ray emission spectroscopy (XES). The possibility of achieving both unidimensional and three-dimensional information on the local structure around the isoelectronic impurities and the structure of the complexes playing on different analysis strategies is explained; the importance of coupling spectroscopy data analysis with state-of-the-art theoretical calculations described in Chapter 4 is stressed. Section 6.4 shows the experimental results and data analysis of different X-ray spectroscopy experiments performed on hydrogenated dilute nitrides, focusing on the local structural information obtained and placing it in its historical context with respect to the results of different characterization techniques reported in Chapters 5 and 7. The complementarity between different technical approaches is emphasized. Section 6.5 draws the main conclusions of the chapter and future perspectives.

6.2 Structural Effects of Hydrogenation of Dilute Nitrides

As is often the case for semiconductors materials, the first-observed effects of hydrogenation on the physical properties of dilute nitrides concerned their optical and electronic properties, largely addressed in Chapters 2 and 3. This is not surprising considering that these properties are extremely important for the final integration of semiconductors in optoelectronic devices. Probably for this reason, the effects of hydrogenation on the structure of dilute nitrides were addressed only a couple of years later. It is worth remarking here that a modification of the emission properties of a material due to a postgrowth treatment does not automatically imply an important modification of the structure: in fact, photoluminescence (PL) and related techniques are not bulk probes, and the emission spectrum of a semiconductor can be ruled by transitions between states related to impurities or defects present at very low concentrations and not representative of the main crystallographic structure. The first XRD[a] pattern on hydrogenated $GaAs_{1-y}N_y$ was taken by some of us in March 2002 at the crystallography laboratory of CNRS in Grenoble, France. The measurement was conceived as sample precharacterization in view of a spectroscopy experiment with synchrotron radiation and quickly performed just to be in time for using the beam time at the synchrotron. In this context, the results obtained and reported below in Fig. 6.1 and Fig. 6.2 were particularly astonishing and perhaps more interesting than the scheduled experiment itself.

The characterized samples were $GaAs_{1-y}N_y$ epilayers (y = 0.0081 and 0.013, layer thickness equal to 300 nm) and

[a]Knowledge of the principles of XRD is assumed to be held by the readers of the present book, and this technique will be not addressed in the next section, along with the spectroscopy ones. Here, we only call back to mind that XRD permits the determination of the lattice parameter of a crystal and refer to master course textbooks (Ashcroft and Mermin, 1976) or dedicated books (Warren, 1990) for details.

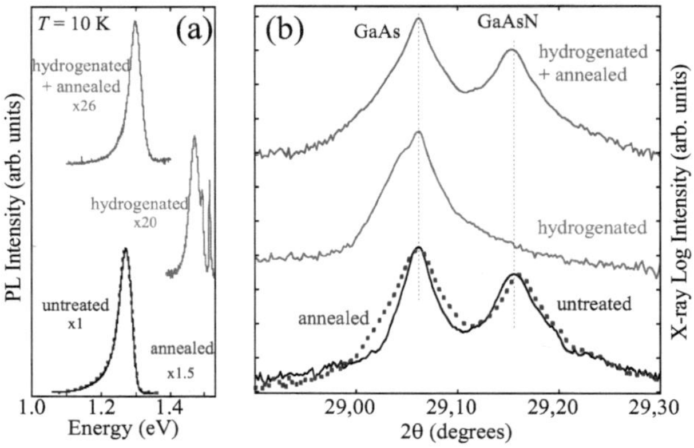

Figure 6.1 (a) PL spectra of a 300 nm thick $GaAs_{0.9919}N_{0.0081}$ epilayer after different postgrowth treatments. Bottom (continuous line): Untreated sample. Middle: The same sample after exposure to a hydrogen dose equal to 3.0×10^{18} ions/cm^2. The narrow bands on the high-energy side of the main PL band are due to carrier recombination from GaAs states. Top: A hydrogenated sample with the same H dose but annealed at 500°C for 30 minutes. The spectrum of a sample subjected to the same heat treatment (temperature 300°C and duration 6 hours) of the hydrogenated sample but in the absence of hydrogen is shown at the bottom by the gray dotted line superimposed to the spectrum of the untreated sample (bottom continuous line). (b) XRD curves for the same samples whose PL spectra are displayed in part (a). The XRD data has been recorded in the vicinity of the [004] crystal plane reflection. The curves have been vertically offset for clarity. Modified from Ref. (Polimeni et al., 2003b).

$In_xGa_{1-x}As_{1_y}N_y$/GaAs QWs ($x = 0.36$, $y = 0.052$ and 0.042, QW thickness equal to 8.0 nm) grown on GaAs substrates by solid-source molecular beam epitaxy. Postgrowth incorporation of atomic hydrogen was obtained as described in Chapter 2. XRD measurements on $GaAs_{1-y}N_y$ were performed in the $\theta - 2\theta$ geometry by exploiting a molybdenum-rotating anode as an X-ray generator and setting the incident wavelength at the Kα1 Mo fluorescence line (0.7092 Å) by means of a Si [111] channel-cut monochromator. The scattered intensity was detected by a NaI(Tl) scintillation detector. The XRD data has been recorded in the vicinity of the (004) crystal plane reflection. Figure 6.1b shows XRD data of a $y = 0.0081$ $GaAs_{1-y}N_y$ epilayer exposed to different postgrowth

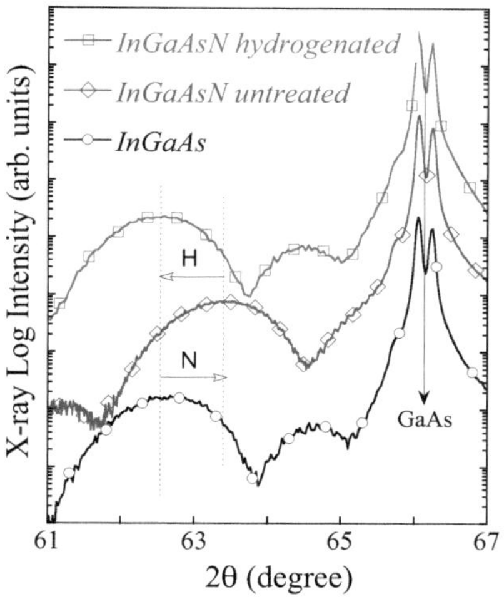

Figure 6.2 XRD from 8.0 nm thick, In-containing QWs. Bottom: $In_{0.36}Ga_{0.64}As$ (no nitrogen). Middle: Untreated $In_{0.36}Ga_{0.64}As_{0.948}N_{0.052}$. Top: The same $In_{0.36}Ga_{0.64}As_{0.948}N_{0.052}$ sample now hydrogenated with 3.0×10^{18} H ions/cm^2. The vertical dashed lines mark the diffraction peaks due to the QW layers. The horizontal arrows depict the effect of N incorporation in $In_{0.36}Ga_{0.64}As$ and H irradiation in $In_{0.36}Ga_{0.64}As_{0.948}N_{0.052}$. The double peak at about 66° is due to the GaAs substrate. The curves have been vertically offset for clarity. Modified from Ref. (Polimeni et al., 2003b).

treatments, whose corresponding PL spectra are shown in Fig. 6.1a. As already shown in Chapter 2, H insertion (middle curve in Fig. 6.1a) and removal (top curve) dramatically modify the material energy gap in a reversible way. In the XRD curve for the as-grown $GaAs_{1-y}N_y$ sample (bottom solid curve), one can observe two diffraction peaks. The higher and lower intensity peaks originate from the GaAs substrate and the $GaAs_{1-y}N_y$ epilayer, respectively. The value of the angular separation between the two peaks allows us to measure the N concentration (Li et al., 2001); the positive angular shift of the N-containing epilayer peak indicates that it has a smaller lattice constant along the growth direction (5.636 Å) than it has in the plane, where it is lattice-matched to the GaAs substrate (5.653 Å).

Remarkably, in the diffraction curve recorded on the hydrogenated $GaAs_{1-y}N_y$ epilayers (middle curve in Fig. 6.1b), the peak associated with the $GaAs_{1-y}N_y$ epilayer completely disappears. At the same time a shoulder can be detected at slightly smaller angles than that of the substrate peak, indicating the presence of compressive strain in the hydrogenated $GaAs_{1-y}N_y$ epilayer. These results indicate a very important lattice relaxation of the out-of-plane lattice constant of the hydrogenated layer, which can even exceed that of the GaAs. Since heat treatment is necessary during the hydrogenation process, to be sure that this does not bring about any structural effect itself, we also measured a sample subjected to heat treatment identical to the one the sample is subjected to during the hydrogenation process: the relative XRD curve (gray dotted line, bottom of Fig. 6.1b) shows no variation with respect to the one of the untreated sample, which parallels the lack of variation in the optical properties shown at the bottom of Fig. 6.1a by the gray dotted curve. Moreover, XRD data recorded on a same piece of a sample that was previously hydrogenated and then annealed until all H was removed (top of Fig. 6.1b) ruled out the hypothesis that the effect can originate from nitrogen diffusion out of the lattice and demonstrated that the large variation of the crystal unit cell size upon H insertion is a reversible process, which goes along with the full recovery of the $GaAs_{1-y}N_y$ optical band gap (Fig. 6.1a top). Similar results were obtained for the $y = 0.013$ samples series.

This result was confirmed by studying the case of a more disordered system under opposite initial strain conditions, that is, the $In_xGa_{1-x}As_{1_y}N_y$ alloy. In this case, the first XRD measurements on thin QWs were performed by using a copper-rotating anode without a monochromator and referred to the Cu Kα1 line (1.54088 Å) for the determination of the lattice parameter;[a] the XRD data was recorded in the vicinity of the [004] crystal plane reflection, as for the case of $GaAs_{1-y}N_y$. In Fig. 6.2 (bottom curve), we show the XRD diffraction curve of a N-free $In_xGa_{1-x}As$ QW ($x = 0.36$, thickness $= 8$ nm): due to the substitution of 36% of the Ga atoms with larger In atoms, the active layer is characterized by compressive strain (out-of-plane lattice parameter $= 5.929$ Å), which can be partially relieved by adding N. This is also shown in Fig. 6.2 (middle curve) where XRD data for an $In_xGa_{1-x}As_{1_y}N_y$ QW having the same

[a]A splitting of the substrate peak due to the Kα2 line is visible in the spectra.

thickness and In concentration as the N-free reference but $y = 0.052$ is reported; in this case the out-of-plane lattice parameter extracted from XRD is equal to 5.868 Å. Hydrogenation of the quaternary $In_xGa_{1-x}As_{1_y}N_y$ QW completely restores the lattice properties of the ternary N-free QW since the position of the active-layer XRD peak (top curve) is indistinguishable from that of the $In_xGa_{1-x}As$ samples. We also checked that the XRD curve of the N-free sample is left unaffected by the same H irradiation (not shown), while, as for the case of $GaAs_{1-y}N_y$, PL data recorded on the hydrogenated $In_xGa_{1-x}As_{1_y}N_y$ QW shows a full recovery of the band gap to the value of the N-free $In_xGa_{1-x}A$ QW (Polimeni et al., 2001a).

These XRD results were published in 2003 (Polimeni et al., 2003b), more than one year later with respect to the date of our first XRD experiment, together with density functional theory (DFT) calculations of the lattice parameter (Chapter 4) and IR absorption spectroscopy results (Chapter 5) in an attempt to provide an explication to the mechanism at the origin of the experimental observation, depending on the theory of the N–H complexes. Meanwhile, analogous results of another XRD precharacterization of the same sample series were published, together with Raman and photomodulated reflectance spectra, by Klar et al. (Klar et al., 2003) during the same year. Later, more systematic high-resolution X-ray diffraction (HRXRD), including in situ studies during thermal annealing, was performed by Berti et al.: The results of such characterization, deeply connected to nuclear reaction analysis in the study of the stoichiometry of nitrogen–hydrogen complexes, will be the subject of Chapter 7.

The XRD results presented in this section were extremely important because they provided the first clue that the optical effects of hydrogen irradiation originate from a large modification of the crystal structure and suggested a massive interaction of H ions with the constituent elements of dilute nitride alloys. They demonstrated, in general, the soundness of an insightful structural analysis of these materials and motivated not only more rigorous crystallographic studies but also an analysis of the structure at the local (atomic) scale, including the X-ray spectroscopy studies presented in the following sections of this chapter.

6.3 X-Ray Spectroscopies of Semiconductors

In this section we provide a brief introduction to three X-ray spectroscopies, X-ray absorption fine structure (XAFS), DAFS, and XES, in order to provide a basis for understanding the results obtained with these advanced methods in the field of dilute nitrides.

6.3.1 *X-ray Absorption Spectroscopy*

Photoelectric absorption is the dominant interaction mechanism between X-rays and matter. An *X-ray absorption spectrum* is a measurement of the absorption coefficient of a sample, $\mu\,(\hbar\omega)$, as a function of the energy of the impinging photons $\hbar\omega$ in the range of the characteristic *absorption edge* due to excitation of strongly bound core electrons of one of the elements of which it is composed. The nomenclature for the various absorption edges is well established; for example, excitation of electrons from 1*s* atomic orbitals corresponds to the K absorption edge.

The presence of *fine structure* in X-ray absorption spectra has been known since nearly 100 years; this fine structure shows up as a modulation of $\mu(\hbar\omega)$ at energies greater that the absorption edge, the relative magnitude of which is a few percent and decays with increasing energy from the edge itself. Two decisive factors have contributed to asserting XAFS as a valuable *structural tool*: the availability of brilliant and broadband synchrotron radiation sources and the development of an approximate but simple and useful point scattering theory. The "modern" era of XAFS is usually traced to the paper by Sayers et al. (Sayers et al., 1971).

The physical origin of XAFS in the framework of scattering theory can be understood quite easily. Consider an atom bound in a molecule or a solid. In the one-electron approximation, the core-level photoelectric absorption process is described in terms of the initial state consisting of the impinging photon of energy $\hbar\omega$ plus the electron in the core atomic orbital characterized by a wavefunction ψ_i with a binding energy E_{B} and a final state consisting of a core hole plus an excited electron characterized by a wavefunction ψ_f with an energy E_f. Depending on the photon energy, the electron can be excited to a discrete and bound final state or to a continuum and unbound final state. The final energy of the photoelectron is dictated by energy conservation

$$E_f = \hbar\omega - E_{\mathrm{B}}. \tag{6.1}$$

In the dipole approximation, the photoelectric absorption cross section is proportional to the square modulus of the matrix element of the scalar product of the photon polarization unit vector and the position operator between the initial and final states

$$\sigma\left(\hbar\omega\right) = 4\pi^2 \alpha\ \hbar\omega \left|\langle\psi_c|\,\hat{\varepsilon}\cdot\vec{r}\ |\psi_f\rangle\right|^2 \rho\left(E_f\right), \tag{6.2}$$

where $\alpha = \frac{e^2}{4\pi\varepsilon_0\hbar c} \cong \frac{1}{137}$ is the fine structure constant, $\hat{\varepsilon}$ is the unit vector defining the direction of the electric field of the exciting X-ray beam, and $\vec{r}$ is the position operator. One important consequence of the dipole matrix element is the selection rules, the most important one being $\Delta\ell = \pm 1$, that is, the orbital angular momentum quantum number must change by 1, an expression of the angular momentum conservation law. Let us consider only continuum final states. If the atom were not embedded in a matrix (e.g., a monoatomic gas) the final-state wavefunction would be, at sufficiently high energies, that of an outgoing spherical wave and no fine structure would result. The relation between the wavenumber ($k = \frac{2\pi}{\lambda}$, with λ the wavelength), and the kinetic energy for such a free electron is

$$k = \frac{\sqrt{2m\,E_f}}{\hbar} = \frac{\sqrt{2m\,\left(\hbar\omega - E_{\mathrm{B}}\right)}}{\hbar}. \tag{6.3}$$

If the atom is bound in a molecule or a solid the photoelectron wavefunction will be modified by the presence of the surrounding atoms' potential; one describes this modification as due to the *scattering* (one or more times) of the photoelectron by the surrounding atoms. The final-state wavefunction is now the outgoing spherical wave plus the scattered waves. Depending on the photoelectron wavenumber and the relative atomic arrangement, the scattering process may lead to an enhancement or a decrease of the amplitude of the final-state wavefunction close to the core-level orbital and thus to an increase or decrease of the cross section. This is an *interference* effect between the outgoing and scattered wavefunctions; the scattered field is the electron wavefunction, while the scattering "objects" are the neighboring atoms. As for any interference effect the position of maxima and minima in reciprocal space depends on the wavelength of the scattered field and on the geometry of the scattering object (e.g., slits, grating, or atomic lattice). In the present case, for a given atomic structure the cross section varies with photon energy, exhibiting characteristic oscillations which are quasi-periodic as a function of the wavenumber. The structural sensitivity stems from the dependence on the relative atomic positions between absorbing ("central") atoms and scattering (neighboring) atoms.

The preceding discussion easily allows us to illustrate the common separation of an XAFS spectrum into an extended and a near-edge region. If the energy of the photoelectron is high it will be only weakly affected by the neighboring atoms' potential; usually, it will be scattered a very limited number of times by the neighboring atoms and single-scattering (SS) processes will be dominant. This is the so-called extended X-ray absorption fine structure, or EXAFS, region. As the photoelectron energy decreases, closer to the absorption edge, photoelectron scattering by the neighboring atoms' potential will become increasingly important and multiple-scattering (MS) processes will give a greater contribution to the fine structure. This spectral region, which extends from the edge to roughly 50 eV above it, is commonly referred to as the X-ray absorption near-edge structure (XANES) or the near-edge X-ray absorption fine structure (NEXAFS). It must be stressed that the distinction between the two regions is largely arbitrary and that there is no difference in the basic physical mechanism that gives rise to the fine structure; however, the analysis methods for EXAFS and XANES are different, and thus it is reasonable to maintain the nomenclature.

In the past 40 years XAFS has evolved into a reliable *tool* that is able to provide a quantitative measurement of the local structure in condensed matter and has been applied to fields ranging from solid-state physics to cultural heritage materials. A number of review papers (Lee et al., 1981) and books (Teo and Joy, 1981; Koningsberger and Prins, 1988; Bunker, 2010) on XAFS have been published, and the reader is referred to them for a thorough description of experimental, theoretical, and data analysis aspects. We note that recently a review of the applications of XAFS to semiconductor heterostructures and nanostructures (Boscherini,

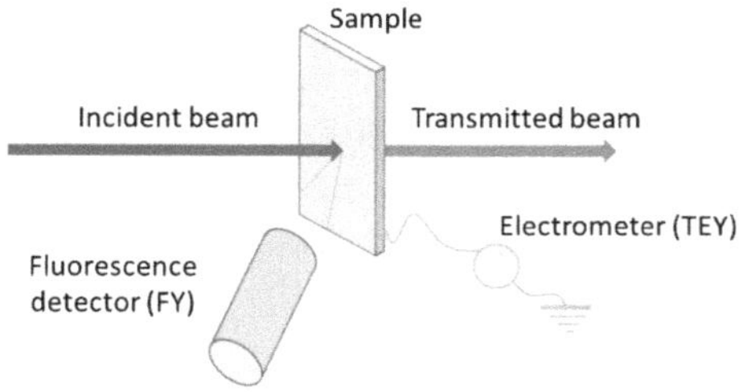

Figure 6.3 Sketch of the different XAFS detection setups: transmission, fluorescence yield (FY), and total electron yield (TEY).

2013), a book describing the full range of applications of XAFS to semiconductors in general (Schnohr and Ridgway, 2014), and a paper describing the use of advanced XAFS analysis applied to semiconductors (D'Acapito, 2011) have been published.

We now briefly describe the experimental setups relevant for XAFS measurements on dilute nitrides (sketched in Fig. 6.3). The objective of any XAFS measurement is to record, as a function of photon energy, a quantity that is directly proportional to the cross section of the atom of interest in the particular environment or phase that is relevant for the experiment. A number of methods have been devised. A transmission measurement of XAFS is the most commonly used in general. The transmission technique can be applied when the atom of interest is not dilute or at or near the surface; its main advantage is that it generally provides data of a very good signal-to-noise ratio, since it relies on the measurement of rather high photon fluxes (10^9–10^{11} photons/sec). In the field of semiconductor heterostructures and nanostructures one is mostly interested in studying the local structure of ultrathin films, nanolayers, or low-dimensional structures on, or near, the surface. Therefore, the transmission mode is mostly used to record spectra on concentrated reference samples but not the samples themselves.

The most commonly used method to record spectra for dilute atomic species is fluorescence yield (FY) (Jaklevic et al., 1977), in which a photon detector is used to record the intensity of the X-ray fluorescence (also known as characteristic radiation) as a function of energy. In the most common experimental geometry used with horizontally polarized X-rays, the detector is placed in the horizontal plane at right angles to the incident beam to minimize the intensity of elastic scattering of the impinging X-ray beam. It can be shown that the FY spectrum is directly proportional to the absorption coefficient of the excited atom for samples of thickness much less than the penetration depth or when the atom of interest is dilute (typically below around 10%).

Many types of detectors for FY mode have been used, but the most common are based on hyperpure Ge diodes or Si drift diodes (SDDs), often in the form of multielement arrays (Oyanagi et al., 1998; Derbyshire et al., 1999), and have an energy resolution in the range of 150–200 eV at 6 keV. Pulse-counting electronics is used, so care must be taken to avoid, or correct, dead-time effects (Ciatto et al., 2004).

Since both the probe and the detected signals are penetrating X-rays, FY is the best choice to study dilute elements in the bulk of materials. This has led to a wide application of XAFS to the study of dopants in semiconductors. The lowest concentration that can be measured in a FY-XAFS experiment ultimately depends on the magnitude of the inelastic-scattering background in the specific experimental geometry. The lowest detectable dose or thickness depends on the absorbing atom, the matrix, and the experimental geometry; XAFS spectra on thicknesses equivalent to 1–0.1 monolayers are nowadays measurable with an adequate signal-to-noise ratio, and even lower concentrations can be measured in specific cases.

Decay of the core hole following photoexcitation can occur also with the emission of an Auger electron. This Auger electron creates a cascade of lower-energy ("secondary") electrons due to multiple ionization and inelastic-scattering processes that can be used to record XAFS spectra. The experimentally simplest method consists in measuring the drain current using an electrometer, which connects the sample to ground, termed total electron yield (TEY). The main peculiarity of all electron detection modes is the surface sensitivity, which is guaranteed by the limited electron mean free path. Hence, these modes are particularly useful for the investigation of the near-surface region or of interfaces. Erbil et al. (Erbil et al., 1988) have reported a detailed study of the depth sensitivity of TEY mode, concluding that it depends on the range of the original Auger electrons and that it generally is in the range of a few 100 to a few 1000 Å.

6.3.2 *EXAFS and XANES*

In this section the basis of XAFS in the framework of MS theory (Schaich, 1973) is briefly outlined with the objective of providing a guide for the reader.

As described above, the photoelectric absorption cross section is directly proportional to the square modulus of the matrix element of the scalar product of the photon polarization unit vector and the position operator between the initial and final states. For a free atom the final state is simply an outgoing spherical wave. For a molecule or condensed matter one must the modification of the final state due to the presence of neighboring atoms, that is, to the scattering of the photoelectron by their potential. Since the principal interaction of the photoelectron is with the core-level electrons of surrounding atoms, the *muffin-tin approximation* for the scattering potential is commonly adopted. This consists of spherically averaging the potential around each atom and adopting a constant interstitial potential in between. In this approximation, the calculation of the absorption cross section is simplified into an MS problem of the final-state wavefunction by a collection of spherically symmetric scattering centers.

Under the previously stated assumptions, it can be demonstrated that an expression for the cross section for a given central atom can be obtained that depends on the scattering amplitude of neighboring atoms (which is related to their atomic number) and on their relative geometric arrangement (i.e., on the local atomic structure). At energies sufficiently far from the absorption edge (roughly 50 eV above it) this expression simplifies into a sum of quasi-sinusoidal terms as a function of the photoelectron wavenumber, which depend rather simply on the one-dimensional radial distribution function; this is the EXAFS region of the X-ray absorption spectrum. Instead, near the absorption edge, the relation between the cross-section and local structure is more complex since it depends on the three-dimensional local structural arrangement; this is the XANES region of the spectrum.

In the EXAFS region one defines the EXAFS spectrum $\chi(k)$ as

$$\chi(k) = \frac{\mu(k) - \mu_0(k)}{\mu_0(k)}, \tag{6.4}$$

where k is the photoelectron wavenumber and $\mu_0(k)$ is the absorption coefficient in the absence of neighboring atoms. It can be shown that the dominant contribution to the EXAFS function is due to so-called SS processes, which involve the central atom and one neighbor (a "back and forth" scattering path consisting of two "legs"). For an isotropic sample (i.e., a liquid, an amorphous solid, or a polycrystal with isotropic orientation of the crystallographic axes) the EXAFS function depends on the local structure, as described by the following equation, written for a central atom taken as reference, the neighboring atoms being grouped into coordination shells labeled by the index j and consisting of N_j identical atoms at the interatomic distance r_j and in the harmonic approximation for thermal motion:

$$\chi_2(k) = S_0^2 \sum_{j=\text{shells}} \frac{N_j f_j(r_j, k)}{kR_j^2} \sin\left[2kr_j + \varphi_j + 2\delta\right] e^{-2k^2\sigma_j^2}, \tag{6.5}$$

where $f_j(r_{j,}k)$ and φ_j are the modulus and argument of the effective scattering amplitude of atoms in shell j and δ is the phase shift due to the central atom,

$$S_0^2 = \left|\left\langle \Psi_{N-1}^i \right|\left| \Psi_{N-1}^f \right\rangle\right|^2 \tag{6.6}$$

is the many-body amplitude reduction factor (Ψ_{N-1}^i and Ψ_{N-1}^f are the many-body wavefunctions of the "passive" electrons before and after excitation of the "active" electron)

$$\sigma_i^2 = \left\langle \left[\hat{r}_{oi} \cdot \left(\vec{u}_0 - \vec{u}_i\right)\right]^2 \right\rangle, \tag{6.7}$$

is the Debye–Waller (DW) factor, which takes into account thermal and structural disorder; in the last expression $\langle \ldots. \rangle$ denotes a configurational average, $\hat{r}_{oi}$ is the unit vector joining the equilibrium positions of atoms 0 and i, and $\vec{u}_i$ is the instantaneous deviation of the position of atom i from its equilibrium position.

If the sample is an oriented single crystal and if, as usual, a linearly polarized photon beam is used, each atomic correlation will contribute to $\chi(k)$ with a weight w_{0i}, which depends on the relative orientation of $\hat{r}_{oi}$, the unit vector joining atoms 0 and i, and the direction of polarization $\hat{\varepsilon}$; a simple expression for this weight can be obtained in the *small atom approximation* in which the spatial extension of the scattering region associated with each atom is neglected with respect to photoelectron wavelength:

$$w_{0i} = 3\ (\hat{\varepsilon} \cdot \hat{r}_{0i})^2 . \tag{6.8}$$

Contributions to $\chi(k)$ due to MS paths have a form similar to Eq. 6.5. It can be shown that the MS contribution of order n due a set of atoms the coordinates of which are identified by a path index $\Gamma = \{\vec{r}_i\}$ and for which the total scattering path length is r_Γ are of the form

$$\chi_n^\Gamma = A_n(k, \Gamma) \sin[kr_\Gamma + \phi_n(k, \Gamma) + 2\delta] \tag{6.9}$$

The total EXAFS function will be the sum of all the SS and MS paths. In principle, there is an infinite number of such paths, but in practice, their number is limited by the rapid damping of the signal for long path lengths (usually paths the length of which is greater than $\sim$ 10 Å have negligible amplitude) and the fact that the amplitude of the MS signal quickly decreases with the order n (usually paths with $n > 4$ can be neglected). MS paths are weak in the EXAFS region but must be taken into account for an accurate structural determination.

Since the EXAFS function is the sum of sine functions, the argument of which is kr_Γ, a Fourier transform (FT) of $\chi(k)$ will exhibit peaks in correspondence to the path length; the FT is usually plotted as a function of the half path length, which, for SS paths, is equal to the interatomic distance.

From the analysis of an EXAFS spectrum the following local structural parameters can be obtained: interatomic distances (R, typically $\pm$ 0.005 Å for the first shell), identity of neighboring atoms and their coordination number (CN, typically $\pm$ 10%), and, provided a Gaussian model for the radial distribution function is appropriate, the DW factor (σ^2, typically $\pm 5 \times 10^{-4}$ Å^2).

EXAFS analysis requires an adequate starting structural model, which can be obtained from prior knowledge of the sample. An up-to-date analysis usually is based upon a simulation of the

EXAFS signal and a subsequent refinement of the local structural parameters via a least squares routine.

In the XANES spectral region it is not possible to express the cross section as a sum of sine functions. It is possible to perform a simulation of the XANES line shape in terms of the local structure in the so-called full multiple-scattering (FMS) approach, which takes account of all scattering paths, a relatively lengthy matter in terms of computer time. The XANES line shape is sensitive both to the details of the scattering potential (because of the low energy of the photoelectron) and to the three-dimensional position of atoms contained within a radius significantly greater than that necessary to reproduce EXAFS. As a consequence of the many parameters that determine the XANES line shape, quantitative analysis of XANES with a structural fitting approach is not generally possible at present. The inadequacy of the muffin-tin approximation for the scattering potential in the XANES region has been highlighted in cases in which the local structure of the absorbing atom has a low symmetry; a code to compute XANES by a numerical method without the muffin-tin approximation has been proposed by Joly (Joly, 2001, 2003).

Since XANES is sensitive to the three-dimensional atomic geometry (while the dominant contributions to EXAFS are of a one-dimensional character) and because XANES spectra are more easily recorded than EXAFS ones, especially for dilute elements, a quantitative interpretation of XANES would be of great interest. In the field of semiconductors, investigations of dilute nitrides were particularly important (Ciatto et al., 2005a) since it was one of the first cases in which, by making a comparison between experimental spectra and ab initio simulations, it was possible to determine the three-dimensional structure of the nitrogen–hydrogen complex responsible for the elimination of the red shift of the band gap in the dilute nitride alloy GaAsN (see Section 6.4.1). In a significant paper, Longa et al. (Longa et al., 2001) have described a structural fitting of XANES spectra for biomolecules; the extension of this method to solids would be a very important step forward in the investigation of dopants and defects in semiconductors.

An interpretation of XANES spectra in terms of *electronic structure* is also possible. In fact, referring to Eq. 6.2, if one neglects the energy dependence of the matrix element in a limited range of the near-edge region, one can see that the absorption cross section is proportional to the density of final states (in the presence of the core hole), selected in angular momentum by the selection rule $\Delta\ell = \pm 1$ and projected on the absorbing atomic species. This kind of interpretation links XANES to electronic structure calculations and will be useful in Section 6.4.3 below.

6.3.3 *Spectroscopy in Diffraction Conditions: DAFS*

6.3.3.1 The basic idea

DAFS refers to a fine structure in the intensity of an XRD peak, as a function of the incident X-ray energy, in the vicinity of an absorption edge. This phenomenon, first observed by Y. Cauchois (Cauchois, 1956) in 1956, finds its physical origin in the proportionality between the imaginary part of the atomic form factor f (Ashcroft and Mermin, 1976; Warren, 1990) and the total atomic cross section. The last is defined as the sum of the atomic cross sections for elastic scattering and absorption:

$$\sigma_{A,\text{total}(E)} = \sigma_{\text{elestic scatt.}} + \sigma_{\text{absorption}} \tag{6.10}$$

If we call f_{Aj} the atomic form factor (also referred to as the atomic scattering factor) for an atom A on a site j, the relationship with $\sigma_{Aj,\text{total}(E)}$ is given (in the forward scattering limit) by the optical theorem (Cohen-Tannoudji et al., 1973)

$$f''_{Aj}(\vec{Q} = 0, E) = \frac{E}{2hcr_0}\sigma_{Aj,\text{total}} \tag{6.11}$$

where r_0 is the classical electron radius and f''_{Aj} is the imaginary part of f_{Aj}:

$$f_A(\vec{Q}, E) = f_{0A}(\vec{Q}) + f'_A(E) + if''_A(E) \tag{6.12}$$

For nonforward scattering the relation 6.11 is valid in the dipole approximation; some studies have recently addressed the issue of the limit field of application for this relationship (Benfatto and Felici, 2001).

The absorption cross section $\sigma_{\text{absorption}}$ is proportional to the absorption coefficient, which shows oscillations in the vicinity of the absorption edges; the amplitude and frequency of these oscillations contain information on the local environment of the absorber (Section 6.3.1). Due to the optical theorem, these oscillatory structures propagate into f''_{Aj} and also into the energy-dependent real part of the atomic form factor f'_A, since it is related to f''_{Aj} via the Kramers–Kronig relationships (Sakurai, 1967):

$$f'(E) = \frac{2}{\pi} P \int_0^{\infty} \frac{E' f''(E')}{E^2 - E'^2} dE' \qquad (6.13)$$

$$f''(E) = -\frac{2E}{\pi} P \int_0^{\infty} \frac{f'(E')}{E^2 - E'^2} dE'$$

P indicates here the principal part of the integral. $f'_A(E)$ and $f''_A(E)$ are called anomalous contributions of the atomic form factor; elsewhere they are very low but become relevant near an absorption edge. Since they enter the expression of the structure factor of the cell F_{cell} and of the diffracted intensity, which is proportional to the square of F_{cell}, the oscillatory XAFS-like behavior propagates in the energy dependence of the Bragg peak amplitude. The same information available from an XAFS analysis can be, in principle, obtained studying the fine structure of the diffracted peak in the vicinity of an absorption edge.

The fundamental idea that aimed at the developing of DAFS was to unify the *chemical selectivity* of XAFS with the *spatial* and *site selectivity* of XRD. By means of DAFS it is possible, in spite of a more elaborate experimental setup and data analysis compared to XAFS, to select the contribution of a particular absorber (choosing the X-ray energy near its absorption edge) embedded in a specific long-range structure (choosing a defined Bragg peak). This allows us, for example, to study the local structure of a particular element present both in the active layer and in the substrate/cap of a thin QW by selecting the Bragg peak of the only active layer, thanks to the mismatch between the lattice parameters. DAFS was exploited to study the local environment of Ga in thin $In_xGa_{1-x}As_{1-y}N_y/GaAs$ QWs (see Section 6.4.2).

6.3.3.2 Data analysis

A step-by-step mathematical description of how spatial and site-selective local structural information can be extracted from a DAFS spectrum is out of the scope of this short section; a first demonstration of the DAFS spectroscopy was advanced by Stragier et al. (Stragier et al., 1992) in 1992, while an explanation of the DAFS data analysis in the extended energy region above an absorption edge (EDAFS) was provided by Proietti et al. (Proietti et al., 1999): we refer to these papers for a more detailed description. Here, we summarize that if we take only the first terms in the expansion of the square modulus of the structure factor (Ashcroft and Mermin, 1976; Warren, 1990) and write the complex fine structure as an expansion over the scattering paths of the photoelectron around the absorbing atom (Stragier et al., 1992), the EDAFS oscillations can be expressed in an EXAFS-like form:

$$\chi_{\vec{Q}}(k) = \sum_j \sum_\Gamma w_j(\vec{Q})\mathrm{Amp}_j^\Gamma(k)\sin[2kR_j^\Gamma + \delta_j^\Gamma(k) + \varphi_0(\vec{Q},k) - \varphi_j(\vec{Q}) - \frac{\pi}{2}] = S_d\left(\frac{|I_{\mathrm{exp}}| - |I_{0\,\mathrm{exp}}|}{|I_{0\,\mathrm{exp}}|}\right) \quad (6.14)$$

where Γ indicates a photoelectron-scattering path, $\mathrm{Amp}_j^\Gamma(k)$ is the amplitude of the scattering process along the path Γ, $\delta_j^\Gamma(k)$ is its phase shift, R_j^Γ is the effective path length, w_j are crystallographic weights, $\varphi_j(\vec{Q}) = \vec{Q}\cdot\vec{r_j}$, and $\varphi_0(\vec{Q},k)$ is the phase of the complex smooth structure factor ($\vec{Q}$ is the scattering vector); the first sum runs over the anomalous atoms in the unit cell and the second on the scattering paths. I_{exp} is the experimental diffraction intensity (Fig. 6.4) corrected for several factors (absorption, polarization, Lorentz, fluorescence background; see Ref. (Renevier et al., 2003)), and I_0 is the smooth atomic background; therefore the right-hand side of Eq. 6.14 represents the background subtracted scaled EDAFS oscillations.

The terms $\varphi_0(\vec{Q},k) - \varphi_j(\vec{Q}) - \frac{\pi}{2}$ and S_d in Eq. 6.14 are the DAFS phase and amplitude correction factors. In the case that the sample crystallographic structure is well known and that the chemical element defying the absorption edge occupies a single site in the cell (single anomalous site), the terms $\varphi_0(\vec{Q},k) - \varphi_A(\vec{Q}) - \frac{\pi}{2}$ and S_d can

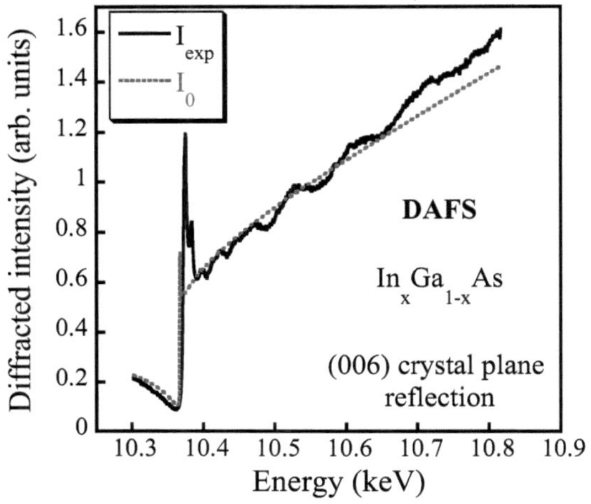

Figure 6.4 Normalized experimental DAFS (I_{exp}) and fit of the smooth part (I_0). Modified from Ref. (Ciatto et al., 2005c).

be evaluated by fitting the smooth DAFS spectrum (Proietti et al., 1999). Provided that these amplitude and phase correction factors are taken into account, the fine structure analysis can be performed according to standard criteria and available theoretical simulation codes for X-ray absorption (Section 6.3.1).

As for the experiment on $In_xGa_{1-x}As$ and $In_xGa_{1-x}As_{1-y}N_y$ QWs at the Ga K-edge described below in Section 6.4.2, we dealt with a noncentrosymmetric zinc-blende structure whose basis consisted of four elements: two cations located in position (0,0,0) and two anions located in position $(\frac{1}{4},\frac{1}{4},\frac{1}{4})$. The only "anomalous" atom of the basis was the Ga (0,0,0), which allowed us to determine $\varphi_0 - \varphi_A$ and S_d by fitting of the smooth part of DAFS, as shown in Fig. 6.4. The chosen reflection was the weak (006) one,[a] where $f_{0A}(\vec{Q})$ for Ga and As almost cancel out: this causes the anomalous effect to be quite large:

[a]The (006) reflection is forbidden for the diamond lattice, as well as all reflections for which the sum of the Miller indices is twice an odd integer (Ashcroft and Mermin, 1976). In the case of the zinc-blende structure, which is similar to the diamond except for the fact that a basis is necessary to account for the presence of chemically different elements, the different atomic form factors of anions and cations preserve the structure factor from completely vanishing. Nevertheless the structure factor is significantly reduced and the (006) is called a "weak" reflection.

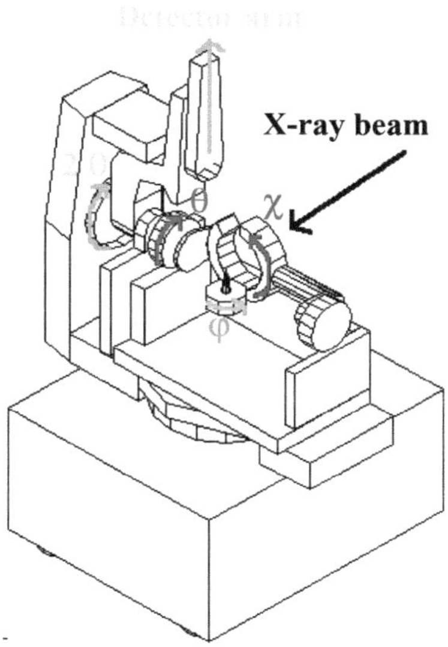

Figure 6.5 Pictorial view of a diffractometer for a DAFS experiment. The figure reports a modified scheme of the diffractometer of the D2AM CRG beamline of the ESRF where the experiment described in Section 6.4.2 has been performed; only the circles of interest are indicated. The sample is located on the tip.

diffracted intensity at the Ga edge is about a factor 5.5 larger than 80 eV below the edge (Fig. 6.4). It is possible to demonstrate that in case of a weak reflection, and in the experimental conditions used absorption, polarization, and Lorentz correction can be neglected. Instead, the fluorescence background, which in the case of weak reflections is comparable to the diffracted intensity, was eliminated by using a crystal analyzer, as described below.

6.3.3.3 The experimental setup

Set up for a DAFS experiment requires a diffractometer with at least four circles like the one sketched in Fig. 6.5. Three Euler circles (θ, φ, χ) are dedicated to sample movements, and the last circle is the vertical motion 2θ of the detector arm (same rotation axis as θ). The angular movements have typically a resolution better than 0.001°. The tracking of the Bragg's peak during an energy scan is obtained by previously determining the θ and 2θ angles for three or four points equally spaced in the energy range of interest and by applying a linear regression. The experiment requires as well a synchrotron beamline equipped with a monochromator, which permits us to continually tune the X-ray energy, X-ray mirrors to reject high-order harmonics, and appropriate beam intensity monitors for normalization of the signal, similar to the setup for XAFS (Section 6.3.1). Detection of the diffraction intensity is achieved either with a photomultiplier (Cyberstar) combined with NaI scintillator crystals or PIN silicon photodiodes, depending on the count rate at play. In the experiment on $In_xGa_{1-x}As_{1-y}N_y$ QWs presented in Section 6.4.2, elimination of the sample fluorescence background, which is a concern when dealing with weak reflections, was achieved by the utilization of a (002) flat graphite single-crystal analyzer. In Fig. 6.6 it is possible to note how the diffraction peak (on the right side) can be separated, after the edge, from the Ga Kβ fluorescence one (on the left side), by an opportune energy calibration of the analyzer. Elimination of the Ga Kα fluorescence is by far less problematic due to the larger energy shift of the related peak with respect to the diffraction one. More details on the DAFS setup can be found in Ref. (Renevier et al., 2003).

6.3.4 *X-Ray Emission Spectroscopy*

XES is a photon-in photon-out experimental method that is able to provide refined information on the electronic structure of condensed matter (Groot and Kotani, 2008). The physical processes involves the excitation of a core-level electron into an unoccupied state followed by the emission of a second photon due to the filling of the core hole; XES is thus a core hole decay spectroscopy, the other being Auger electron spectroscopy. The electron filling the core hole may originate from higher-energy core orbital or from the valence band (VB) (in this case the term "valence band XES" is used).

If the energy of the impinging photon is sufficiently far from the absorption threshold the core electron is excited into a high-energy continuum state and one speaks of "normal XES," or NXES. In this case, when the one-electron approximation is valid, excitation and decay processes can be thought of as independent (absorption

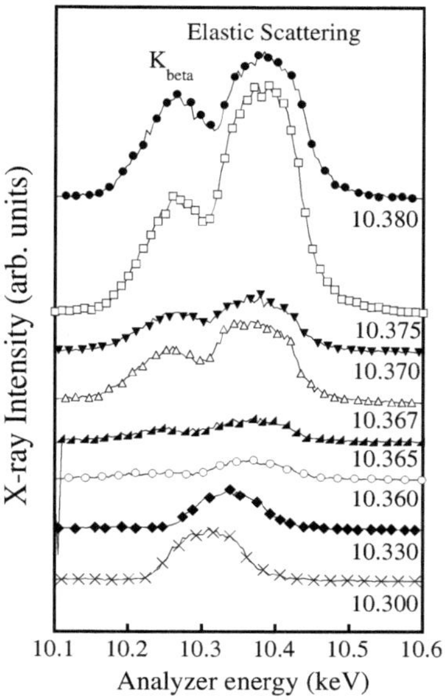

Figure 6.6 Energy scans performed with the crystal analyzer on the diffracted intensity from an $In_xGa_{1-x}As_{1-y}N_y$ sample for different values of the incident beam energy (reported in the legend): after the Ga edge (E = 10.367 KeV) the Ga Kβ fluorescence radiation can be separated from diffraction (elastic scattering). Reprinted with permission from Ref. (Ciatto et al., 2005c). Copyright 2005 by the American Physical Society. http://prb.aps.org/abstract/PRB/v72/i8/e085322.

followed by emission) and the emission line shape is essentially independent from the excitation energy; when the emitted photon is related to a transition between core levels this is the case of the common "X-ray fluorescence" or "characteristic radiation." If the energy of the impinging photon is close to the absorption threshold, instead, absorption and emission must be treated coherently as a second-order quantum process and the term "resonant XES" (RXES) is used; to stress that this can also be considered an inelastic-scattering process the acronym RIXS for resonant inelastic X-ray scattering is also used. In this case the emission line shape depends significantly on the excitation energy. RXES is particularly useful to study strongly correlated electron phenomena and materials, for example, oxides. Both NXES and RXES share the elemental selectivity of XAFS since they are based on core-level excitation. Since it is a photon-in photon-out technique XES is usually quite bulk sensitive, making it complementary to photoemission as a probe of electronic states.

From the experimental point of view XES requires a photon-out detector with sufficiently high-energy resolution to analyze the emission spectrum and, clearly, with sufficient solid-angle acceptance to provide acceptable count rates. In the soft X-rays (Nordgren et al., 1989; Ghiringhelli et al., 1998; Hague et al., 2005; Zangrando et al., 2004) a grating must be used as a dispersive element, and this is often coupled to some kind of parallel photon detector, for example, a charge-coupled device (CCD) camera. In the hard X-ray range, instead, Bragg diffraction from crystals is used, often in the Rowland geometry (Glatzel and Bergmann, 2005). In both energy ranges a further requirement is a high-brilliance and tunable source for excitation; since the XES cross section is rather low, XES apparata are hosted on undulator beamlines at synchrotron radiation facilities. Because the technique is rather "photon hungry," XES (and especially RXES) has developed more recently than XAFS with the availability of very high-brilliance beamlines and sophisticated experimental setups.

The interaction Hamiltonian necessary in the theoretical description of XES involves both linear and quadratic terms in the vector potential. The XES differential cross section is in general given by the Kramers–Heisenberg formula, which describes, in the electric dipole approximation, inelastic scattering by an atomic system (Bransden and Joachain, 2003):

$$\frac{d\sigma}{d\Omega} = r_0^2 \omega\omega'^3 \times \left| m \sum_n \left[\frac{(\hat{\varepsilon}' \cdot \vec{r}_{fn})(\hat{\varepsilon} \cdot \vec{r}_{ni})}{(E_n^0 - E_i^0 - \hbar\omega + i\Gamma_n)} + \frac{(\hat{\varepsilon} \cdot \vec{r}_{fn})(\hat{\varepsilon}' \cdot \vec{r}_{ni})}{(E_n^0 - E_i^0 + \hbar\omega' + i\Gamma_n)} \right] \right|^2 \quad (6.15)$$

where Ω is the solid angle, ω and ω' are the angular frequencies, $\hat{\varepsilon}$ and $\hat{\varepsilon}'$ are the polarization unit vectors of the incoming and outgoing photons, respectively, i and f identify the initial and final atomic states, $\vec{r}_{\alpha\beta}$ is the matrix element of the position operator between atomic states α and β, the sum is over all intermediate atomic states identified by the label n, Γ_n is the lifetime of state n, m is the electron mass, and r_0 is the classical electron radius. When the incident photon energy is close to or above the absorption threshold as in NXES or RXES, $E_n^0 \cong E_i^0 + \hbar\omega$ and the first term dominates the cross section.

In the case of NXES and in the one-electron approximation it can be proved (Groot and Kotani, 2008; Ma, 1994; Jimenez-Mier et al., 1999) that the cross section is directly proportional to the site and symmetry selected density of states (DOS) of the occupied intermediate states; symmetry selection is according to the dipole selection rule $\Delta\ell = \pm 1$ (thus, for a 1s hole, the relevant DOS is that of p character) and site selection is provided by the localization of the initial state wavefunction on the excited atomic species. For example, for K-edge VB NXES the emission line shape will probe the p-type VB DOS on the excited atom.

6.4 Experimental Results and Simulations

6.4.1 *Local Structure of N–H Complexes in Dilute Nitrides*

It is rather rare to find a situation in condensed matter where the local symmetry around a dopant or impurity is so deeply (and reversibly) modified by postgrowth treatment as it happens for hydrogenation of dilute nitrides. Even if the structural variations monitored by XRD (described in Section 6.2) and first IR absorption measurement (Chapter 5) already preannounced an important effect, it is worth remarking that a characteristic of fluorescence-mode XANES (Section 6.3.2) is that of probing all the possible local atomic configurations around the absorber (N in this case) present in the sample, weighing the contribution of each one to the average spectrum with its relative abundance. Often, when a defect forms in a semiconductor, the number of atoms involved in the defect corresponds to a small fraction of the total atoms; therefore one would expect the XANES (or, in general, XAFS) signal to come from a combination of the contributions from the "defective" atoms and the "well behaving" ones, with the second contribution ruling the signal. In this context, the results of the first N K-edge XANES experiment, shown in Fig. 6.7, were even more surprising than the XRD ones illustrated in Section 6.2.

The XANES experiment was performed on $GaAs_{1-y}N_y$ (y = 0.014) epilayers grown by solid-source molecular beam epitaxy on GaAs(001) substrates, with a typical thickness of 300 nm. Postgrowth atomic hydrogen irradiation was executed, as described in Chapter 2. The XANES measurements were performed on the branch line of the ALOISA beamline (Floreano et al., 1999) of the ELETTRA synchrotron radiation facility in Trieste, Italy. The energy dependence of the N absorption cross section was monitored by recording the intensity of N Kα fluorescence photons using a windowless hyperpure Ge detector (see Section 6.3.1).

In Fig. 6.7 we report the first N K-edge XANES spectra taken on $GaAs_{1-y}N_y$ before (bottom curve) and after (upper curve) hydrogenation: the line shape of the two signals is remarkably different and testifies to a dramatic change of the local environment around the N atoms in the alloys driven by hydrogen irradiation. As it happens for PL and XRD, an appropriate thermal annealing of the

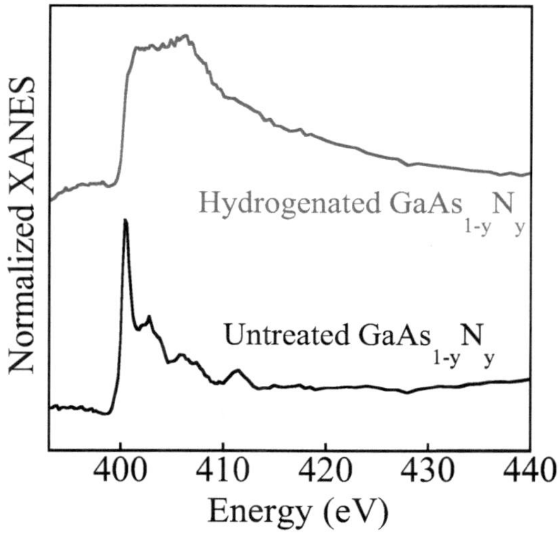

Figure 6.7 N K-edge XANES spectra obtained on untreated (bottom curve) and hydrogenated (top curve) $GaAs_{0.986}N_{0.014}$/GaAs epilayers.

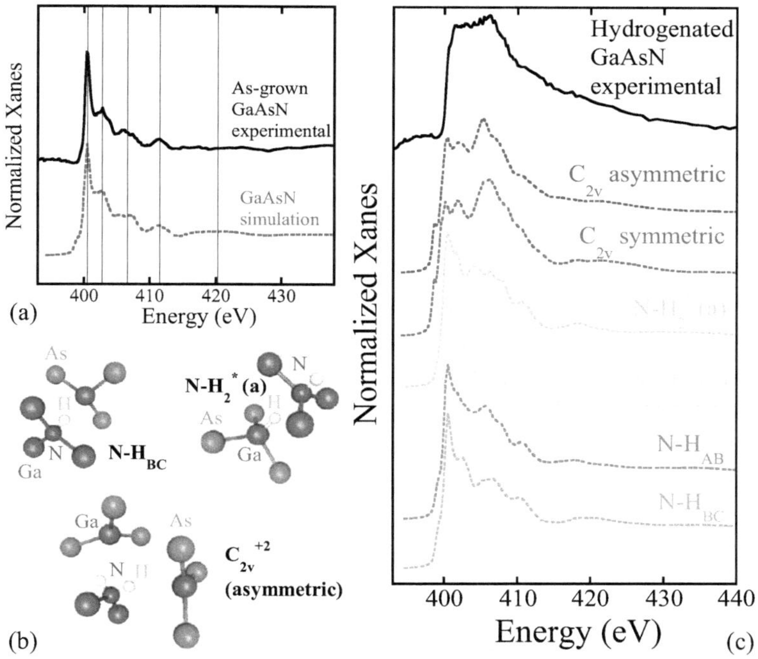

Figure 6.8 (a) Experimental and simulated XANES for the as-grown $GaAs_{0.986}N_{0.014}$ sample before exposure to hydrogen; (b) sketch of the most historically relevant N–H complexes; (c) experimental N K-edge XANES for the hydrogenated $GaAs_{0.986}N_{0.014}$ sample and corresponding FMS simulations based on the atomic coordinates calculated via DFT in presence of the most stable monohydrogen- and dihydrogen–nitrogen complexes. Modified from Ref. (Ciatto et al., 2005a).

hydrogenated sample restores a signal identical to the one of the untreated epilayer (not shown).

This result was obtained in May 2004; at that time there was a controversy in literature, since all theories predicted that an in-line dihydrogen–nitrogen complex, named H_2^* (a), was responsible for hydrogen-related effects in dilute nitrides. In the H_2^* (a) complex (introduced in Chapter 4 and sketched here in Fig. 6.8b), two strong Ga–H_{BC} and N–H_{AB} bonds are formed, which involve, respectively, a H_{BC} atom in bond-centered (BC) position between the Ga and N atoms and a H_{AB} atom in the opposite position with respect to the same nitrogen atom (antibonding position, AB) (Janotti et al., 2002b). However, a new paper published in January 2004 (Jiang et al., 2004) and already addressed in detail in Chapter 5, reported IR absorption measurements defying this hypothesis: In fact, if on the one hand the local vibrational modes related to the Ga–N bond disappeared in GaAsN after H irradiation (in agreement with the H_2^* (a) hypothesis), on the other hand no evidence for a Ga–H_{BC} mode was found, contrary to the expectations of the H_2^* model. In that study, two N–H-stretching modes due to the same N–H complex were detected, instead. More or less in the same period, DFT calculations of the defect energetics (see Chapter 4) proposed that two alternative dihydrogen–nitrogen complexes may form in dilute nitrides and phosphides (Amore Bonapasta et al., 2003, 2004). These complexes consist of two H_{BC} atoms bound to a single N atom, the first with C_{2v} symmetry and the second with quasi-C_{2v} symmetry, since one of the H_{BC} ions is located off the N–Ga axes. The second complex, sketched and indicated as "asymmetric" C_{2v} in Fig. 6.8b is preferred to the first one from the energetics standpoint.[a] Even if the C_{2v} complex was predicted to lead to a recovery of the band-gap energy and was in qualitatively agreement with IR absorption measurements, on the other hand, contrary to the H_2^*, it was not able to account for the lattice parameter changes shown in Fig. 6.1b being the predicted lattice parameter recovery only partial. This left open the question about the nature of the N–H complexes formed in hydrogenated dilute nitrides.

We decided to exploit the sensitivity of XANES to the three-dimensional structure of defects to address this issue and simulated the spectra under the hypothesis of formation of the most stable N–H complexes predicted by the different theoretical works quoted above. The XANES spectra were simulated in a real-space FMS approach using the FEFF code (Ankudinov et al., 1998); input atomic coordinates for the N–H complex were taken from the DFT calculations described in Chapter 4, repeating 8 times a 64-atom supercell and translating the positions in order to drive a N atom to the center of the cell. The clusters exploited in FMS calculations consisted of

[a]Note that the asymmetric C_{2v} complex was also referred to as "C_{1h}" or "canted C_{2v}" in other works as well as in Chapter 5.

512 atoms for GaAsN, 520 for monohydrogen complexes, and 528 atoms for dihydrogen complexes. The use of such large clusters is mandatory to reproduce the region of the absorption spectrum close to the edge, where the photoelectron mean free path can reach a few tens of Å and, as a consequence, probe the "middle range" structure around the N absorber. Figure 6.8a shows a comparison of the fluorescence-detected nitrogen K-edge XANES experimental spectrum for the as-grown $GaAs_{0.986}N_{0.014}$ sample (continuous line) with a simulation based on the atomic coordinates obtained by DFT calculations for a H-free alloy (dotted line). The excellent correspondence between the experimental and simulated spectra in the case of this well-known structure demonstrates the soundness of our method, which can be consequently used to investigated unknown structures, such as the one of hydrogenated $GaAs_{1-y}N_y$ by comparison between the experimental data and the different simulations.

Figure 6.8c shows the experimental nitrogen-edge XANES for hydrogenated $GaAs_{0.986}N_{0.014}$ along with simulations for the most interesting N–H complexes proposed by the theory, some of which are sketched in Fig. 6.8b. Sizable variations among the simulated XANES line shapes for the different N–H complexes can be observed. It is straightforward that only the C_{2v}, both in the symmetric and the asymmetric configuration, reproduce the main features of the experimental spectrum: for these complexes, the XANES signal does not decrease suddenly after the first maximum (the so-called white line) as it does for all other configurations but rather exhibits an extra peak between 405 eV and 408 eV similar to what is observed experimentally. In contrast, the H_2^* models show large discrepancies with and cannot account for the experimental XANES spectrum: this simple comparison allowed us to discard this model, which was believed for a long time to be the actual one due to its high stability and its effects on the crystal unit cell. Monohydrogen–nitrogen complexes (N–H_{BC}, N–H_{AB}) have a smaller effect on XANES and produce a signal that is more similar to the one of the untreated sample: they can be also discarded on the basis of the comparison with the experimental signal of the hydrogenated $GaAs_{1-y}N_y$.

These results, published in 2005 (Ciatto et al., 2005a), provided firm evidence of the predominant formation of complexes of

C_{2v} type in hydrogenated $GaAs_{0.986}N_{0.014}$. Moreover, due to the relatively bulk sensitivity of fluorescence XANES and its capacity to inherently probe and weight all the atomic geometries present in the sample, our findings demonstrated that virtually any single N atom incorporated in the alloy is involved in the formation of the N–H defect. In fact, due to the very different XANES line shapes predicted for the case of H-free $GaAs_{1-y}N_y$ and C_{2v} complexes, if a relevant fraction of N atoms had their environment unaffected by H irradiation, this would immediately result in the measured spectrum of the hydrogenated sample. Quite the reverse, the spectrum of hydrogenated $GaAs_{0.986}N_{0.014}$ shown in Figs. 6.7 and 6.8c is well reproduced by the simulation of the C_{2v} structure only, and there is no trace of any nonhydrogenated component (in particular, of the sharp white line peak at around 400.5 eV). This represents a completely different situation from the usual behavior of hydrogen in semiconductors and that of defects in general: instead of defects we should rather talk about a new crystallographic structure characterized by its own physical properties.

We point out that the present method is highly complementary to IR absorption spectroscopy and allows a more direct and quantitative comparison between experiment and simulation: a full comparison of the experimental spectrum (energy position of the spectral features, their relative intensities, and width) with simulations is indeed possible, which renders an accidental agreement obtained with a wrong model extremely improbable. Moreover, the method avoids interpretative problems of IR absorption due to the varying cross sections of vibrational modes. Straight after publication of our XANES work, two different groups published calculations (Fowler et al., 2005; Du et al., 2005) showing that a canted or asymmetric C_{2v} complex similar to the one described above has vibrational properties consistent with the IR experiment, supporting the XANES results. The controversy about the total recovery of the lattice parameter, which was not possible with the doubly charged C_{2v} complexes used in our simulations, was solved within the hypothesis of similar neutral C_{2v} complexes, able to recover the GaAs lattice constant. The justification for C_{2v} complexes canting and neutralization has been already discussed in Chapter 5.

In the following years, several theoretical and experimental investigations based on alternative techniques attempted to refine the structure of the C_{2v} defects, addressing in particular the issues of the possible presence of additional satellite N atoms around the C_{2v} and their effect on compressive strain. In particular, nuclear reaction analysis (NRA)/channeling investigations coupled with HRXRD (which will be the subject of Chapter 7) suggested the formation of C_{2v} complexes with a deuterium satellite after deuterium (D) irradiation[a] of $GaAs_{1-y}N_y$ (i.e., deuterium/nitrogen ratio $= 3$) and the possibility of removing the satellite via a soft thermal annealing (Berti et al., 2007). In the same period, Buyanova et al. (Buyanova et al., 2007), on the basis of secondary ion mass spectroscopy (SIMS) and PL, suggested that the H/N ratio in $GaAs_{1-y}N_y$ and $GaP_{1-y}N_y$ should be larger (from 3 to 5) and the formation of complexes involving the presence of up to five H atoms correlated with a single nitrogen. DFT calculations detailed in Chapter 4 demonstrated the stability of structures involving additional H "satellite" atoms in the proximity of the C_{2v} defect, which were predicted to produce compressive strain in good agreement with that measured by HRXRD (Berti et al., 2007). However, such calculations predicted that the final step of the hydrogenation process should involve two H satellites (i.e., four H atoms in total) with an expected H/N ratio of 4 (Amore Bonapasta et al., 2007), larger than the value determined by NRA.

Our second XANES experiment (Ciatto et al., 2009) was conceived to examine and validate these recent results on the structure of the N–H(D) complex, essentially the charge state of the complexes, the possibility of H(D) satellites, and some minor issues as the effect of the tetragonal distortion of the unit cell due to the epitaxial growth on GaAs (Woicik et al., 1997). All these hypotheses could in fact a priori affect the XANES line shape since they imply some modifications of the local structure around the N absorber. In the case of formation of neutral C_{2v} complexes (C^0_{2v}, sketched in Fig. 6.9), two dangling bonds are induced on the Ga atoms involved in the

---

[a] Hydrogen was substituted by its isotope deuterium to reduce systematic error in the NRA measurements. It was formerly verified that deuterium modifies electronic and structural properties of dilute nitrides in an identical way as hydrogen.

broken Ga–N bond so that these Ga atoms interact forming a weak Ga–Ga bond, and their interatomic distance (3.24 Å) is sensibly shorter than the distances between the other Ga atoms. In contrast, in the case of the doubly charged C_{2v} complex used in the simulation previously reported in Fig. 6.8c (C_{2v}^{+2}), the same Ga–Ga distance is much longer (5.02 Å) since the Ga atoms that carried the dangling bond have lost their unpaired electron and cannot form any Ga–Ga bond (Du et al., 2005; Amore Bonapasta et al., 2007). Moreover, the angle formed by the same two Ga atoms and centered on N is very different for the two charge states, being 61° for C_{2v}^{0} and 89° for C_{2v}^{+2}. In case of the neutral C_{2v} complex with one H satellite (C_{2v}-H^0, also sketched in Fig. 6.9), the distance between the two Ga atoms involved in the breaking of Ga–N bonds (3.92 Å), the distances

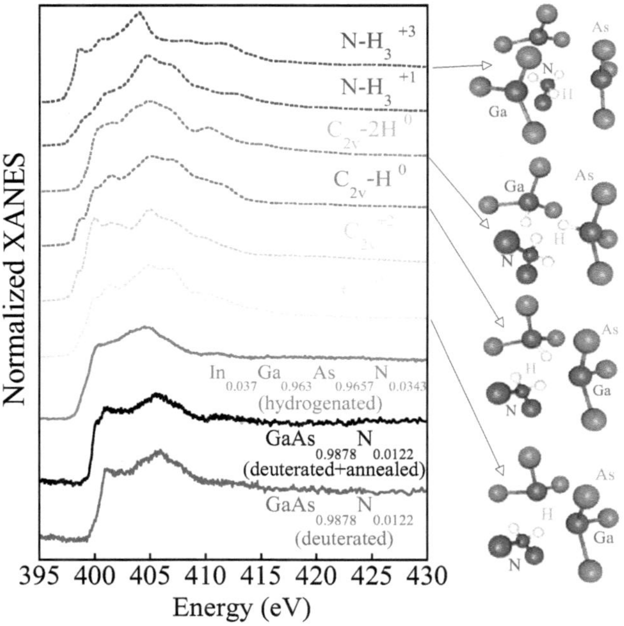

Figure 6.9 N K-edge XANES simulations for selected N-H complexes having an asymmetric C_{2v} core and for N-H_3 complexes (dotted line) compared to three experimental spectra (continous line, at the bottom) for deuterated/hydrogenated epilayers. Some of the complexes are sketched at the right side of the figure. Modified from Ref. (Ciatto et al., 2009).

between N and the same two Ga atoms, and the N-centered angle formed by these two Ga atoms and the N one (72°) are closer to the values for C_{2v}^{0}; however, the two structures still differ for the presence of the additional H atom. As for the neutral complex with two satellite H atoms (C_{2v}–$2H^0$), the calculated distance between the two Ga atoms involved in the breaking of the Ga–N bond (4.26 Å), along with the distances between such Ga atoms and the N one, is longer than in the case of C_{2v}–H^0, due to the H saturation of the two dangling bonds induced on Ga; in contrast the value of the angle centered on N (74°) is close to the value predicted for C_{2v}–H^0.

The XANES spectra, reported in Fig. 6.9, were recorded on two 223 nm thick $GaAs_{1-y}N_y$ epilayers ($y = 0.0122$) grown on a (001) GaAs substrate and then deuterated at fluence of 5×10^{18} ions/cm^2, as described in Chapter 2. After deuteration, one of the two epilayers has been softly annealed at 250°C for 13 hours. We also measured a 149 nm thick $In_xGa_{1-x}As_{1_y}N_y$/Gas epilayer ($x = 0.037, y = 0.0343$) hydrogenated at H fluence $= 5.8 \times 10^{18}$ ions/cm^2. Optimization of the experimental setup allowed us to obtain spectra with an improved signal-to-noise ratio with respect to the first data shown in Fig. 6.7 and 6.8; however, the XANES line shape of the previous experiment for the hydrogenated (deuterated) sample is confirmed and does not depends on the N concentration. The spectra for the deuterated $GaAs_{1-y}N_y$ epilayer, the deuterated + annealed one, and the hydrogenated $In_xGa_{1-x}As_{1_y}N_y$ look quite similar and differ for a few details in the near-edge region only.

Figure 6.9 also shows FMS simulations of the N K-edge XANES performed in presence of neutral asymmetric C_{2v} complexes (C_{2v}^{0}), neutral asymmetric C_{2v} complexes decorated with one H satellite (C_{2v}–H^0), and two H satellites (C_{2v}–$2H^0$), along with the previously published simulation for the doubly charged complex (C_{2v}^{+2}) and simulations for complexes with three H atoms bound to the same N one in the +1 and +3 charge states (N–H_3^{+1} and N–H_3^{+3}, see sketch in the figure). To mimic as close as possible the reality, we took into account the effect of the tetragonal distortion of the unit cell due to the pseudomorphic growth on the substrate. Several charge states were explored for all the complexes under study, but only the most stable ones for each configuration are reported here. Like for the previous analysis, FMS simulations are based on calculation of atomic supercells around a central N atom performed by using DFT (Amore Bonapasta et al., 2007). At a first inspection, the line shapes of all simulations for C_{2v} complexes look rather similar, regardless of the presence of satellites and the charge state. Specifically, if we compare the simulation of the C^0_{2v} complex with the experimental data, we note that the line shape and energy position of the main spectral features are very well reproduced. Looking more attentively, one can note that the simulation for such a neutral complex mimics the experimental data even better than the one based on the C^{+2}_{2v} complex formerly introduced, in particular for what concerns the two peaks at roughly 400 eV and 405 eV, the relative amplitude of which is inverted in case of the doubly charged complex. This is in agreement with XRD data (Section 6.2), since neutral complexes are predicted to cause an expansion of the alloy lattice parameter. However, fixing the charge state at a neutral level, the smaller variation of interatomic distances and angles observed when satellites are included (detailed above) are not sufficient to produce important changes in XANES. An attempt to associate some minor modifications of the three experimental spectra to specific satellite configurations is described in our work (Ciatto et al., 2009); nonetheless the only safe conclusion we can infer from these results is that the C^0_{2v}, C_{2v}–H^0, and C_{2v}–$2H^0$ configurations are all compatible with the experimental data. If we consider other configurations yielding a D/N ratio = 3, in particular N–H complexes with three H atoms bound to the same N (N–H_3), the corresponding FMS simulations (the two topmost curves in Fig. 6.9[a]) are rather at variance with the experimental spectra. This allows us to discard such configurations, in agreement with theoretical results that predict that complexes with three N–H bonds are metastable or unstable (see Chapter 4).

To conclude, if on the one hand XANES is able to give firm evidence that the core of the N–H defect in $GaAs_{1-y}N_y$ and $In_xGa_{1-x}As_{1_y}N_y$ is a C_{2v} complex, most likely in the neutral charge state, on the other hand alternative experimental techniques such as IR, NRA, and HRXRD (see Chapters 5 and 7) can better assist

[a]We show here only the simulations for the +1 and +3 charge, with the simulation for the +2 state being very similar to that for the +3 one.

the theoretical calculations in addressing the issue of satellite configuration and satellite removal upon annealing.

6.4.2 *Effects of Hydrogenation on the Interatomic Distances*

In Section 6.3 we anticipated that, if XANES provides information on the three-dimensional local configuration around the absorber atom, EXAFS and DAFS can measure very precisely (with an uncertainly of the order of 0.01 Å) the interatomic bond lengths between the absorber and the first few neighbors shells. Before 2005, little information was available on the effects of hydrogenation on the bond lengths; however, an evaluation of the interatomic distances in as-grown and hydrogenated dilute nitrides is of paramount importance since not all the N–H complexes introduced in Chapter 4 are predicted to bring about the structural relaxation observed by XRD (Section 6.2). Hence, we decided to complete the XANES experiments described in the previous section with an experimental determination of the cation–As bond length in $GaAs_{1-y}N_y$, $In_xGa_{1-x}As_{1-y}N_y$, and reference $In_xGa_{1-x}As$ samples.

We investigated three series of $In_xGa_{1-x}As_{1-y}N_y$ single QWs: each series consisted of a N-free sample, an as-deposited $In_xGa_{1-x}As_{1-y}N_y$, and an hydrogenated one. The indium concentration, x, was in the range from 0.32 to 0.38; nitrogen concentration, y, from 0.027 to 0.052; and the well width from 6 nm to 8 nm. All QWs had a 100 nm thick GaAs capping layer and buffers of the order of 100 nm. We measured also three 300 nm hydrogenated thick $GaAs_{1-y}N_y$ epilayers and the respective nonhydrogenated references with a N concentration y ranging from 0.008 to 0.027. All samples have been grown on a GaAs(001) substrate by solid-source molecular beam epitaxy. Postgrowth treatment by atomic hydrogen irradiation was obtained as described in Chapter 2; the hydrogen dose was 3.0×10^{18} ions/cm^2. PL (Baldassarri Höger von Högersthal et al., 2001) and XRD (Polimeni et al., 2003b) characterization of the samples was in agreement with the effects described in Section 6.2.

As for $In_xGa_{1-x}As_{1-y}N_y$ QWs, we measured the Ga–As bond lengths by DAFS at the Ga K-edge and the In–As bond lengths by EXAFS at the In K-edge. The spatial selectivity of DAFS (see

Section 6.3.3) was mandatory for Ga K-edge measurements because of the presence of Ga atoms in the substrate, the small thickness of the QWs (less than 10 nm), and the presence of a cap layer. $GaAs_{1-y}N_y$ thick epilayers (300 nm) were investigated by Ga K-edge EXAFS. DAFS experiments were undertaken at the D2AM CRG beamline (BM02) of the European Synchrotron Radiation Facility (ESRF), in the setup described in Section 6.3.3.3, measuring the fine structure of the diffracted intensity beyond the Ga K-edge at the weak (006) reflection. XAFS experiments at the In and Ga K-edges were performed at the Gilda CRG beamline (BM08) of the ESRF. In the case of In K-edge measurements on thin $In_xGa_{1-x}As_{1-y}N_y$ QWs, the In K_α fluorescence signal was monitored using a 13-element hyperpure Ge detector. In case of Ga K-edge measurements on hydrogenated and untreated $GaAs_{1-y}N_y$ thick epilayers, the energy dependence of the X-ray absorption coefficient at the Ga K-edge was recorded by measuring the drain current from the sample (TEY mode) (Ciatto et al., 2005b). In fact, on the basis of a TEY depth-dependence model (Erbil et al., 1988), it is possible to estimate that the contribution of absorption events taking place in the substrate to the total electron current is lower than 3–4% in our experiment (see Section 6.3.1 for a description of the different XAFS setups).

Figure 6.10a shows the EDAFS oscillations ($\frac{|I_{\text{exp}}|-|I_{0\ \text{exp}}|}{|I_{0\ \text{exp}}|}$) for the three studied QWs samples, extracted from raw spectra similar to the one reported in Fig. 6.4. EDAFS spectra were Fourier-transformed to the R-space, a procedure that permits us to separate the signal frequencies related to the different coordination shells, and then the first shell contribution was filtered in the R-space and the corresponding k-signal was obtained by inverse Fourier transform. The spectra obtained in this way, plotted in Fig. 6.10b, permit us to obtain information on the nearest neighbors (NNs) of the central Ga atoms and to extract the bond length values from fits based on GaAs and GaN model clusters by means of dedicated software tools (Ankudinov et al., 1998; Newville et al., 1995).[a] Fits performed on the first shell Fourier filtered signals are reported in Fig. 6.10b (open circles), superposed to the experimental data. Analysis of EXAFS data was performed in a similar way, except

[a]Of course, applying the phase and amplitude corrections introduced in Section 6.2.

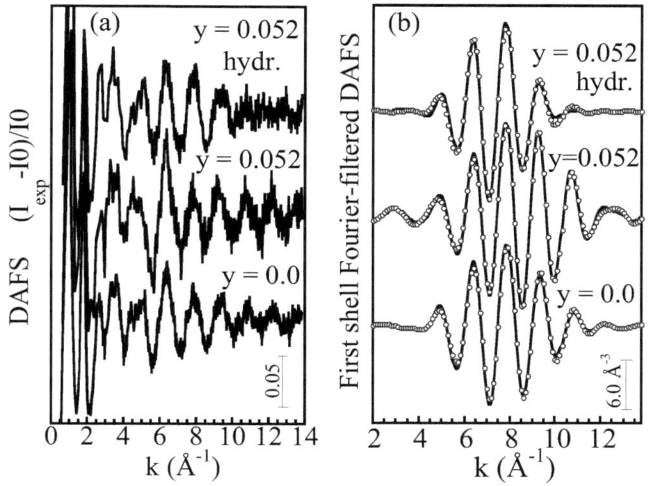

Figure 6.10 (a) Background subtracted Ga K-edge DAFS spectra for the three investigated $In_xGa_{1-x}As_{1-y}N_y$ QWs, In concentration (x) $= 0.38$ for all QWs, N concentration (y) is reported above each spectrum, the top curve shows the spectrum for the hydrogenated sample; (b) fits on the first shell filter (k^3 weighted): continuous line = data, open circle = fit. Modified from Ref. (Ciatto et al., 2005c).

for the fact that three coordination shells instead of one were modeled and fitted, thanks to the better signal-to-noise ratio of the experimental spectra. A detailed description of the data analysis procedure can be found in a paper published in 2005 (Ciatto et al., 2005c); in this section, we only summarize the most important results obtained.

In Fig. 6.11 we report the difference between the experimental values of the cation–As bond lengths for $In_xGa_{1-x}As_{1-y}N_y$ QWs and the theoretical expectations; the Ga–As distances are determined by DAFS, while the In–As distances are determined by EXAFS. Theoretical cation–anion distances (R_{A-B}) are calculated with the equation

$$R_{A-B} = R_{AB} + \delta_{\text{alloying}} + \delta_{\text{strain}} \tag{6.16}$$

where R_{AB} is the sum of the covalent radii of the two atoms (Pauling, 1967) and δ_{alloying} is a correction term expressing the fact that the bond length value is affected by the presence of a third

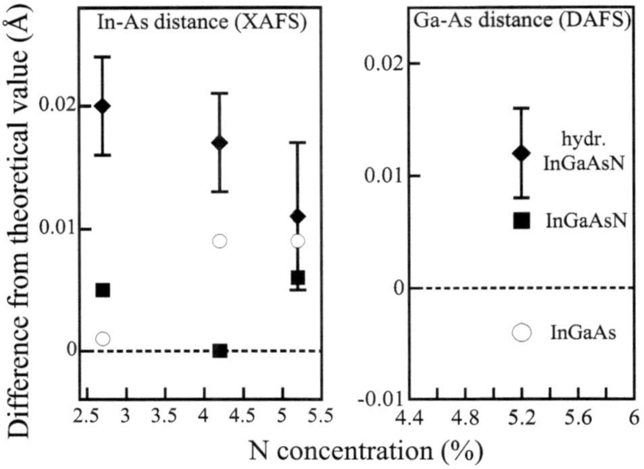

Figure 6.11 Difference between experimental and theoretical determination of the cation–As distances; open circles = $In_xGa_{1-x}As$, squares = $In_xGa_{1-x}As_{1-y}N_y$, diamonds = hydrogenated $In_xGa_{1-x}As_{1-y}N_y$; errors = 1σ (for the sake of clarity error bars are reported only for hydrogenated samples). Reprinted with permission from Ref. (Ciatto et al., 2005c). Copyright 2005 by the American Physical Society. http://prb.aps.org/abstract/PRB/v72/i8/e085322.

(possibly fourth) atom with a different radius. The second term (strain correction), which is absent in bulk alloys, comes from the propagation to the local scale of the tetragonal distortion of the unit cell due to the epilayer–substrate misfit. δ_{alloying} is here calculated via a valence force field approach (Cai and Thorpe, 1992), while δ_{strain} is obtained with a formula applicable to the strained zinc-blende structure (Woicik et al., 1997; Tormen et al., 1999) when the lattice parameter is available from XRD (Polimeni et al., 2003b). As for hydrogenated samples the effect of H incorporation on the δ_{alloying} term is neglected,[a] while the effect on δ_{strain} is taken into account (since experimental XRD data presented in Section 6.2 is used for calculating δ_{strain}). It is evident from the figure that the difference from the expected values is systematically larger in hydrogenated QWs than in the other samples, both for the In–As and for the Ga–

[a]i.e., we did not assume a priori that H had a specific lattice location, nor the existence of N–H complexes since this was still unclear at the time of this experiment.

As bond length, and that it always exceeds zero (dotted line) at a 1σ confidence level in the case of hydrogenated samples. Assuming that the modeling of the strain component δ_{strain} is correct (as verified for a variety of III–V semiconductors), this means that hydrogen affects the alloying component of the distances and has more than just an influence on the strain. These results provided yet more evidence, at a different length scale than XRD, that a relatively high fraction of H atoms incorporates in the crystal lattice of dilute nitrides, sufficient to sensibly modify the interatomic distances. The effect of hydrogenation on the $\delta_{alloying}$ term of the cation–As distance can be justified in the frame of an overall expansion of the structure that affects also the bonds that are not directly involved in the formation of the N–H complex (Ga–As and In–As). We point out that, due to the weakness of the In–N signal, we had to fix the In–N distance to the theoretical value in the fit; therefore we were not able to measure it in these samples.

Results presented in this section are therefore in agreement with dihydrogen–nitrogen complex models (Chapter 4) predicting a relaxation of the lattice parameter, such as the $H_2^*(N)$ (Sanna and Fiorentini, 2004) or the neutral variant of the C_{2v} complex (Du et al., 2005), which well simulates the N K-XANES data reported in the previous section, but not with monohydride $H(BC_N^+)$ complex. In the latter case, the presence of Ga dangling bonds causes a Ga–H interaction, which favors a contraction of the unit cell (Amore Bonapasta and Filippone, 2003).

As for the Ga–As bond length in untreated and hydrogenated $GaAs_{1-y}N_y$/GaAs epilayers, measured by TEY EXAFS, the smaller N concentration with respect to the QWs forbids us to draw the conclusion that hydrogenation brings about an increase of the alloying component of the Ga–As bond length; as a matter of fact only in the sample characterized by the highest N concentration the theoretical estimation of the bond length is close to the lower limit of the 1σ experimental error bar (Ciatto et al., 2005c). Despite the smallness of hydrogenation effects on the bond lengths in $GaAs_{1-y}N_y$, we show in this section an intriguing effect observed on the distances distribution. In Fig. 6.12 we show the Fourier transforms of the EXAFS signal in the region of the first atomic shell peak (which is essentially related to the Ga–As coordination with the

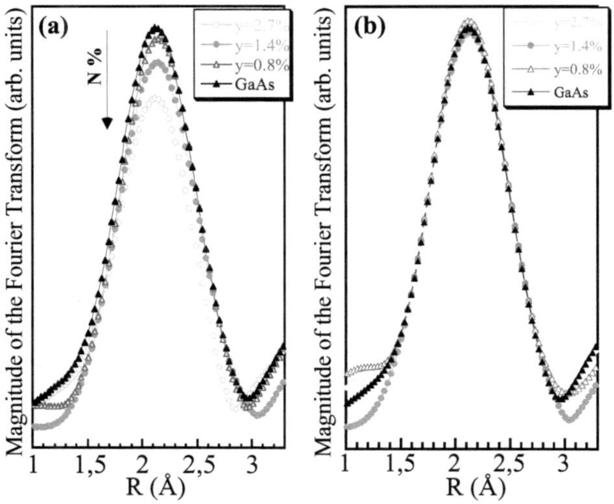

Figure 6.12 Magnitude of the first shell peak of the Fourier transform for (a) as-deposited epilayers and (b) hydrogenated $GaAs_{1-y}N_y$ epilayers. Note that the amplitude decreases with y in the untreated samples but not in hydrogenated ones. Modified from Ref. (Ciatto et al., 2005c).

concentrations at play) for untreated and hydrogenated epilayers. For untreated epilayers (Fig. 6.12a), the amplitude of the peak gradually diminishes with N concentration (the arrow indicates increasing N concentrations); hydrogenation reverses this tendency, by restoring the amplitude value of GaAs, independently of the concentration (Fig. 6.12b). The decrease in the Fourier transform amplitude of as-grown samples is related to the static disorder induced by N incorporation in $GaAs_{1-y}N_y$ (which affects the Ga–As Debye–Waller factor, defined in Section 6.3.2). Such large amplitude variations cannot come from the contribution of a Ga–N first shell signal, due to the very low N concentration in these epilayers (Ciatto et al., 2005b).

We propose that this effect is related to the unique characteristics of the $GaAs_{1-y}N_y$ alloy where As and N have very different electronegativities in the Pauling scale, 2.18 and 3.04, respectively. In turn, a strong charge transfer toward the N anion characterizes the Ga–N bond, while this charge transfer is nearly absent in the Ga–As bond. At the present low N concentrations, the $GaAs_{1-y}N_y$ alloy is mainly formed by Ga-centered As_4N_0 and As_3N_1 clusters, with Ga–As bond lengths slightly longer in the N-containing clusters where the charge is more localized on the nitrogen atom. Ga K-edge EXAFS measurements cannot separate the different Ga–As bond lengths in the two types of clusters, because their values are too close. However, this local disorder generates the damping effect on the bond length distribution shown in Fig. 6.12a. On the other hand, in hydrogenated $GaAs_{1-y}N_y$, the formation of C_{2v} complexes requires breaking of the ionic Ga–N bond, so the aforementioned effect on the bond length distribution due the strong charge transfer is supposed to be lost, in agreement with our observations.

6.4.3 *Probing the Local Electronic Structure of Nitrogen in Dilute Nitrides*

The introduction to XANES and NXES given above (Sections 6.3.2 and 6.3.4) suggests that a combination of the two techniques should be able to provide a novel probe of the symmetry-selected local electronic structure of both occupied and unoccupied electronic states projected on a particular atomic species. We recall that, in the one-electron approximation, assuming a weakly varying matrix element the XANES line shape is directly proportional to the site and symmetry-selected unoccupied (conduction band, CB) DOS, while the valence NXES line shape is proportional to the likewise selected occupied (VB) DOS. It must be mentioned that the final state in the two experiments is different: in XANES a core hole is present, while it isn't in NXES.

In the field of dilute nitrides the combination of XANES and VB NXES at the N K-edge can provide, in fact, precious information on the local electronic DOS and its variation with hydrogenation; this local information is complementary to what can be obtained from more standard probes of the electronic structure, such as PL (Chapter 2), which probe the average properties rather than those specific to the dilute species responsible for the modification of the physical properties.

Recently, Amidani et al. (2014) have reported such a joint study, which will be now summarized. The XANES measurements have been performed as described in detail above (Section 6.3.2). Of importance here is the careful energy calibration of the spectra, which was performed by recording in parallel the signal of N_2 gas (Ciatto et al., 2007). N K-edge valence NXES spectra were recorded on the BACH beamline of the ELETTRA facility. Spectra were recorded with a variable line spacing grating analyzer and a CCD camera detector (Zangrando et al., 2004); energy calibration of the impinging beam was performed using photoemission from a Au sample and that of the emitted beam using the elastic scattering peak. Here we will report spectra of a GaAsN sample with 1.3% N concentration, the growth conditions of which have been described in Section 6.4.1; it is particularly relevant that NXES spectra on such dilute samples could be recorded thanks to the high brilliance of the beamline and the dedicated experimental apparatus. As a reference, XANES and NXES spectra of GaN will also be reported.

Thanks to the reliable energy calibration and using as a reference the N1*s* core level the XANES and NXES spectra can be plotted on a common energy scale with respect to the valence band maximum (VBM); a systematic error is possible due to the presence of a core hole in the final state of XANES, but this will affect equally all spectra so that differences among different samples and due to sample treatment are expected to be significant.

In Fig. 6.13 we show NXES and XANES spectra of GaN (red) and pristine GaAsN (blue); the energy scale is referred to the VBM, determined as an extrapolation to zero of the leading edge of the high-energy part of the spectrum. Clearly, the NXES spectra (proportional to the *p*-projected VB DOS on the N atom) of the two samples are quite similar. This is related to the similar NN environment of a N in the two samples. On the basis of electronic structure calculations and some previous studies (Strocov et al., 2004) we can state that the lower-energy peak is due to hybridization between N 2*p* and Ga 4*s* atomic orbitals, while the higher-energy one is due to hybridization between N 2*p* and Ga 4*p* ones. In the same figure we also report the XANES spectra, which are, instead, quite different. The first evident difference is the greater energy separation between the VB and the CB, a quantity we may call the "local energy gap"; the considerably higher local energy gap of GaN reflects the higher value of the actual energy gap. The second

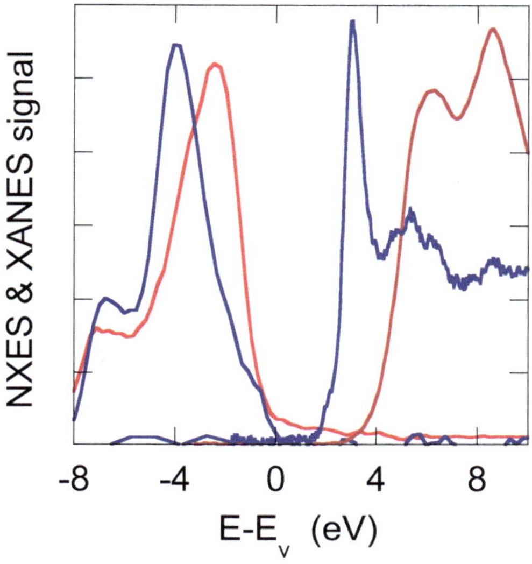

Figure 6.13 NXES and XANES spectra of GaN (red) and pristine GaAsN (blue); the energy scale is referred to the VBM. The normalization of the ordinate scale is arbitrary. Modified from Ref. (Amidani et al., 2014).

striking difference is the presence of a strong resonance at the onset of the XANES, that is at the bottom of the CB, which falls well within the local band gap of GaN. This peak is due to a high DOS localized on N and due to the interaction between N atomic orbitals and the GaAs CB (Mattila et al., 1999); this assignment will be confirmed by hydrogenation studies.

In Fig. 6.14 we report a comparison of N K-edge VB NXES and XANES of the pristine (red) and hydrogenated (green) GaAsN sample. Analysis of these spectra allows us to draw important conclusions on the electronic structure of the N–H complex formed. Firstly, the VB DOS of the complex has a wider band width with respect to the local N DOS of GaAsN, a lower contribution of N $2p$– Ga $4p$ states and a higher one of N $2p$– Ga $4s$ ones. Secondly, and most strikingly, it is clear that the opening of the band gap induced by H irradiation is due to the removal of the high DOS peak at the bottom of the CB of GaAsN. This result provides a "local view" of the modification of the electronic structure, which conclusively

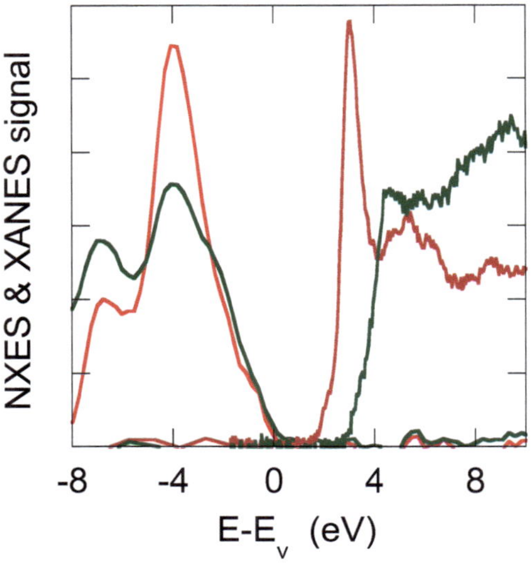

Figure 6.14 NXES and XANES spectra of pristine (red) and hydrogenated (green) GaAsN; the energy scale is referred to the VBM. The normalization of the ordinate scale is arbitrary. Modified from Ref. Amidani et al. (2014).

demonstrates that the high DOS peak at the bottom of the GaAsN CB is due to localized states induced by N.

6.5 Conclusions

X-ray and synchrotron radiation–based investigations of dilute nitrides have played a central role in the understanding of the atomic-scale structural mechanisms at the basis of the modification of the physical properties driven by hydrogenation. Laboratory XRD provided the first clue of large modifications of the crystal structure upon hydrogen irradiation of $GaAs_{1-y}N_y$ and $In_xGa_{1-x}As_{1_y}N_y$ and stimulated finer characterizations at different length scales.

The utilization of synchrotron radiation X-ray spectroscopies (XAFS, DAFS, and XES), to which a large part of the present chapter has been dedicated, represents a rather innovating approach in the analysis of the atomic structure of defect complexes in semiconductors (Ciatto, 2014), highly complementary to more established techniques such as IR absorption spectroscopy discussed in the previous chapter.

Nitrogen K-edge XANES, in particular, thanks to the enhanced sensitivity to the three-dimensional configuration of the neighbors around the absorber atom, provided a firm and direct experimental evidence that a N–H complex with a C_{2v}-like symmetry core was the most abundant species in hydrogenated dilute nitrides, result which was confirmed by a suite of experimental and theoretical approaches. A fundamental point in the successful application of XANES to the present scientific case was the possibility to simulate the experimental data based on state-of-the-art calculations of the atomic positions in the defect complexes described in Chapter 4, while the combination with the complementary structural tools described in Chapters 5 and 7 allowed a refinement of the N–H complex with determination of the number of H satellites around the C_{2v} core.

Finally, the coupling of XANES and XES has very recently started to set up a bridge between the real space and local electronic structure of N centers in dilute nitrides.

Acknowledgments

We acknowledge excellent support by the staff of the different beamlines where the relevant synchrotron radiation experiments described in this chapter were carried out, in particular L. Floreano (Aloisa Beamline, ELETTRA), F. d'Acapito (Gilda CRG Beamline, ESRF), and H. Renevier (D2AM CRG Beamline, ESRF). We are grateful to all students and collaborators who assisted us in the experimental work at different stages of this research: R. Carboni, M. Malvestuto, L. Amidani, M. G. Proietti, and P. Fons. We are also grateful to L. Ortega (CNRS, Grenoble) for technical support during the first XRD characterization described in the chapter and to A. Forchel (University of Wurzburg), L. Grenouillet (CEA, Grenoble), S. Rubini (TASC-CNR), and F. Martelli (IMM-CNR) for providing the different samples analyzed in our experiments. We would like to acknowledge as well the coauthors of this book, the collaboration with whom provided us with constant and fundamental feedback in the interpretation of our data. Measurements at ELETTRA were supported in part by the European Community Research Infrastructure Action under the FP6 "Structuring the European Research Area" program. Measurements at the ESRF were funded through Project No. HS2020 (public user program) and by the INFM. Gianluca Ciatto is grateful to the ESRF and to SOLEIL synchrotron for allowing him to carry on this fascinating research activity in parallel with the instrumental one.

Chapter 7

Stoichiometry of Nitrogen–Hydrogen Complexes in Dilute Nitrides

Marina Berti, Gabriele Bisognin, and Davide De Salvador
Dipartimento di Fisica e Astronomia "G. Galilei," Università di Padova, via Marzolo 8, 35100 Padova, Italy
marina.berti@unipd.it

The electronic and structural properties of dilute nitrides under hydrogenation were deeply described in the previous chapters. To close the loop and better fix the structural counterparts of electronic and optical effects, in this chapter we will describe the use of two additional experimental methodologies, high-resolution X-ray diffraction (HRXRD) and nuclear reaction analysis (NRA). Both of these are experimental methods largely used in material science that can give fundamental and specific contributions to the present topic. By exploiting these techniques and by using deuterium instead of its hydrogen isotope we will show that deuteration proceeds layer by layer and the incorporation of further deuterium is possible only after the N–D complexes have been formed. A 3D–N complex is the final product of the deuteration in all dilute nitrides. The study of the thermal dissolution reveals that the 3D–N complex converts into a 2D–N complex still capable of N passivation. Further annealing

Hydrogenated Dilute Nitride Semiconductors: Theory, Properties, and Applications
Edited by Gianluca Ciatto

ISBN 978-981-4463-45-4 (Hardcover), 978-981-4463-46-1 (eBook)
www.panstanford.com

treatments induce the total loss of deuterium and the recovery of the pristine GaAsN properties.

7.1 Introduction

The electronic and structural properties of hydrogenated dilute nitrides were deeply described in the previous chapters. Different complementary methodologies were used to understand many aspects of the material.

It has been shown that all the effects associated with N in GaAs and GaP are dramatically affected by H irradiation, which restores in a controllable manner the pristine GaAs and GaP electronic properties, for example, band-gap energy, electron effective mass, and gyromagnetic factor (Chapters 2 and 3). Ab initio calculations, the study of vibrational properties, and the investigation of the local structure by synchrotron radiation techniques (Chapters 4, 5, and 6) allowed us to have an "atomistic snapshot" of the N–H candidate structures that can explain the peculiar properties of the material under hydrogenation.

To close the loop and better fix the structural counterparts of electronic and optical effects, in this chapter we will describe the use of two additional experimental methodologies, high-resolution X-ray diffraction (HRXRD) and ion beam analysis (IBA) and among them nuclear reaction analysis (NRA). All of these are experimental methods largely used in materials science that can give fundamental and specific contributions to the structural counterpart of the present topic.

By the use of HRXRD it has been shown that as a consequence of hydrogen irradiation and of the formation of H–N complexes, the interplanar distance of the films undergoes an expansion that reverses the type of the strain from the tensile one of the as-grown material into a compressive one of the treated layer. In this way, the presence of compressive strain was used as a structural fingerprint of the H–N complex formation. These findings were exploited also to trace a qualitative picture of the evolution of the H–N complex formation in the dilute nitrides lattice as a function of increasing H irradiation fluence.

Moreover, the HRXRD measurements were used to quantitatively determine the compressive strain to be associated with these H–N complexes, providing fundamental information to be merged with the results of theoretical calculations. Finally, thanks to the ability to study the evolution of the strain in H–N complexes as a function of both temperature and time, HRXRD was used during the annealing of hydrogenated materials. As a result, it has been found that the compressive strain originally found in the hydrogenated films disappears by overcoming two different temperature windows strongly connected with the energy barriers involved in the dissolution of the pristine H–N complexes. The application of HRXRD during thermal annealing allowed us not only to trace a scenario of the N–H complex formation and dissolution but also to obtain ad hoc samples with a particular and stable strain level that can be in principle associated with a particular population of H–N defects. These samples open the route to the quantitative determination of the [H]/[N] ratio by the use of NRA. In fact, by applying the appropriate nuclear reactions, it was possible to extract this ratio, not only just after the formation of the H–N complexes, but also at each dissolution step accurately determined by HRXRD. The application of NRA was mandatory since it was the only way to detect both very low amounts of light elements in heavy matrixes and their lattice location by the use of the channeling phenomenon.

Besides HRXRD, in this chapter we will discuss the application of selected nuclear reactions to the quantification of both deuterium and nitrogen. Hydrogen was substituted by its deuterium isotope, not only to allow the use of NRA, but also to avoid the problem of spurious contaminations that can, in principle, introduce an artifact in the quantification of H. Moreover NRA was used not only to determine the whole amount of D and N in the layers but also to extract D depth profiles with a resolution of about 50 nm. This resolution is quite different from that achieved by secondary ion mass spectrometry (SIMS) (about 10 nm), but as we will show in the following section, it is good enough to be applied to the phenomena under investigation. The depth profiles ability of NRA was of fundamental importance since it enabled us to establish that a not negligible amount of D is introduced in dilute nitrides, even if it is not bound to N. This fact allowed the quantitative determination of the right [D]/[N] stoichiometry.

This synergy of HRXRD and NRA and the whole set of the data collected with these techniques finally enabled us to state that the complexes formed in dilute nitrides after hydrogenation are made by three D atoms bound to a single N atom and that they populate the host film in a layer-by-layer way. Moreover we have found that after thermal annealing at 235°C, a D atom is lost, while the full desorption of the other D atoms bound to N occurs at 315°C. The energy barriers for the dissolution of the D–N complexes are 1.77 eV and 1.89 eV, respectively.

Finally, it was determined that the remaining D not bound to N is completely removed after a more aggressive thermal treatment involving a higher temperature of 603°C.

In the following section first some fundamental ideas on HRXRD and IBA techniques will be given and experimental results will be presented and discussed.

7.2 High-Resolution X-Ray Diffraction

The phenomenon of X-ray scattering from matter is the core of a set of experimental techniques that allow the detection of the structural properties of both amorphous and crystalline materials. HRXRD is an experimental technique particularly suitable when the material under investigation is a crystal of very high structural quality (i.e., the atoms of the sample are arranged in a periodic lattice with a very low content of extended defects). In this case, considering Bragg's law (Ashcroft and Mermin, 1976), it clearly appears that every type of phenomenon inducing a modification of the crystalline lattice interplanar distance (d) can be easily detected by the use of HRXRD.

When interested in the study of semiconductor heterostructures, the first cause of a d change is the epitaxial growth of a film on a substrate with a different lattice parameter. If the thickness of the film is under the critical value (People and Bean, 1985), the film lattice is able to inherit the lattice parameter of the substrate in the direction parallel to the substrate surface. As a consequence, the film undergoes a variation of the lattice parameter in the

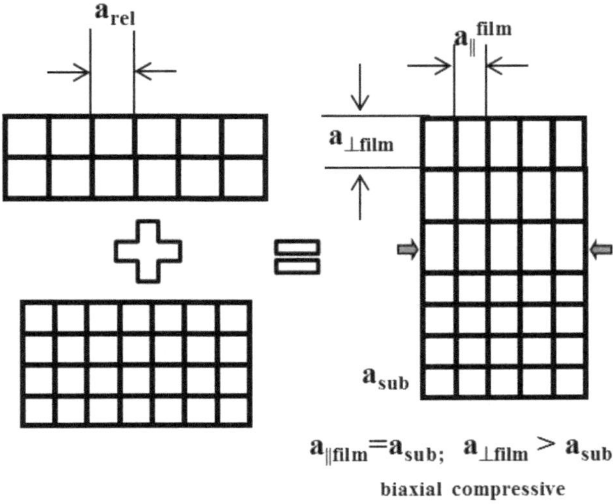

Figure 7.1 Epitaxy of a film with a lattice parameter larger than that of the substrate.

direction perpendicular to the substrate surface, according to a phenomenon well known as the Poisson effect (Landau and Lifshitz, 1986) (Fig. 7.1). The type of the interplanar distance change is governed by the sign of the perpendicular lattice mismatch $m_{\perp} = (a_{\perp\ film} - a_{sub})/a_{sub}$ (where $a_{\perp\ film}$ is the film lattice constant in the perpendicular direction and a_{sub} is the lattice parameter of the substrate), that is, the film will undergo an extension if $m_{\perp} > 0$, differently, a shortening. In addition to the perpendicular lattice mismatch $m_{\perp}$, it is useful to introduce another meaningful quantity, the perpendicular strain $\varepsilon_{\perp} = (a_{\perp\ film} - a_{rel})/a_{rel}$, where a_{rel} is the lattice parameter of the bare film material (i.e., of the material free from epitaxial constraints) (Holý et al., 1999).

The simplest way to induce a modification in the lattice parameter of a crystal is to introduce an impurity in the host lattice of the crystal itself. By the substitution of a certain percentage of the host-type atoms with another one provided of a different covalent radius, it is usually possible to tune the lattice parameter of the so-formed alloy. As an example, the introduction of N atoms into a GaAs lattice originates a reduction of the GaAs lattice parameter

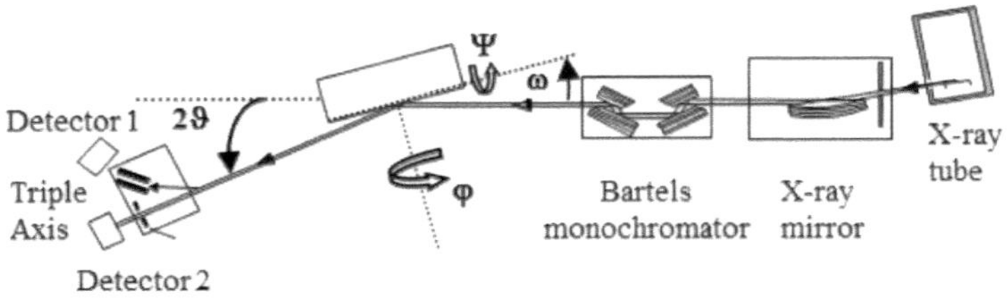

Figure 7.2 Schematic representation of the diffractometer used in this work. Ψ and φ represent the rotation degrees around the axes parallel and normal to the sample holder surface, respectively. The accuracy of ω and 2θ is of 10^{-3} degrees.

(Bisognin et al., 2004). On the contrary, the introduction of In causes the opposite effect (Bisognin et al., 2004). Consequently, GaAsN alloys exploit a negative $m_{\perp}$ (interplanar distance contraction) when epitaxially grown on GaAs, while InGaAs is just the opposite.

The HRXRD measurements of this work were carried out by means of a laboratory apparatus with a "conventional" X-ray tube source with 8 keV energy. In particular the system is a Philips X'Pert PRO MRD diffractometer (see Fig. 7.2) equipped with a Bartels Ge (220) four-crystal monochromator and a parabolic mirror, using a channel-cut Ge (220) analyzer before the proportional Xe methane detector (triple-axis configuration with an angular acceptance of 12 arcsec). It is important to underline that, thanks to the use of the parabolic mirror, the photon flux impinging on the sample is raised up to 2.6×10^7 photons/second. In this way the collection time of a spectrum with a good signal-to-noise ratio is shortened, allowing, in the specific case of the samples studied in this chapter, to record HRXRD measurements in five minutes. Moreover, the presence of an Anton Paar DHS 900 hot stage, a commercial stage particularly equipped for heating the samples, allowed us to perform HRXRD measurements during the annealing of the material in a N_2 (inhert) atmosphere. In this way, the result of both the high photon flux and the heating sample holder enabled us to study by fast (few minutes) and in situ HRXRD measurements the effects of the thermal treatment on the samples under investigation. Particular care was devoted to preserving the Bragg diffraction conditions by checking the alignment of the sample every time a new measurement was collected. Moreover, the sample temperature was also carefully checked by comparing the lattice parameter of the thick (0.3 mm) substrate dilatation with literature values, finding a good agreement between the nominal and the hot-stage temperature (3–4 K). Sample contamination during the annealing was prevented by placing a Tantalum foil between the sample and the clips fixing it on the hot stage (Bisognin et al., 2008). The typical result of an HRXRD measurement performed with a monochromatic beam is a Bragg peak located at the angular value corresponding to the interplanar distance of the reflecting family planes. In the case of a thick crystal (several microns) no additional features characterize the Bragg peak (Holý et al., 1999).

When the sample is made by a thin (some hundreds of nanometers) film epitaxially grown on a thick substrate, an additional peak originating from the film will appear (Fig. 7.3). Moreover, due to the finite thickness of the film, this peak will be accompanied by periodic interference fringes. From the angular distance between the peak of the substrate and that of the film, the value of $m_\perp$ can be obtained, while from the period of the interference fringes the thickness of the film can be extracted. Starting from the measurement of $m_\perp$, by exploiting elasticity theory, the values of $\varepsilon_\perp$ is finally obtained (Holý et al., 1999). Furthermore, when the trend of the lattice parameter as a function of the crystal composition is known, it is possible to determine the percentage of dopants and impurities substitutionally hosted in the crystal (Vegard, 1921). The extraction of both the $\varepsilon_\perp$ and the film thickness by means of peak analysis and the distance between the fringes, respectively, correctly works only in the case that the perpendicular strain is constant inside the film. However, in practice, the strain distribution can vary and the reconstruction of the strain profile requires the reproduction of the whole HRXRD spectrum, depending on the well-assessed X-ray diffraction dynamical theory (Holý et al., 1999). To this aim simulation or fitting codes able to extract strain depth profiles with the resolution of a few nanometers were developed and are nowadays commercially available (Wormington et al., 1999).

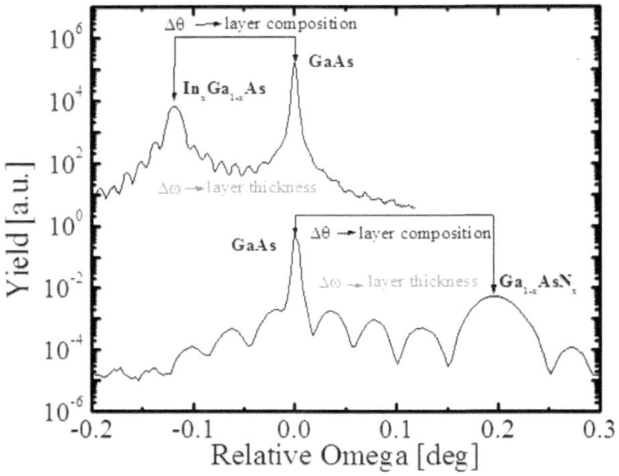

Figure 7.3 RC of a compressively strained layer of InGaAs epitaxially grown on GaAs (upper curve) and of a GaNAs tensile strained layer grown on a GaAs substrate (lower curve). From the period of the interference fringes ($\Delta\omega$) the thickness of the film can be deduced. Moreover, from the angular distance between the Bragg peaks ($\Delta\theta$) the perpendicular mismatch and layer composition can be deduced.

7.3 Ion Beam Analysis Techniques

The techniques based on IBA used in this work encompass both Rutherford back scattering spectrometry (RBS) and NRA. Their use allows the absolute quantification of very low amounts of selected elements, exploiting the comparison with a calibrated standard and/or with a well-established cross section. The physical phenomenon underlying IBA is a collision process: from the study of the collision parameters, information about the materials under investigation can be obtained. The experimental apparatus for RBS and NRA is essentially the same. A collimated beam of particles with energy ranging from 1 MeV to 3.5 MeV, in most cases, strikes the target, and the light products of the reaction are revealed by a solid-state detector (see Fig. 7.4). RBS essentially uses the large-angle elastic scattering described by E. Rutherford in 1911 (Rutherford, 1911). An energetic beam of light particles (usually protons or alpha particles) hits a target. The atoms in the targets are supposed to

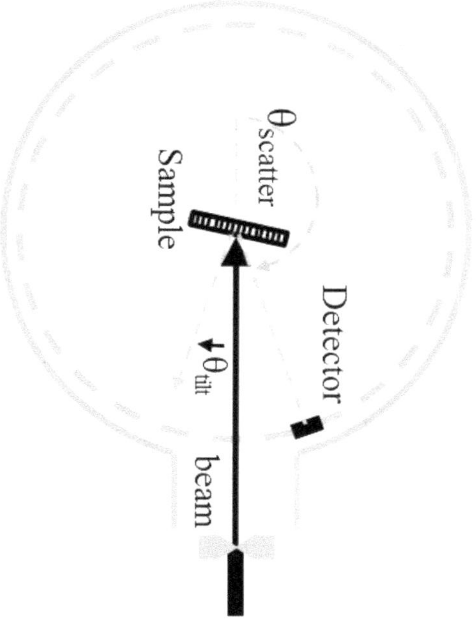

Figure 7.4 Schematic picture of the RBS/NRA measurement chamber. The scattering angle and tilt angle are reported. The sample is usually mounted on a goniometric system to allow for orientation under channeling conditions.

be at rest, and the particles of the incoming beam are elastically deflected by the Coulomb barrier of the target atoms. The collision can be described in terms of classical mechanics, and the energy of the elastically scattered light particles is easily related to the mass of the heavy target atom. The differential cross section of the process, which is related to the probability of a scattering event in the solid angle, is analytical and depends on the scattering angle and on the square of the atomic number (Z) of the target atoms. Moreover it is inversely proportional to the square of the beam energy. Thus the scattering probability of heavy elements will be much higher than that of light elements. This makes RBS a good technique for measuring medium-high Z targets, while it is not applicable for measuring light elements in heavy matrices.

NRA uses a beam made by particles that, thanks to their energy, are able to overcome the Coulomb barrier of the target atom, promoting the formation of an excited atom that gains the fundamental state through the production of different types of particles. The energies supplied by most accelerators used for IBA techniques (< 3 MeV) allow us to overcome the Coulomb barrier of most light nuclei, up to $Z = 15 \div 20$. When the barrier is overcome a composite unstable nucleus forms in most of the cases; the nucleus can decay in different ways with different probabilities. Of course the decay properties of the composite nucleus strictly depend on the isotopes type; therefore NRA is sensitive to the isotopes of an atomic species.

As an example (this will be useful in the following section) the notation $^{14}N(d,p)^{15}N$ means that a deuteron beam (d) hits the ^{14}N atoms in a target, and an excited ^{16}O is formed that decays in ^{15}N trough the emission of a proton (p). After the reaction the excited nucleus (^{16}O in our example) can be in the ground state or in an excited state. According to this the light product (p in our example) can exit the reaction with different energies: due to energy conservation, the highest-possible energy for the light product corresponds to the ground-state case for the produced nuclei. In this case the light product will be indicated with a 0 subscript (p_0 in our example); reaction channels involving exited nuclei are indicated with a progressive number, and the light products have lower and lower energies (p_1, p_2, etc.). Usually the composite nucleus does not decay only in a single light product. The same deuteron beam can induce the $^{14}N(d,\alpha)^{12}C$ reaction in which the excited ^{16}O decays trough the emission of an alpha particle; also the emitted alpha particles can have different energies. In summary, provided that the energy of the analyzing beam is high enough to overcome the Coulomb barrier of the target atoms, different interfering products with different probabilities (cross sections) could be detected. Moreover the cross sections for a specific product are, in general, strongly angle dependent. In most nuclear reactions used to detect light elements the energies of the light products are largely higher than the beam energy. This is essentially due to the mass defect of the products with respect to the reagents. The energy loss (dE/dx) of the ejected particles when they exit the target is negligible. Then the possibility of depth-profiling the elements, typical of RBS, is reduced by a large majority when dealing with NRA, and the available information is in most cases an integral quantification of the dose (areal density in at/cm^2) of the isotope under investigation. The cross sections of many nuclear reactions used for material characterization were studied in half of the sixties mainly by the pioneering work of G. Amsel. Among the reactions of interest in our work, he studied ^{14}N(d,p)^{15}N and ^{14}N(d,α)^{12}C (Amsel and Davies, 1969). Both of these reactions can be used to quantify N by using a deuteron beam with an energy of about 1 MeV. In the same energy range the D(d,p)T reaction can be induced. This reaction has been deeply studied as it is produced in many radioactive decays, but it had almost never been used to quantify deuterium. The D(d,p)T reaction has in principle two disadvantages: a) it has practically no depth resolution, and b) the same species constitute the analyzing beam and the atoms under study. The latter fact can, in principle, induce a D self-contamination of the sample. On the other hand, the cross section is relatively high and self-contamination can be easily controlled. Moreover it is possible to measure directly on the same spectrum the N and D amounts, reducing the systematic error in evaluating the [D]/[N] ratio. To check the possibility of simultaneously measuring N and D, an accurate study of the ^{14}N(d,p)^{15}N and ^{14}N(d,α)^{12}C reactions was performed in the energy range (900$\div$1200) keV, finding that the most suitable reaction is ^{14}N(d,α)^{12}C. In fact its cross section ($\sigma_{\alpha 1}$) related to the first excited level is almost constant ($\pm$ 2%) in the (1100$\div$1250 keV) energy interval (Bisognin et al., 2004). The same holds for the D(d,p)T nuclear reaction allowing a simultaneous measurement of nitrogen and deuterium. Figure 7.5 shows the spectra recorded on two GaAsN samples with (solid line) and without deuteration (dashed line). Many peaks from different reaction products are visible: ^{16}O(d,p$_0$)^{17}O, ^{16}O(d,p$_1$)^{17}O, ^{14}N(d,p$_i$)^{15}N, and ^{12}C(d,p)^{13}C. But the two peaks of interest appear separated from the others, allowing an accurate measure. The N atomic dose is derived from the integrals of the corresponding signals and by comparing it with that of a calibrated standard. In the case of deuterium a calibrated standard was not available. The D concentration was derived from the N standard, taking into account the ratio of D and N cross sections (Berti

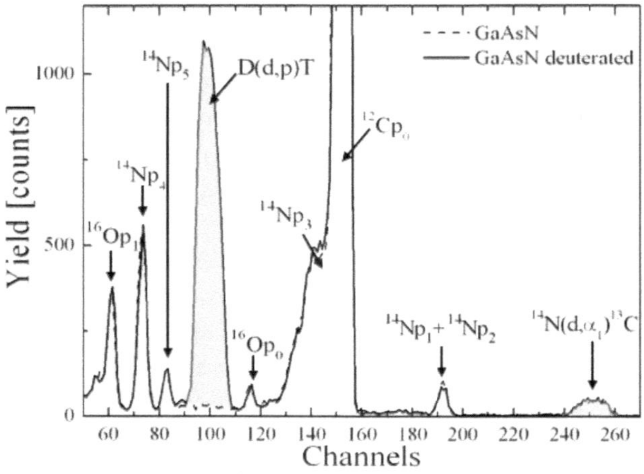

Figure 7.5 Reaction products from deuterium-induced nuclear reaction. Beam energy = 1200 keV. The peaks of interest for the analysis are evidenced.

et al., 2008; Bisognin et al., 2004). The overall procedure allows us to determine the nitrogen concentration with an accuracy of 3.5% and the deuterium concentration with an accuracy of less than 10%.

As already mentioned, deuterium-induced reactions usually do not allow us to get the concentration profiles of the elements under study. The reason is twofold: The energy lost into the sample by the analyzing beam is very low, and the reaction products have high energy so that they lose only a small fraction of this energy while traversing the sample toward the detector. This is a strong limitation for deuterium characterization, while it is not important for nitrogen, for which, in these samples, HRXRD showed constant depth–strain profiles, which means constant concentration profiles. To overcome this difficulty we used the less conventional ^{3}He(d,p)^{4}He reaction (Moller and Besembacher, 1980; Bosch and Hale, 1992; Dieumegard et al., 1979; Berti et al., 2007), which allowed us to reach a best depth resolution of about 50 nm, thanks to the relatively high energy loss of the incoming beam. If the sample under investigation is a crystal, both RBS and NRA techniques can exploit the advantage of the channeling phenomenon. Due to the periodic arrangement of the atoms into a crystalline lattice, a crystal shows some empty channels along particular directions. The density and position of the atoms into these channels depend on the structure of the crystal under investigation. The atoms of the lattice rows generate a potential within the channels that can steer inside the crystal the beam impinging on the crystal surface in a direction parallel to that of the crystal rows (Tesmer and Nastasi, 1995). In this channeling condition the scattering probability between the beam and the crystal atoms is strongly lowered (up to 2 orders of magnitude). On the contrary, defects or impurities located off-lattice inside the crystal give rise to an increase in the scattering yield.

The use of IBA techniques in channeling configuration requires a beam with low divergence and the possibility to rotate the sample holder in order to align the beam to the crystallographic direction of interest. The angular precision required for the goniometric sample holder is typically of 10^{-2} degrees. All the cross sections measured for nuclear reactions and those derived for Rutherford scattering assume that the atoms in the sample are randomly located (i.e., amorphous material). By comparing spectra recorded in channeling conditions and in random conditions (crystal rows or planes not aligned with the beam) quantitative information on the atomic organization of the crystal and on the location of defects can be extracted. In particular, once the beam has been aligned with a selected crystal direction, it is possible to rotate the sample holder in steps of a few tenths of degrees, recording at each step a spectrum or a part of a spectrum. This sequence of spectra is called "channeling dip": in fact when the beam is aligned with the crystal direction the yield of the spectrum is minimum and increases at increasing misalignments. Defects or misplacements of the atoms of the row enhance the scattering (or reaction) probability, producing features in the dip that can easily be compared with those of an unperturbed crystal.

7.4 Results and Discussion

As pointed out in the previous chapters, hydrogen irradiation of dilute nitrides leads to the formation of N–H complexes that strongly

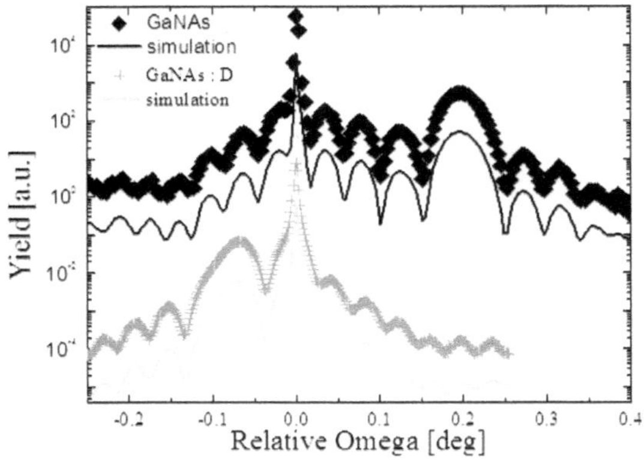

Figure 7.6 RCs of a 110 nm thick $GaAs_{0.986}N_{0.014}$ film covered by a 35 nm GaAs cap layer both before and after the deuteration process (3×10^{18} D/cm^2) (black diamonds and gray crosses, respectively). The continuous lines are the simulations.

affect both the optical and the structural properties of the original materials. In Fig. 7.6 the rocking curves (RCs) of a 110 nm thick $GaAs_{0.986}N_{0.014}$ film covered by a 35 nm GaAs cap layer both before and after the deuteration process (3×10^{18} D/cm^2) (black diamonds and gray crosses, respectively) are reported. The tensile strain due to substitutional N is not only neutralized but even converted into a compressive one, as testified by the shift of the Bragg peak of the epilayer from positive to negative angles. Moreover, the presence of interference fringes in both RCs suggests the pseudomorphicity of the epilayers, as verified by means of reciprocal space maps. This allows us to extract the lattice parameter of the film only by the measure of the out-of-plane lattice parameter by means of the (004) reflection, obtaining the related quantities (i.e., the relaxed lattice parameter and the perpendicular strain) by the use of elasticity theory.

To quantify the extent of the change in the $\varepsilon_{\perp}$ strain profiles (see Fig. 7.7) were extracted by means of RC simulations (continuous lines of Fig. 7.6). It is interesting to note that the strain profiles are flat both before and after the D irradiation process, confirming the strong coupling between N and D atoms and pointing out the

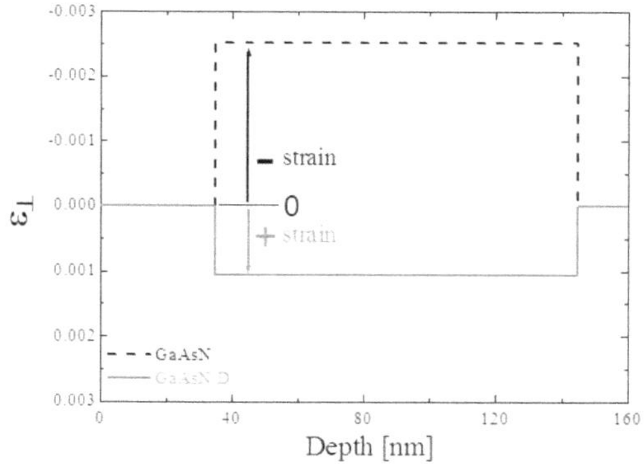

Figure 7.7 Strain profiles extracted by the simulations of the RCs of Fig. 7.6.

reliability of HRXRD to monitor the structural changes induced by the formation of N–D complexes in dilute nitrides.

Due to its sensitivity to the change in the lattice parameter induced by the formation of N–D complexes, HRXRD was also used to monitor the strain evolution as a function of the D irradiation fluence. In Fig. 7.8 the RCs of a 280 nm thick $GaAs_{0.9873}N_{0.0127}$ epilayer irradiated with 7×10^{17} D/cm^2 and 12×10^{17} D/cm^2 are reported (gray and black lines, respectively). For comparison, also the RC of the as-grown sample is displayed in the same plot (light-gray, thick line).

It is evident that after the first D irradiation, a Bragg peak appears also at negative angles, while the original peak undergoes a broadening. These facts mean that the original tensile film is converted into a bilayer structure, part of which hosting the N–D complexes induces compressive strain, while the remaining part is keeping on the strain of the pristine structure. Raising up the D fluence we can observe a sharpening of the peak located at negative angles, testifying that the thickness hosting the N–D complexes is increased. From the RC simulations (not reported here), the thickness of the deuterium-containing layers is measured, finding 130 ± 10 and 160 ± 20 nm for the lower and higher D fluences. NRA measurements with ^{14}N(d,α)^{12}C and D(d,p)T allowed us to measure the total nitrogen and deuterium content in these samples; the results are reported in Table 7.1, together with the thickness

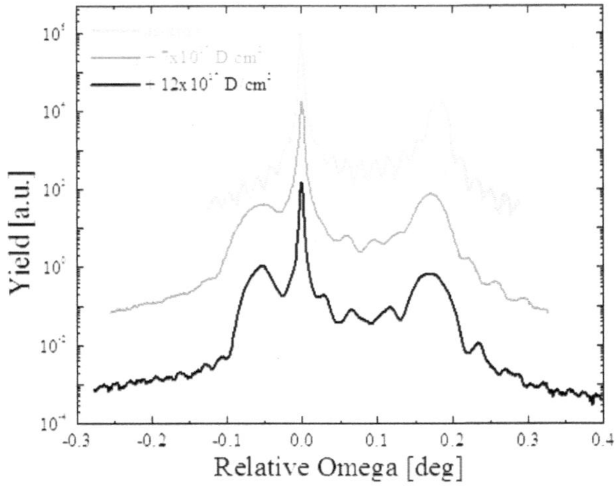

Figure 7.8 RCs of a 280 nm thick $GaAs_{0.9873}N_{0.0127}$ sample irradiated with different deuterium fluences.

Table 7.1 Nitrogen and deuterium doses, as derived from NRA for two samples irradiated with the deuterium fluences reported in the first column. The thickness of the passivated layer, as derived from HRXRD is also given. The [D]/[N] ratio has been computed by assuming that deuterium is incorporated only into the passivated layer

Irr. Fluence (ions/cm^2)	N (10^{15} at/cm^2)	D (10^{15} at/cm^2)	Pass. Thick. (nm)	[D]/[N]
7.0 10^{17}	7.7 ± 0.3	10.6 ± 0.7	130 ± 10	2.76 ± 0.26
12.0 10^{17}	7.7 ± 0.3	16.4 ± 1.1	160 ± 20	3.82 ± 0.35

of the passivated layer, as derived from HRXRD. By assuming that deuterium is distributed only into the passivated layer the [D]/[N] ratio can be derived, and it is also shown in Table 7.1.

A value ranging from 3 to about 4 was found also in other samples with different N concentrations and increasing deuterium irradiation fluences. The question was, does this ratio mean that at least 3 hydrogen atoms are necessary to passivate nitrogen? And why does the ratio fluctuate from 3 to 4? The answer was given

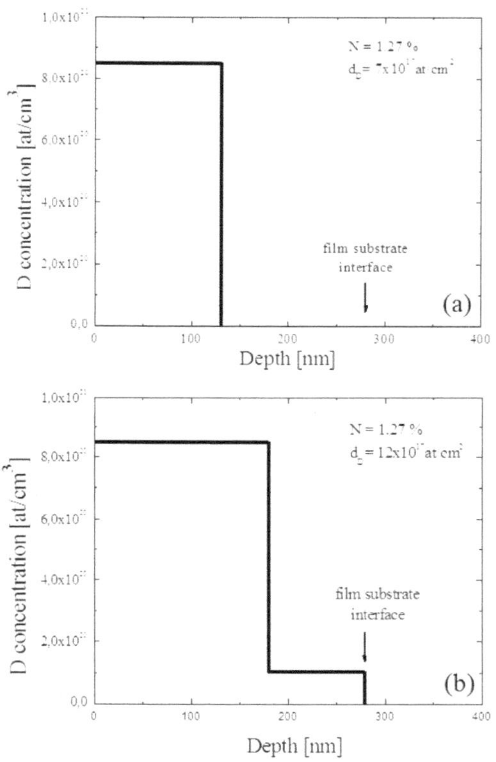

Figure 7.9 Deuterium concentration profiles derived from the ^{3}He(d,p)^{4}He nuclear reaction. The samples are the same as in Fig. 7.8.

by NRA, exploiting the ^{3}He(d,p)^{4}He reaction, which gave for the samples reported in Table 7.1 the concentration profiles of Fig. 7.9.

It is clear that deuterium incorporation proceeds layer by layer once a critical concentration ([D]/[N] $=$ 3) has been reached. When the layer has been almost totally deuterated additional deuterium can be incorporated possibly weakly bound to nitrogen. This additional deuterium could be responsible for the [D]/[N] ratio greater than 3 found in some samples.

So at least three hydrogen atoms are necessary to passivate nitrogen, but could two also be enough?

We addressed the issue of how many D atoms are required to passivate a single nitrogen by performing HRXRD measurements during annealing of fully passivated epilayers. Since the full passivation of N is accompanied by compressive strain, its lowering promoted by thermal annealing can indicate dissolution of the N–D complexes and give information about the thermal stability of the N–D complexes. Exploiting the sample holder allowing HRXRD measurements in a N_2 atmosphere during the heating of the sample, we progressively raised the temperature of the sample, monitoring possible shifts of the Bragg peak of the layer. No significant peak shift occurs up to about 235°C. At this temperature the peak starts to move toward positive angles (i.e., the reduction of compressive strain), and we followed the movement till the saturation of the phenomenon was reached. In Fig. 7.10a some selected RCs collected during 69 hours of annealing are reported. It is clear that the peak shift saturates when the strain of the film passes from compressive to slightly tensile values. Since the strain value gained at $T = 235°\text{C}$ is stable, we can conclude that the original complex dissolution is stopped and that the film hosts a particular type of N–D aggregate, presumably simpler with respect to the original one.

Then the temperature was raised up to 315°C, and a second strain variation was observed (Fig. 7.10b). In fact, at this temperature, the strain of the pristine GaNAs lattice was almost completely restored after six hours of annealing. In Fig. 7.10 both the (004) and (224) reciprocal space maps are reported (right panels), which testify the absence of strain relaxation of the epilayer (both the film and the substrate reciprocal space nodes are perfectly aligned along the y axis). Finally, full tensile strain recovery was gained with a more aggressive annealing performed at 603°C.

These findings confirmed the reversibility of the deuteration process and indicate the existence of at least two different dissolution stages of the N–D complexes formed after D incorporation.

More insights on the energy barrier involved in these two dissolution steps come from the analysis of the RCs' evolution as a function of time. By simulating the HRXRD measurements, we could extract the relaxed lattice parameter of the film, which in both annealing steps shows a trend easily fitted by a single exponential decay (Fig. 7.11). By appropriate modeling (Bisognin et al., 2008) the

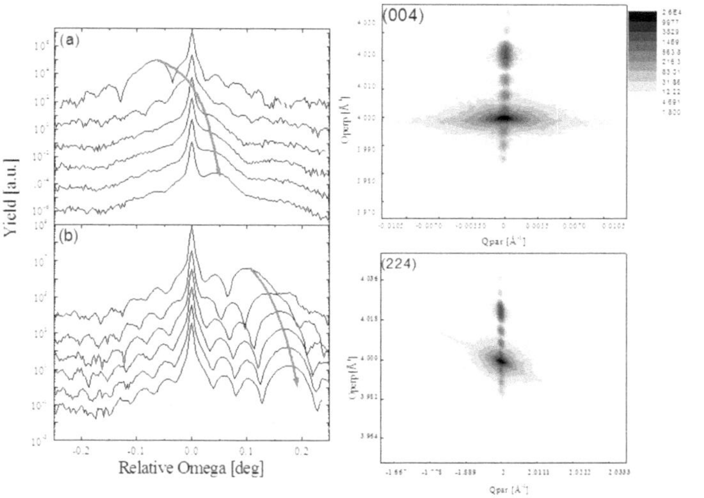

Figure 7.10 (a) Evolution of the RCs of a 110 nm thick $GaAs_{0.986}N_{0.014}$ film covered by a 35 nm GaAs cap layer irradiated with 3×10^{18} D/cm^2 annealed at 235°C. Each RC differs in time by about 14 hours from the previous one; (b) evolution of the RCs related to the same sample of (a) at $T = 315$°C. Each RC differs in time by 1 hour from the previous one. The dashed line represents the as-grown sample. (004) and (224) reciprocal space maps related to the sample at the end of the 315°C annealing are also reported.

energy barriers of the two dissolution steps are determined: 1.77 eV and 1.89 eV, respectively. Photoluminescence, infrared absorption, and EXAFS measurements, reported in the previous chapters, took into account the results on the stoichiometry of the GaAsN (D) complexes, leading to a complete description of the formation and dissolution process of D–N complexes. The deuteration proceeds layer by layer, and the further incorporation of deuterium is possible only after the N–D complexes have been formed. A 3D–N complex is the final product of the deuteration in all dilute nitrides. The study of the thermal dissolution showed that the 3D–N complex converts into a 2D–N complex still capable of N passivation. Further annealing treatments induce the total loss of deuterium and the recovery of the pristine GaAsN properties.

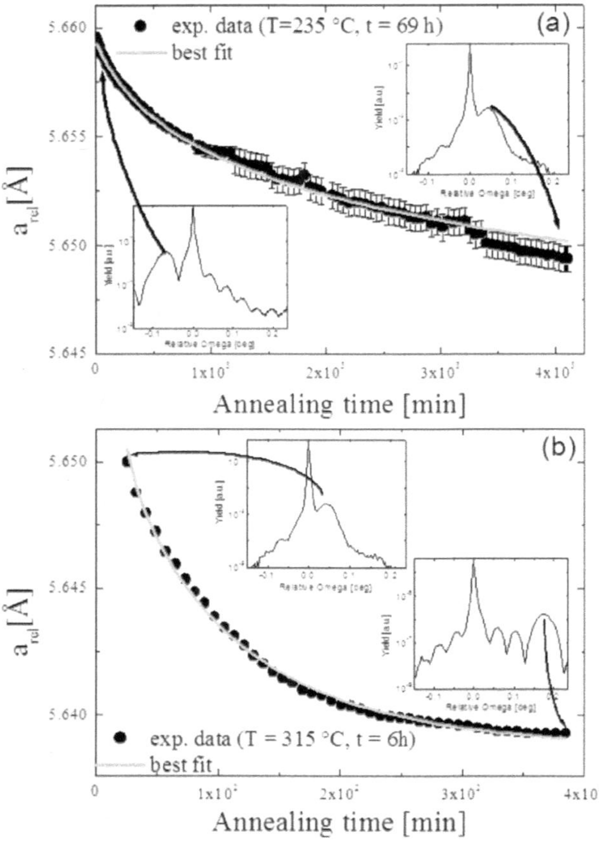

Figure 7.11 Evolution of the relaxed lattice parameter as a function of the annealing time for $T = 235°C$ (a) and for $T = 315°C$ (b). The best fits are also reported (gray lines), together with the RCs related to the starting and ending points of both annealing processes.

Chapter 8

Technological Applications of Hydrogenated Dilute Nitrides and Perspectives

Rinaldo Trotta[a] and Marco Felici[b]
[a] *Institute of Semiconductor and Solid State Physics, Johannes Kepler University Linz, Altenbergerstrasse 69 Linz, A-4040, Austria*
[b] *Dipartimento di Fisica, Sapienza Università di Roma, Piazzale Aldo Moro 2, 00185 Roma, Italy*
rinaldo.trotta@jku.at, marco.felici@roma1.infn.it

The capability to control and manipulate semiconductor materials at the nanoscale has opened up new frontiers in the promising fields of *nanoelectronics* and *nanophotonics*. However, the strict requirements set by advanced quantum technologies keep spurring the search for smart fabrication methods, offering the means to precisely control the position and electronic properties of individual nanostructures. This chapter describes an original nanofabrication approach based on the *spatially selective hydrogenation* of dilute nitride semiconductors. The strong N–H interaction in dilute nitrides turns H diffusion into a multiple trapping process, resulting in the formation of extremely sharp heterointerfaces between H-free and H-containing regions. This finding—combined with the striking effects of hydrogen on the electronic, optical, electrical, and

Hydrogenated Dilute Nitride Semiconductors: Theory, Properties, and Applications
Edited by Gianluca Ciatto

ISBN 978-981-4463-45-4 (Hardcover), 978-981-4463-46-1 (eBook)
www.panstanford.com

structural properties of dilute nitrides—allows for the fabrication of a novel class of site-controlled micro- and nanostructures, whose properties meet a wide set of technological applications, ranging from nanophotonics to integrated optical and electrical circuits. A number of examples are described in the following sections, and future directions are discussed.

8.1 Introduction

In the last decade, the investigation of light-matter coupling phenomena in the solid state has pushed its boundaries beyond the limit represented by the wavelength of light ($\sim$1 μm for visible/near-infrared light). This intriguing possibility was made available by the development of techniques allowing for the integration of nanometer-sized light emitters—mainly quantum dots (QDs) (Yoshie et al., 2004; Hennessy et al., 2007) but also quantum wires (QWRs) (Atlasov et al., 2010) and radiative defects (e.g., nitrogen-vacancy [NV] centers in diamond (Englund et al., 2010))—with optical cavities of (sub-) micrometric dimensions, such as micropillars (Bayer et al., 1998; Peter et al., 2005; Reithmaier et al., 2004) and photonic crystal (PhC) microcavities (Yoshie et al., 2004; Hennessy et al., 2007; Atlasov et al., 2010; Baba et al., 1999; Akahane et al., 2003; Noda, 2006; Song et al., 2005; Kuramochi et al., 2006; Akahane et al., 2005). This has recently allowed for the observation of strong coupling phenomena (such as vacuum Rabi splitting (Khitrova et al., 2006)) in several QD–microcavity integrated systems (Yoshie et al., 2004; Hennessy et al., 2007; Peter et al., 2005; Reithmaier et al., 2004), and it carries great promise for the realization of numerous optoelectronic applications.

In spite of these early accomplishments, however, the field of nanophotonics is still in its infancy. The departure from this pioneering phase will require the emergence of novel methods for the design and fabrication of advanced photonic devices, offering the possibility to control both the electromagnetic field distribution of the confined cavity modes (Englund et al., 2005; Srinivasan and Painter, 2003; Felici et al., 2010) and the spatial/spectral position of the integrated light emitters. The strive to reach the latter goal, in particular, is the main driving force behind current, massive efforts to control the physical properties of *semiconductor QDs*.

Often referred to as "artificial atoms," QDs exhibit electronic and optical properties that are a direct consequence of their zero-dimensional nature (Bimberg et al., 1999). In particular, semiconductor QDs—which can be easily integrated into conventional optoelectronic devices (such as light-emitting diodes [LEDs] (Salter et al., 2010))—can emit single (Baier et al., 2004) and entangled (Hudson et al., 2007) photons "on demand," in response to an external trigger. This feature, not shared by other sources of nonclassical light, has pushed the potential reach of nanophotonics toward the promising fields of quantum information and quantum communication. The fulfillment of the strict requirements set by advanced quantum information applications, however, asks for the development of new, original QD fabrication technologies, offering the possibility to control not just the spectral and spatial positions of individual nanostructures but rather the complete set of their electronic properties, that is, the exciton fine structure splitting and the biexciton binding energy (Trotta et al., 2012a,c). In spite of the significant efforts that have been recently dedicated to the fabrication of self-assembled QD systems possessing a high degree of order and finely tunable emission properties (Kiravittaya et al., 2006; Huggenberger et al., 2011; Hughes et al., 2011; Skiba-Szymanska et al., 2011; Yuan et al., 2011), the establishment of nanophotonics as a truly viable technology will likely require the emergence of novel nanofabrication approaches, combining lithography, epitaxial growth, and postprocessing steps in a smart way to deliver the required precision and dot quality.

In the following sections, we will describe an innovative method for the fabrication of site-controlled nanostructures on the basis of the hydrogen-induced *passivation* of N atoms in dilute nitride (GaAsN, InGaAsN, GaPN...) semiconductors. As discussed throughout this volume, in these systems H forms stable N–2H–H complexes (Wen et al., 2010), which wipe out the effects of the presence of N in the lattice. In turn, this leads to profound—and fully controllable—modifications of the electronic (e.g., band-gap energy (Polimeni et al., 2001a; Baldassarri Höger von Högersthal et al., 2001; Polimeni et al., 2003a), transport (Masia et al., 2006),

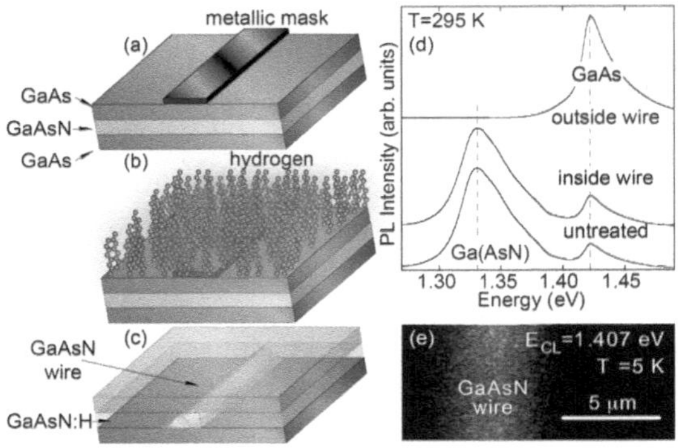

Figure 8.1 (a–c) Sample processing steps entailed by in-plane band-gap engineering (BGE). (a) A metallic mask is deposited on the sample's surface (e.g., a GaAsN quantum well [QW]). (b) Hydrogenation is performed on the masked sample. (c) Following the formation of the designed GaAsN/GaAsN:H structure (e.g., a *wire*), the mask is removed. (d) Microphotoluminescence (micro-PL) spectra of a selectively hydrogenated $GaAs_{1-x}N_x$ sample ($T = 295$ K) in the regions outside (top) and inside (middle) a 3 μm wide wire, fabricated following the steps outlined in (a–c). The micro-PL spectrum before hydrogen irradiation is also shown for reference purposes (bottommost line). (e) Cathodoluminescence (CL), mapping ($T = 5$ K, beam current $I_b = 1$ nA, beam energy $E_b = 20$ keV) of a 3 μm wire. The energy of the detected photons, $E_{CL} = 1.407$ eV, matches the CL peak of the untreated $GaAs_{1-x}N_x$ sample at $T = 5$ K. Panels (d and e) adapted from Ref. (Felici et al., 2006a).

and spin properties (Pettinari et al., 2006)), as well as optical (refractive index (Geddo et al., 2007)), structural (lattice constant (Berti et al., 2007), and ordering (Ciatto et al., 2005c; Felici et al., 2006b)) characteristics of these materials (Henini, 2005), which are of interest in many areas ranging from telecommunication optoelectronics (Diehl et al., 2007) to THz sources (Ignatov et al., 2006). In particular, H binding to N atoms in $GaAs_{1-x}N_x$ leads to an increase in the band-gap energy of the N-containing material ($\sim$1.33 eV for $x = 1\%$ at $T = 5$ K) up to the value it has in GaAs (1.52 eV at 5 K). Therefore, by allowing H incorporation only in selected regions of the sample—for example, by deposition of H-opaque metallic masks prior to hydrogenation, as sketched in Fig. 8.1—it is possible to attain a spatially controlled modulation of the band-gap energy in the growth plane. Felici *et al.* first demonstrated the feasibility of this approach in 2006 (Felici et al., 2006a) by depositing photolithographically defined masks on the surface of a GaAsN sample. Following H irradiation, pristine (i.e., H-free) GaAsN wires of micrometric dimensions were left underneath the masks, whereas the remainder of the sample was hydrogenated fully. The fabricated structures were characterized by micro-PL spectroscopy, and their emission spectrum was indeed found to be virtually identical to that of untreated GaAsN (see Fig. 8.1d, relative to a 3 μm wire). The quite distinctive emission spectrum of the masked areas with respect to their surroundings was also exploited to obtain nicely contrasted CL images of the fabricated structures (see Fig. 8.1e) by setting the energy of the detected photons, $E_{\text{CL}} = 1.407$ eV, to match that of the GaAsN emission peak (Felici et al., 2006a).

Having established the possibility of using this method—which will be referred to as *in-plane BGE* in the remainder of this book—to tailor the in-plane potential profile of dilute nitride materials, successive refinements of the selective-hydrogenation protocol were aimed at the realization of potential wells that would be small enough to result in the establishment of *quantum confinement* in the fabricated (nano)structures. These efforts reached their culmination in 2011, when the realization of *site-controlled QDs* by in-plane BGE was first reported by Trotta et al. (Trotta et al., 2011) (see also Section 8.3.2).

While the switch to *electron beam lithography*—which allows for defining the H-opaque metallic masks with a precision of a few nanometers—represented one of the main reasons for the achievement of this goal, a crucial role was also played by the systematic investigation of the effects of the hydrogenation conditions—most notably of the hydrogenation *temperature*—on the *diffusion profile of H* in the material. Thanks to the information gathered during this study, whose most salient results are summarized in the next section, the forefront of the H diffusion profile in the material can indeed be made to be exceptionally sharp (less than 5 nm/decade for hydrogenation performed at 200°C; see Ref. (Trotta et al., 2011) and Section 8.2), thus effectively enabling a nanometer-level control of the carrier-confining potential defined by in-plane BGE.

8.2 Controlling Hydrogen Diffusion in Dilute Nitrides: A New Ground for Nanotechnology

In the previous section, we have seen that a spatial control of H irradiation (and of H-induced effects) can be obtained at the micrometer scale using a simple lithography step preceding H treatment. In the perspective of nanophotonics applications based on hydrogenated dilute nitrides, a natural question arises: Is it possible to push this approach down to the nanometer scale length? As briefly mentioned above, the positive answer to this question passes through the investigation of *H diffusion in dilute nitrides*, which is the focus of the present section.

The kinetics of H in crystalline semiconductors depends strongly on the type of interactions H experiences while traveling through the host matrix and is a rather complex matter to deal with. H is indeed one of the most chemically active atoms present in nature. It can exist in a number of different forms (i.e., atomic or molecular, free or bound to a defect impurity), finally leading to an exceptionally large number of effects in semiconductors (Pearton et al., 1992). In the following section, we restrict our discussion to the particular case of dilute nitrides (specifically GaAsN), once more emphasizing the unique properties of this material system when compared to more "conventional" semiconductors.

Standard diffusion of an impurity into a semiconductor matrix is governed by Fick's law, which in one dimension—and for a diffusion coefficient (D_{dif}) independent of the spatial coordinate, z—takes the following mathematical form:

$$\frac{\partial n_{\text{H}}}{\partial t} = D_{\text{dif}} \frac{\partial^2 n_{\text{H}}}{\partial z^2}, \tag{8.1}$$

where n_{H} is the impurity concentration. In the case of a semi-infinite medium extending from $z = 0$ to $z = \infty$ and with a constant impurity concentration, n_{H0}, at $z = 0$, the solution of Eq. 8.1 is the *complementary error function*, shown in Fig. 8.2a for different diffusion times. This equation is commonly used in semiconductor research (Pearton et al., 1992)—even for the modeling of electronic devices (Kennedy and O'Brien, 1965)—in spite of the fact that it neglects completely the retardation of the impurity diffusion caused

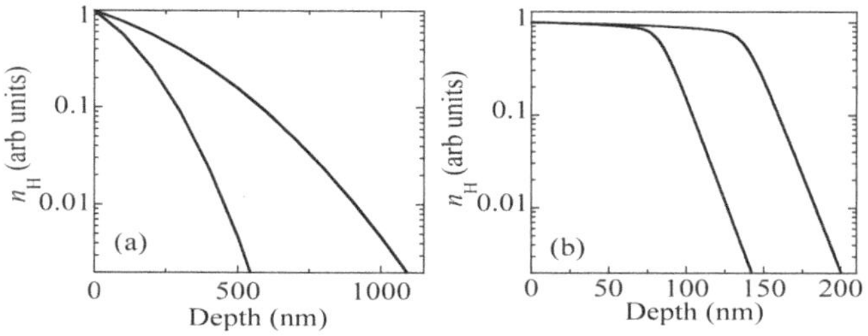

Figure 8.2 (a) Solution of Fick's diffusion equation (see Eq. 8.1) for a constant concentration of the impurity at the surface and for different diffusion times. (b) Solution of Eqs. 8.3, 8.4, and 8.5, valid for hydrogen diffusion in the presence of multiple trapping processes. The different curves correspond to different diffusion times.

by defect trapping, by the formation of impurity clusters, and by the presence of electrostatic potentials (for the case of charged impurities) (Pearton et al., 1992). Through the application of Eq. 8.1, all "retardation effects" are actually incorporated in D_{dif}, which eventually becomes an *effective diffusion coefficient* for the specific impurity in the specific semiconductor. This approximation partially explains the large number of diffusion coefficients that can be found in the literature for H in GaAs (Zavada et al., 1985; Raisanen et al., 1988; Morrow, 1989) and—as long as that the impurity profile retains the "Fick-like" shape shown in Fig. 8.2a—it is indeed valid for most practical applications. However, when retardation effects become more and more important, diffusion is profoundly modified and Fick's equation no longer holds. Generally speaking, in those cases Eq. 8.1 has to be modified in order to include the possible chemical reactions that allow mobile and immobile species involved in the diffusion to transform into each other. The kinetics of impurity diffusion is then governed by the *diffusion reaction* equation, which in its simplest formulation can be written as (Kolmogorov et al., 1937)

$$\frac{\partial n_{\mathrm{H}}}{\partial t} - D_{\mathrm{dif}}\frac{\partial^2 n_{\mathrm{H}}}{\partial z^2} = \mathrm{R}(n_{\mathrm{H}}), \qquad (8.2)$$

where $\mathrm{R}(n_{\mathrm{H}})$ accounts for all the possible reactions taking place during diffusion and in the general case depends on the spatial and temporal evolution of n_{H}. This term complicates the problem considerably, as it requires the knowledge of the main "retardation effects" and of their explicit mathematical form. In the following section, we derive R for H diffusion in dilute nitrides, considering the H–N interaction as the main "retardation effect."

On the basis of the studies reported in the previous chapters (Berti et al., 2007; Bisognin et al., 2006), we know that the interaction of H with N leads to the formation of N–2H–H complexes, where (*i*) two H atoms bind to a single N atom, forming a very stable N–2H complex responsible for the electronic passivation of nitrogen and (*ii*) a third H atom is more weakly trapped nearby passivated N atoms, causing an additional compressive strain in the GaAsN layer. As described in Chapter 7, structural and optical measurements on deuterated GaAsN samples subjected to different thermal annealing treatments showed that the third D (or, equivalently, H) atom is removed from its site at 250°C, with an activation energy equal to $E_a = 1.9$ eV (Bisognin et al., 2007, 2008). Upon removal of this "extra" D (H) atom the compressive strain disappears, while N atoms continue to be electronically passivated. A higher activation energy ($\sim$2.2 eV) characterizes, instead, the dissolution of the N–2D (N–2H) complexes, and the concomitant recovery of the electronic properties (e.g., the band gap) of pristine GaAsN. These findings suggest that H (or D) kinetics in GaAsN is characterized by a *multiple trapping process* (Borenstein et al., 1993), wherein capture ceases as soon as at least three H atoms bind to one N atom.

To model H diffusion in dilute nitrides we must therefore modify Fick's equation, introducing a *reaction term* that accounts for multiple trapping processes (Trotta et al., 2008, 2009a, 2012b). We first treat the GaAsN epilayer as a semi-infinite medium, occupying the $z \geq 0$ space. Let n_{H} be the concentration of free (i.e., unbound) H atoms and n_{NiH} be the concentration of N atoms with $i = 0, 1, 2, 3$ H atoms bonded such that $N_{\mathrm{N}} = \sum_{i=0,3} n_{\mathrm{NiH}}(z, t)$, where N_{N} is the initial concentration of nitrogen. The evolution of the only mobile species is regulated by the following equation for $x \geq 0$ and $t \geq 0$:

$$\begin{aligned}\frac{\partial n_{\mathrm{H}}}{\partial t} - D_{\mathrm{dif}}\frac{\partial^2 n_{\mathrm{H}}}{\partial z^2} &= R_N^H \\ &= -k\left(n_{\mathrm{H}}n_{\mathrm{N}} + n_{\mathrm{H}}n_{\mathrm{NH}} + n_{\mathrm{H}}n_{\mathrm{N2H}}\right) \qquad (8.3) \\ &\quad + \gamma_a n_{\mathrm{N3H}} + \gamma_b n_{\mathrm{N2H}} + \gamma_c n_{\mathrm{NH}}\end{aligned}$$

The reaction term—composed of several subterms—now appears explicitly on the right side of Eq. 8.3. The first term represents the capture of one H atom by N atoms having $i = 0,1,2$ H atoms already bonded (note that $n_{N0H} = n_N$). The same capture rate coefficient ($k = 4\pi r D_{dif}$, where r is an effective capture radius (Morrow, 1989)) is used for each trapping step. The second, third, and fourth terms account for the thermal release of H atoms and can be written as (Morrow, 1989) $\gamma_m = \nu_m e^{-E_m/k_B T}$, where ν_m is the attempt frequency and E_m (with $m = a, b, c$) is the activation energy for the release of one H atom from the particular complex considered. According to previous studies (Berti et al., 2007; Bisognin et al., 2008), we effectively assume $\gamma_a \neq \gamma_b = \gamma_c$. Before attempting to find a numerical solution of this equation, we have to consider the temporal evolution of the immobile species, that is

$$\begin{aligned}
\frac{\partial n_N}{\partial t} &= -k n_H n_N + \gamma_c n_{NH} \\
\frac{\partial n_{NH}}{\partial t} &= k n_H n_N - \gamma_c n_{NH} - k n_H n_{NH} + \gamma_b n_{N2H} \\
\frac{\partial n_{N2H}}{\partial t} &= k n_H n_{NH} - k n_H n_{N2H} + \gamma_a n_{N3H} - \gamma_b n_{N2H} \\
\frac{\partial n_{N3H}}{\partial t} &= k n_H n_{N2H} - \gamma_a n_{N3H}
\end{aligned} \tag{8.4}$$

Equations 8.3 and 8.4 form a system of five coupled partial differential equations, which can be solved numerically with the following initial and boundary conditions: $n_{NiH}(z, 0) = 0$ (for $i = 1,2,3$ and $z > 0$) and $n_N(z, 0) = N_N$ for $z > 0$, $n_H(0, t) = const$ (set by the current of impinging H^+ ions), $n_N(0, t) = 0$, and $n_H(\infty, t) = n_{NH}(\infty, t) = n_{N2H}(\infty, t) = n_{N3H}(\infty, t) = 0$. Finally the total hydrogen depth profile is given by

$$n_H^{tot}(z) = n_H + \sum_{i=1,3} i n_{NiH}(z) \tag{8.5}$$

Figure 8.2b shows, as an example, the hydrogen profiles obtained solving Eqs. 8.3, 8.4, and 8.5 for two different diffusion times. By looking at the marked differences between these profiles and those displayed in Fig. 8.1a, relative to conventional Fick-like diffusion, it is easy to appraise the importance of including the reaction term in Eq. 8.2. First of all, (*i*) a clear plateau develops close to the sample

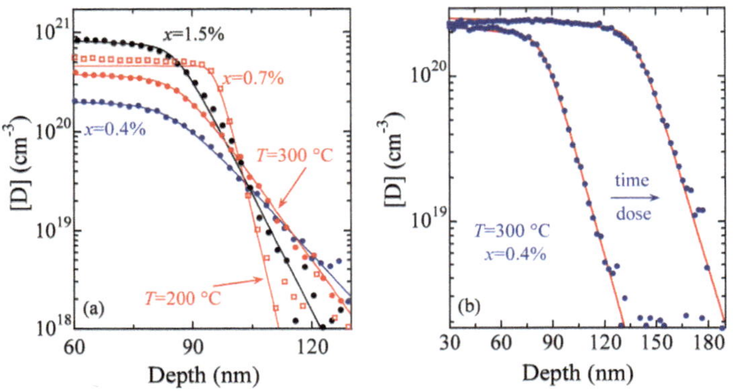

Figure 8.3 (a) SIMS depth profiles of deuterium for $GaAs_{1-x}N_x$ samples with different x (different colors) and irradiated at different temperatures (circles and squares), along with the simulations obtained using Eqs. 8.3, 8.4, and 8.5 (solid lines). The origin of the abscissa is set at the sample surface. (b) SIMS depth profiles of deuterium for a $GaAs_{1-x}N_x$ samples with $x = 0.4\%$ $T_D = 300°C$ and treated with two different D doses (full circles). The results of the simulations obtained using Eqs. 8.3, 8.4, and 8.5 are also reported.

surface, followed by a very steep forefront; second, (*ii*), the profile rigidly shifts toward the inner part of the sample when different diffusion times are employed.

To verify the model described above, we investigated the kinetics of *deuterium* (D) in GaAsN performing secondary ion mass spectrometry (SIMS) on several samples having a different nitrogen content ($0.09\% \leq x \leq 1.5\%$). Deuterium was used instead of hydrogen due to the higher sensitivity of SIMS to D; it should be noted, however, that D and H induce the same modifications in the electronic and structural properties of dilute nitrides. To determine the complete impurity profile, the samples were only partially deuterated. Moreover, the samples were irradiated at several temperatures ($200°C \leq T \leq 350°C$). Figure 8.3a shows selected diffusion profiles of GaAsN samples (symbols), along with the results of the simulations obtained using Eqs. 8.3, 8.4, and 8.5 (solid lines). For the details on SIMS measurements and the specific values of the parameters used we refer the reader to specific papers (Trotta et al., 2008, 2009a). Here, we would like to discuss qualitatively the main results of this study. The experimental data confirms that hydrogen (deuterium) diffusion is well described by Eqs. 8.3, 8.4, and 8.5, that is, by multiple trapping processes. The measured deuterium concentration indeed displays a plateau, followed by an exponential-like decay; moreover, it shifts rigidly toward the inner region of the sample when different deuterium doses (diffusion times) are used (see Fig. 8.3b). Interestingly, the sample-averaged stoichiometric ratio in the plateau region, $q = [D]/[N]$, gives 2.7 for $T_D = 300°C$ and varies little with x. This indicates that the same kind of complexes—characterized by three D atoms for each N impurity—form for all concentrations, thus supporting the assumptions of the model. Finally, the most striking feature visible in Fig. 8.3 is the exceptionally sharp deuterium forefront obtained in samples with high nitrogen concentration and/or irradiated at low temperature. In particular, the deuterium forefront can be as sharp as 5 nm/decade for $T_D = 200°C$ and $x = 0.7\%$. This important result answers the question posed at the beginning of the section and represents the first experimental evidence that spatially selective hydrogenation can be used to tailor the physical properties of dilute nitrides on a scale of a few nanometers, thus paving the way to the realization of nanostructures.

In this context, it is worth extending the theoretical analysis developed in this section to two dimensions and to include the presence of a metallic mask that impedes H diffusion in selected regions of the sample (the extension of this model to three dimensions is straightforward, as we will see in the following section). The mask is assumed to be an impenetrable barrier for hydrogen (i.e., the hydrogen flux beneath the mask is zero), and we solve Eqs. 8.3, 8.4, and 8.5 with the finite-element method using the same initial and boundary conditions described above. Figure 8.4 shows the results of such calculation for 200 nm GaAsN epilayers with $x = 0$ (panel a) and $x = 0.9\%$ (panel b), treated at 200°C and covered with a 200 nm wide wire. The data refers to a section containing the growth axis of the sample (the section is perpendicular to the wire direction) and shows the total hydrogen concentration (see Eq. 8.5; the parameters employed in the calculations are the same used for obtaining the nanostructures that will be described in the following sections).

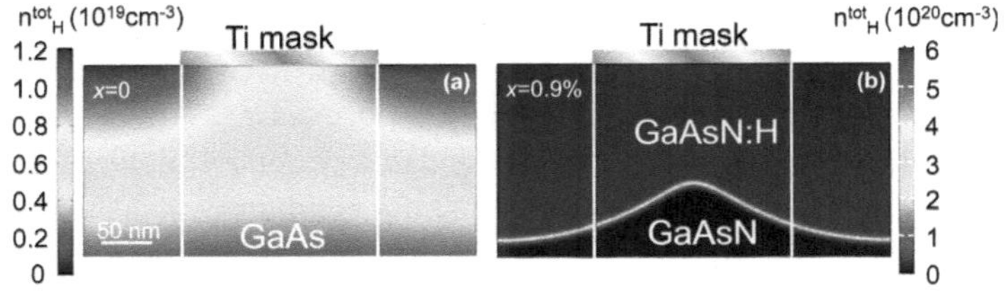

Figure 8.4 2D distribution (false color scale) of hydrogen atoms in GaAsN epilayers, as obtained by solving Eqs. 8.3, 8.4, and 8.5 in the presence of a 200 nm wide metallic mask opaque to hydrogen. (a) Simulated H distribution for $x = 0$ (i.e., GaAs) and $T_D = 200°C$. (b) Simulated H distributions for $x = 0.9\%$ and $T_D = 200°C$. The same parameters were used for obtaining the two distributions. Adapted from Ref. (Trotta et al., 2012b).

This figure highlights how the presence of N atoms in patterned GaAsN samples modifies dramatically the 2D distribution of hydrogen, resulting in a H diffusion forefront much sharper than in the $x = 0$ case (see Fig. 8.4a). As discussed in the next section, such an extremely sharp boundary of the H distribution—stemming from a careful control of the hydrogenation time and of the experimental conditions—is what actually allows for defining nanometer-sized regions by in-plane BGE, ultimately resulting in the fabrication of site-controlled QDs.

8.3 In-Plane Band-Gap Engineering: Site-Controlled Semiconductor Nanostructures

Before reporting on the first nanostructures based on in-plane BGE, we would like to point out that the extremely sharp H diffusion forefront, due to the extremely high number of traps—that is, N impurities—present in the sample, is not a unique property of GaAsN. For instance, hydrogen diffusion in Zn-doped GaAs leads to similar results when a very high concentration of Zn is used, and a similar behavior can also be found in silicon heavily doped with B (Pearton et al., 1992). However, dilute nitrides (and GaAsN in particular) are the only class of materials for which such an ultrasharp H diffusion profile is combined with the possibility of using H to neutralize the effects of N on the structural and electronic properties—including, most notably, the band gap—of the alloy. It is exactly this combination that renders hydrogenated dilute nitrides so interesting for nanotechnology. In the perspective of future applications, it is also worth noting that the simple method described in the previous sections does not require cutting-edge lithographic apparatuses, since H diffusion and sample geometry are what ultimately determine the actual nanostructure size, which can be ultimately smaller than the metallic pattern deposited on the sample surface. Because of these features, in-plane BGE can be rightfully placed at the forefront of the recently very active research field (see, e.g., Refs. (Baier et al., 2004; Kiravittaya et al., 2006; Huggenberger et al., 2011; Hughes et al., 2011; Skiba-Szymanska et al., 2011; Yuan et al., 2011; Hartmann et al., 2000; Baier et al., 2006; Leifer et al., 2007; Felici et al., 2009; Mohan et al., 2010; Surrente et al., 2009; Kapon et al., 1989; Nagamune et al., 1992; Vouilloz et al., 1997; Biasiol and Kapon, 1998)) aiming at the realization of the high-quality, site-controlled nanostructures that will be required by future nanophotonics applications.

8.3.1 *First Evidence of Carrier Quantum Confinement in GaAsN/GaAsN:H Nanostructures*

In this section, we describe how 2D heterostructures can be realized by selective hydrogenation of a bulk (3D) system, that is, of a 200 nm thick $GaAs_{1-x}N_x$ epilayer ($x = 0.9\%$) (Trotta et al., 2010). The sample in question was patterned with ensembles of Ti wires via electron beam lithography and metal evaporation. The wire widths (as determined by a scanning electron microscope) were $w = 80$ nm, 175 nm, and 485 nm, and the wire-to-wire distance was 5 μm. The sample was hydrogenated at 300°C, with a hydrogen dose equal to $1.5{\cdot}10^{18}$ ions/cm^2. Figure 8.5a–c shows the calculated distributions of electronically active N atoms (N with no bound hydrogen atoms) for different wire widths. In the simulations, the H ion current impinging on the sample surface, the duration of the H exposure, and the irradiation temperature were set to the experimental values, while D_{dif} and r (see Eq. 8.3) were adjusted to make sure that the H diffusion forefront reached the GaAsN/GaAs interface (i.e., that the GaAsN epilayer was *completely passivated*) outside the Ti mask, as

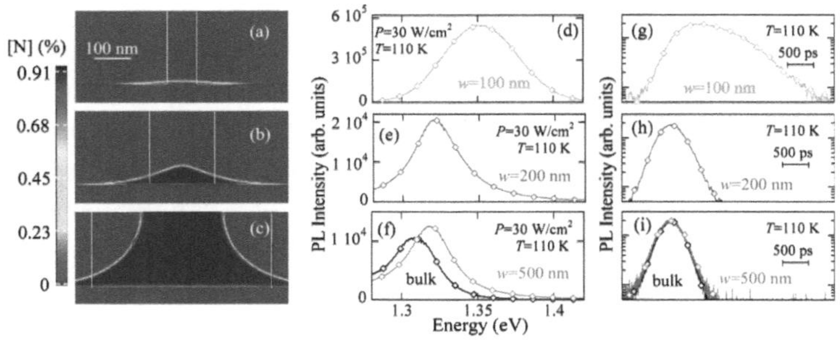

Figure 8.5 (a–c) Posthydrogenation distribution of electronically active N atoms in a sample section containing the growth axis, as obtained by solving Eqs. 8.3, 8.4, and 8.5 in the presence of Ti wires of different widths deposited on the sample surface. Panels (a), (b), and (c) are the simulations corresponding to wires having a width equal to $w = 485$ nm, 175 nm, and 80 nm, respectively. (d–f) PL spectra at $T = 110$ K recorded for wires having different widths, w. The spectrum of the as-grown sample is also reported. (g–i) Time-resolved micro-PL measurements at $T = 10$ K recorded for single wires having different widths, w. The curve relative to the as-grown sample is also reported. To compare carrier dynamics consistently, different power densities were used to construct this plot. Adapted from Ref. (Trotta et al., 2010).

verified *a posteriori* by PL measurements. The figure highlights that the size and shape of the resulting GaAsN/GaAsN:H heterostructures change rapidly with w, thus causing a marked variation of the potential profile to which carriers are subjected. In particular, for the 80 nm wide wire, a QW having lateral size of ~100 nm and a height of ~10 nm (along the growth axis) is formed. Carrier quantum confinement is therefore expected for this structure, at least in the direction of the growth axis.

To verify this possibility we investigated the optical properties of the different wires by means of PL and time-resolved micro-PL spectroscopy. Figure 8.5d–f shows the PL spectra at $T = 110$ K of the ensembles of GaAsN/GaAsN:H wires after hydrogenation and Ti removal. The temperature was chosen to highlight the free-exciton recombination with respect to localized states, while maintaining an intense, narrow PL signal. For the 485 nm and 200 nm wide wires, the PL spectra resemble the spectrum of the as-grown (not hydrogenated) sample (see black line in Fig. 8.5f) but for a slight increase of the integrated PL intensity and for a small blue shift of the free-exciton recombination, the latter most probably caused by the strain related to the lattice expansion of the barrier (Bisognin et al., 2006; Trotta et al., 2009b). In the spectrum of the smallest wires ($w = 80$ nm), on the other hand, a sizeable ($\sim$30 meV) blue shift of the free-exciton emission energy and a pronounced increase of the integrated PL intensity (by a factor of 50) can be clearly observed. Both of these two features hint at the presence of quantum confinement in the wires, since (*i*) the blue shift of the free-exciton recombination nicely matches the quantum confinement energy estimated for a 10 nm thick GaAsN (with $x = 0.9\%$) QW sandwiched between GaAs barriers and (*ii*) the increase of the integrated PL intensity for a constant number of photoexcited carriers points to an enhanced radiative efficiency, which is typical of low-dimensional systems—when compared to bulk materials—and is mainly caused by an increased overlap of the electron–hole wavefunction.

To further confirm the occurrence of carrier quantum confinement in the 80 nm wires, time-resolved micro-PL spectroscopy was employed. Figure 8.5g–i shows the time evolution of the emission at $T = 10$ K for different single wires and for the unpatterned sample. The detection energy is set at the maximum of the free-exciton band, as measured by continuous-wave (CW) micro-PL. To maintain a constant density of photogenerated carriers, different laser power densities were used for different wire widths (see Ref. (Trotta et al., 2010)). The comparison between the largest wires ($w = 485$ nm and 200 nm) and the as-grown sample mimics the result obtained with a CW laser: no appreciable difference can be observed, and the three curves almost overlap. The behavior of the $w = 80$ nm wire, on the other hand, is considerably different, that is, an overall longer decay with a pronounced plateau is observed. This plateau is the fingerprint of band-filling effects, which usually take place in low-dimensional systems because of a reduced density of states (Zwiller et al., 1999). In fact, carrier replenishment of the ground state from higher-energy states is slowed down by the breaking up of the bulk continuum levels into discrete levels, at least in one dimension in our case (Trotta et al., 2010). The experimental evidence reported in this section confirms that the band-gap profile of the GaAsN/GaAsN:H heterostructures fabricated by in-plane BGE

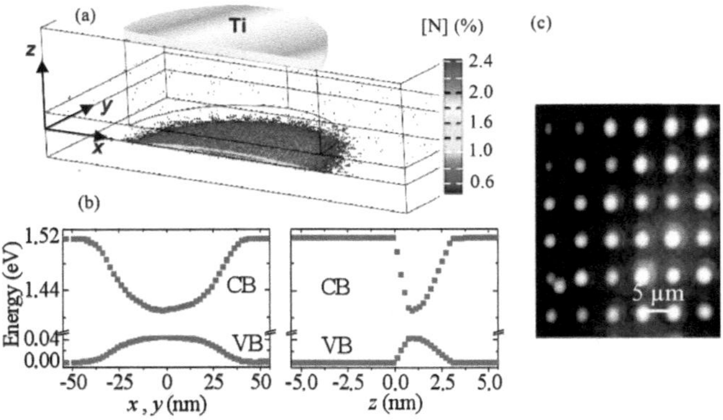

Figure 8.6 (a) Calculation of the 3D distribution of N atoms not passivated by H (namely, N atoms without bound H atoms) in a 6 nm thick $GaAs_{0.989}N_{0.011}$/GaAs QW (thickness of GaAs capping layer = 30 nm). Hydrogen is irradiated over the entire sample surface but for the region covered by a H-opaque Ti mask (diameter d = 80 nm). (b) Energy band diagrams in the conduction band (CB) and valence band (VB), as obtained by performing scans along the main axes of the QD (the scan along the x axis was performed at a height $z = 1$ nm). The diagrams were derived from the relation between the effective N concentration (i.e., the density of N atoms with no bound H atoms) and the $GaAs_{1-x}N_x$ band-gap energy. The CB offset was set to 0.7. The axis origin (0,0,0) is set in the dot center at the GaAsN/GaAs buffer interface. (c) Micro-PL mapping of the d = 80 nm dot emission (an 850 nm high-pass filter was used to cut the GaAs emission). The lateral size of the QD emission spot is limited by diffraction and by carrier diffusion. Adapted from Ref. (Trotta et al., 2011).

can be arbitrarily shaped with nanometer precision. As we will see in the next section, such precision can be exploited for the fabrication of nanostructures with even lower dimensionality.

8.3.2 *GaAsN/GaAsN:H Site-Controlled Quantum Dots*

For the fabrication of GaAsN/GaAsN:H *QDs* the carrier-confining potential, defined through the controlled diffusion of hydrogen in the sample, has to be constrained in all three spatial dimensions. Figure 8.6a shows the results of a finite-element calculation of the distribution of electrically active N atoms (N atoms with no bound hydrogen) obtained after the hydrogenation of a 6 nm thick

$GaAs_{1-x}N_x$/GaAs QW ($x = 1.1\%$, thickness of GaAs capping layer = 30 nm), whose sample surface was covered with a metallic dot having a diameter of 80 nm (Trotta et al., 2011). The calculation was performed by extending the model presented in the previous section to three dimensions and by applying the same boundary and initial conditions used in real experiments: nitrogen concentration $x = 1.1\%$, hydrogen irradiation temperature $T_H = 200°C$, hydrogen ion current = 30 μA/cm^2, and exposure time $t = 420$ seconds. The resulting nanostructure (see Fig. 8.6a) is expected to behave like a quantum box with the hybrid band structure diagram depicted in Fig. 8.6b. To check for the occurrence of 3D quantum confinement in such nanostructures, we have thus fabricated ensembles of Ti dots (dot-to-dot separation equal to 5 μm) with the diameter d ranging between 500 nm and 80 nm. Figure 8.6c shows a micro-PL mapping of an ensemble of $d = 80$ nm dots after hydrogenation and Ti removal. The image was recorded using an 850 nm low-energy-pass filter to reject the emission from the GaAs-like regions of the material (cap layer and hydrogenated barriers). The mapping demonstrates an excellent control over the dot position, coupled to a rather uniform emission efficiency from the array. Figure 8.7 summarizes the emission characteristics of the heterostructures obtained using Ti dots of different diameters. As shown in panel (a), for large values of d the micro-PL spectrum of the region corresponding to a single Ti dot is very similar to that of the untreated GaAsN QW. As long as $d \geq 200$ nm, indeed, the spectrum is composed of a multitude of sharp peaks associated with the localized states originating from compositional fluctuations of the GaAsN alloy, and the center of mass of the emission does not show any sizeable changes. This indicates that the GaAs-like barriers created by H irradiation do not sizably confine carriers, nor do they influence the ground-state energy of the GaAsN dots. However, the situation changes dramatically for the $d = 80$ nm structure (topmost spectra in Fig. 8.7a). For this dot (containing approximately 1000 N atoms), the micro-PL spectrum consists of a single excitonic line positioned in a spectral region where the untreated QW does not emit.

The observed energy blue shift of this line (~40 meV with respect to the center of mass of the QW spectrum) is consistent with

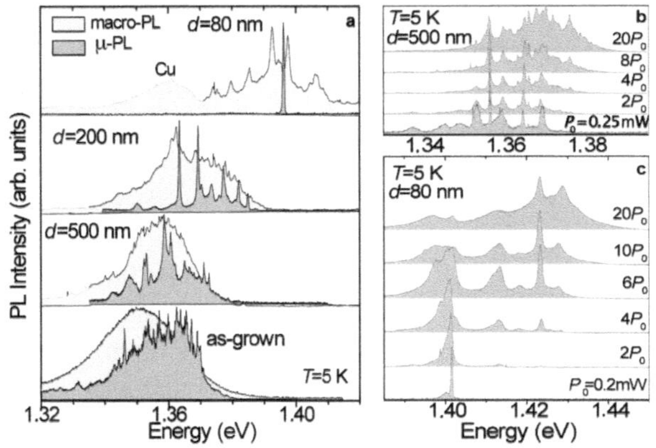

Figure 8.7 (a) Low-temperature macro-PL (brighter areas) and micro-PL (darker areas) spectra recorded on the different structures described in the text. The bottommost spectra refer to the as-grown QW. The band indicated by Cu in the topmost macro-PL spectrum is due to a recombination involving Cu deep levels in the substrate (the recombination at 1.325 eV is the LO-phonon replica of the Cu band) (Pavesi and Guzzi, 1994). (b, c) Low-temperature ($T = 5$ K) power dependence of the micro-PL spectra of single dots formed after the hydrogenation of samples having the top surface patterned with a Ti dot with the diameter d equal to (b) 500 nm or (c) 80 nm. Note the different energy scales in the two panels. Adapted from Ref. (Trotta et al., 2011).

an estimate of the 3D confinement energy obtained by treating the dot potential (see Fig. 8.6b) as a decoupled square box with finite-height barriers. Furthermore, important features of the $d = 80$ nm dot emission can be understood by analyzing its evolution with increasing excitation power P (see Fig. 8.7c). As already discussed, at the lowest power a single line, likely due to the QD ground-state exciton, dominates the spectrum. With increasing P, the intensity of the low-energy components increases more rapidly than that of the ground state. These additional lines can be attributed to charged/multiple-exciton recombinations, as also found in other QD-like systems. By further increasing the photogenerated carrier density, other distinct states can be observed at higher energy, while the intensity of the lowest-lying recombination lines becomes progressively saturated. This kind of behavior has also been widely reported in many zero-dimensional systems and ascribed to the population of excited states. It is interesting to compare this behavior with the dependence on P found in the case of the $d =$ 500 nm dot, shown in Fig. 8.7b. In this case, indeed, different lines appear continuously with increasing P, forming a quasi-continuum without clear evidence of quantum-state discreteness.

8.4 Light Polarization Control via Strain Engineering

As discussed several times through this volume, the surprising effects of H irradiation are not restricted to the *electronic* properties of dilute nitrides. The reopening of the band gap is indeed accompanied by an astonishing expansion of the GaAsN cubic cell (Polimeni et al., 2003b; Bisognin et al., 2006; Berti et al., 2007) that, in turn, causes a reversal of the *strain status* of GaAsN epilayers from tensile to compressive. When combined with the masked-hydrogenation procedure described in the previous sections, this effect opens the way to the unprecedented possibility of tailoring the *strain fields* in the growth plane of the alloy. As it turns out, such an anisotropic in-plane strain modulation allows for a tight control of the polarization properties of the light emitted along the crystal growth direction (Trotta et al., 2009b). In this section we show how to achieve such control, and we explain the origin of the observed polarization anisotropy (Trotta et al., 2009b). This effect has a high potential interest for applications in the field of data transmission, especially in the light of the possibility of embedding dilute nitrides in GaAs-based vertical-cavity surface-emitting laser (VCSEL) structures (Henini, 2005).

The investigation presented here was based on 200 nm thick $GaAs_{1-x}N_x$ epilayers having N concentrations ranging between $x =$ 0.4% and 0.9%. Electron beam lithography was employed to define ensembles of 50 nm thick titanium dots and wires with varying lateral size. Figure 8.8a shows the low-temperature ($T = 10$ K), polarized PL spectra of ensembles of GaAsN wires (having $x =$ 0.4% and 0.9%) after hydrogenation ($T_H = 300°C$) and removal of the Ti mask. The wires were 500 nm wide, separated by 5 μm and directed along the [110] direction of the crystal. Solid and dashed lines refer to light polarized perpendicular and parallel to

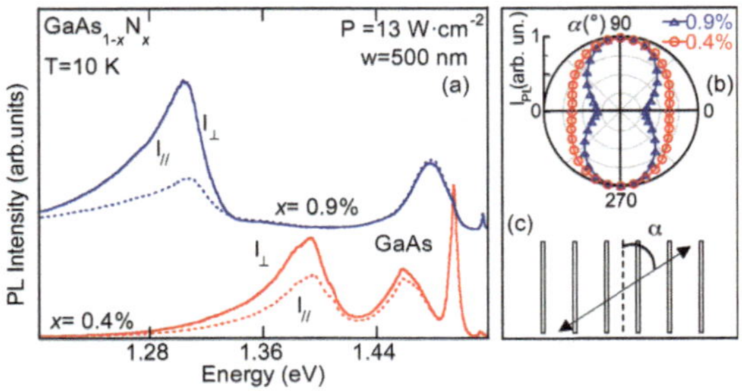

Figure 8.8 (a) $T = 10$ K polarized PL spectra of two wire ensembles ($w = 500$ nm) fabricated on 200 nm thick GaAsN epilayers having $x = 0.9\%$ (upper blue spectra) and $x = 0.4\%$ (lower red spectra), respectively. Continuous (dashed) lines refer to light polarized perpendicular (parallel) to the wire direction. (b) Polar plot of the normalized PL intensity (I_{PL}) as a function of the angle α formed between the polarization analyzer and the wire axis; see sketch in (c). Panel (a) adapted from Ref. (Trotta et al., 2009b).

the wire axis, respectively. The light emitted from the GaAsN wires (low-energy emission band) is strongly polarized perpendicularly to the wire axis, with a polarization degree—defined as $\rho = (I_{\|} - I_{\perp})/(I_{\|} + I_{\perp})$—as large as $\rho = 40\%$ for the sample with $x = 0.9\%$. This preferential polarization direction can be observed more clearly in Fig. 8.8b, where the normalized PL intensity is recorded as a function of the angle, α, the polarization analyzer (composed by a linear polarizer and a rotating half lambda waveplate placed in front of the spectrometer) forms with the wire axis. The reported optical anisotropy is absent in the as-grown sample and, more interestingly, in ensembles of circular dots having different diameters (ranging from 5 μm to 500 nm). These findings suggest that the polarization effect is caused by the in-plane structural anisotropy resulting from the formation of the GaAsN wire.

To get better insight into this effect, the properties of *isolated wires* were investigated. Figure 8.9a shows the micro-PL intensity imaging ($T = 77$ K) of $GaAs_{0.991}N_{0.009}$ wires having widths $w = 5.0$, 2.0, 1.0, 0.5, and 10 μm, from left to right. Brighter (darker) regions correspond to a higher (lower) intensity of the GaAsN PL signal (the emission from GaAsN:H was rejected with a filter). Figure 8.9b shows a map of the polarization degree ρ recorded on the same

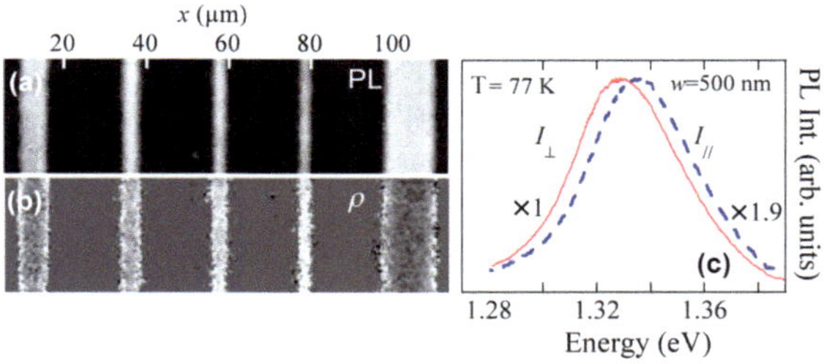

Figure 8.9 (a) PL imaging of GaAsN wires having different widths (white corresponds to maximal intensity). (b) Spatial map of the PL polarization degree ρ on the same area shown in (a); see Fig. 8.8. White indicates maximum ρ. Note the higher value of ρ at the wire borders. (c) Normalized PL spectra of the wire with $w = 500$ nm ($T = 77$ K) for light polarized parallel (dashed blue line) and perpendicular (continuous red line) to the wire axis. Note the relative shift of the peaks. Adapted from Ref. (Trotta et al., 2009b).

sample zone. For $w \geq 2$ μm, the ρ mapping clearly singles out the wire boundaries as the main source of polarization anisotropy. For smaller wires, that is, closer interfaces, ρ increases progressively. Finally, Fig. 8.9c shows that the two polarization-resolved micro-PL spectra recorded—for polarization parallel and perpendicular to the wire—at the center of the $w = 500$ nm wire present a clear energy splitting. Taking into account the H-induced lattice expansion of the hydrogenated barriers, these results allow for ascribing the observed optical anisotropy to the modified *strain status* of the GaAsN layer. More specifically, we expect the lattice expansion taking place in the GaAsN:H barriers to alter the built-in tensile stress of the GaAsN epilayer only in the direction *perpendicular* to the wire axis, while leaving the strain *along the wire* almost unaffected. This in-plane symmetry breaking triggers the activation of polarization selection rules for the two topmost VBs (heavy and light holes) and modifies their built-in energy splitting.

To support this picture, finite-element calculations in the framework of thermoelasticity were performed. We modeled the strain distribution by considering the deformation induced on a $GaAs_{1-x}N_x$ wire by the thermal expansion of the adjacent $GaAs_{1-x}N_x$:H barriers (Trotta et al., 2009b). The extent of such thermal expansion was set to account for the experimental values of

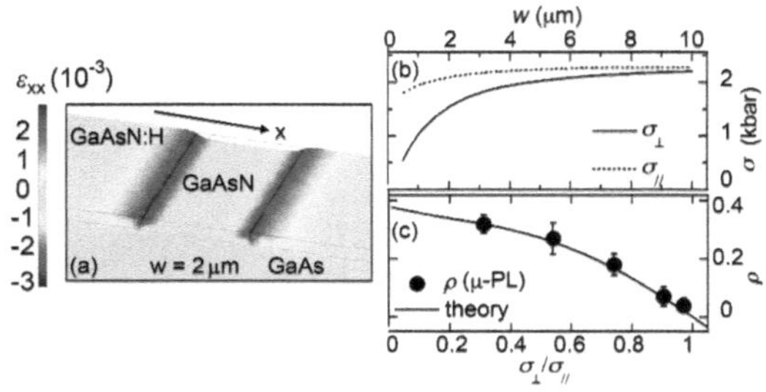

Figure 8.10 (a) 3D distribution of the strain field ε_{xx}, calculated along the direction perpendicular to the axis of a $w = 2$ μm GaAsN wire embedded between hydrogenated GaAsN barriers. The structure deformation is exaggerated by a factor of 100. (b) Wire width dependence of the calculated stress perpendicular (solid line) and parallel (dotted line) to the wire axis. The stress along the direction perpendicular to the wire axis is calculated at the wire center. (c) Dependence of ρ (measured at the wire center and at $T = 77$ K) on $\sigma_{\perp}/\sigma_{\parallel}$ (symbols). The solid line is a theoretical prediction combining finite-element simulations and optical selection rules. Adapted from Ref. (Trotta et al., 2009b).

the compressive (tensile) strains measured in unpatterned samples (Bisognin et al., 2006). Figure 8.10a shows the 3D distribution of the strain fields, ε_{xx}, calculated in a 2 μm-GaAsN wire ($x = 0.9\%$) laterally delimited by hydrogenated barriers. The calculation accurately reproduces the main features of the investigated structure, that is, the in-plane structural anisotropy of the strain fields, the barrier expansion against the wire, and the pronounced value of strain at the wire border. Furthermore, Fig. 8.10b shows that the built-in stress in the direction perpendicular to the wire axis decreases on going from the largest to the smallest wire, while the stress along the wire axis remains almost constant, in complete agreement with our simple picture.

The ability to correctly reproduce the in-plane structural anisotropy of GaAsN/GaAsN:H wires within our simple model represents the first step toward the full understanding of the *polarization properties* of these heterostructures. To achieve this final goal, the *selection rules* for the radiative transitions involved in the wire emission must also be taken into account. When a *biaxial anisotropic stress* is applied along the [110] direction, there is a mixing of the two topmost VBs, which in the [110] representation can be written as (Trotta et al., 2009b; Pollak and Cardona, 1968; Rammohan et al., 1995)

$$\begin{aligned} u_{v1} &= c_a \left|3/2; 3/2\right\rangle_{110} + c_b \left|3/2; -1/2\right\rangle_{110} \\ u_{v2} &= -c_b \left|3/2; 3/2\right\rangle_{110} + c_a \left|3/2; -1/2\right\rangle_{110} \end{aligned}. \quad (8.6)$$

The c_a and c_b coefficients depend on the elastic constants, the uniaxial deformation potentials, and, most importantly, the ratio $\sigma_\perp/\sigma_{//}$ between the stress components perpendicular and parallel to the wire axis (see Ref. (Rammohan et al., 1995) for the explicit formula). The in-plane structural anisotropy of the sample enters directly in the character of the valence band states and, as a consequence, in the matrix elements that account for their interaction with the CB states, u_c. In turn, these matrix elements depend on the orientation of the polarization vector of the emitted light, $\varepsilon^{||,\perp}$, with respect to the wire axis:

$$M_i^{||,\perp} \propto \left|< u_c | \, \varepsilon^{||,\perp} \cdot \vec{p} \mid u_{vi} >\right|^2 \quad (8.7)$$

(with $i = 1, 2$; see Ref. (Pollak and Cardona, 1968) for the complete expressions). For a quantitative analysis of the PL spectra, the matrix elements have to be weighted by the corresponding reduced density of states (proportional to $\mu_i^{3/2}$, where μ_i is the exciton reduced mass) and by the relative thermal population of the VB states, $e^{-\delta E_{12}/(k_B T)}$, where δE_{12} is the calculated energy separation between u_{v1} and u_{v2} (the latter also depends on $\sigma_\perp/\sigma_{||}$). Finally, one finds $I_{||,\perp} \propto [\mu_1^{3/2} M_1^{||,\perp} + \mu_2^{3/2} M_2^{||,\perp} e^{-\delta E_{12}/(k_B T)}]$. Figure 8.10c shows the dependence of ρ on $\sigma_\perp/\sigma_{||}$ for $x = 0.9\%$. A very good agreement is found between the model (with $\mu_2/\mu_1 = 1.2$ being the only adjustable parameter) and the polarization degree measured at the wire center at 77 K.

To summarize, in this section we have shown that the polarization properties of the light emitted by $GaAs_{1-x}N_x$ microstructures can be controlled by modulating the lattice expansion of hydrogenated $GaAs_{1-x}N_x$. This result mirrors the band-gap modulation, demonstrating that the structural properties of dilute nitride alloys can also be modified on a submicrometer scale by spatially selective hydrogenation.

8.5 H-Assisted Tailoring of Pre-Existing, Site-Controlled InGaAsN Nanostructures

As mentioned a few times in the previous sections, the last decade has witnessed the birth of a new, very active research field aimed at the development of novel approaches to the fabrication of site-controlled nanostructures. Arguably, the most successful of these approaches is the one developed by Kapon et al. at Ecole Polytechnique Fédérale de Lausanne (EPFL), Switzerland. On the basis of metal–organic chemical vapor deposition (MOCVD) of the proper sequence of epitaxial layers (e.g., GaAs/InGaAs/GaAs) on a GaAs substrate patterned with linear (*V-grooves*) or tetrahedral (*pyramids*) recesses, this approach guarantees the formation of a single nanostructure—*QWRs* for V-grooves, *QDs* for pyramids—at the bottom of each recess (Kapon et al., 1989; Nagamune et al., 1992; Vouilloz et al., 1997; Biasiol and Kapon, 1998; Hartmann et al., 2000; Baier et al., 2004, 2006; Leifer et al., 2007; Felici et al., 2009; Surrente et al., 2009; Mohan et al., 2010). As demonstrated by a series of recent publications, these QDs (Gallo et al., 2008; Calic et al., 2011) and QWRs (Atlasov et al., 2007, 2008, 2009, 2010, 2011) are extremely well suited for the integration with PhC cavities, mainly due to their near-perfect site control (positioning precision better than 20 nm) and to an excellent uniformity. As regard the latter point, particular mention should be made of the standard deviation of just 0.6 meV, recently estimated for the emission energy distribution of the QDs grown with this method (Mohan et al., 2010).

The recent pairing of this nanofabrication method to the large *band-gap bowing* typical of dilute nitride materials (O'Reilly et al., 2009) allowed for growing high-quality InGaAsN QWRs (Carron et al., 2011a)—and, prospectively, QDs (Carron et al., 2011b)—that emit at $\sim$1.3 μm at room temperature and are thus naturally suited for the integration with nanophotonic devices operating in the telecommunications window. Shortly afterward (Felici et al., 2012a), the 1D character of the carrier wavefunction in InGaAsN V-groove QWRs was confirmed by measuring the anisotropy of the diamagnetic shift of the QWR emission for different orientations of the applied magnetic field, B (see Fig. 8.11) (Nagamune et al., 1992). In the same work, the possibility to passivate N impurities in

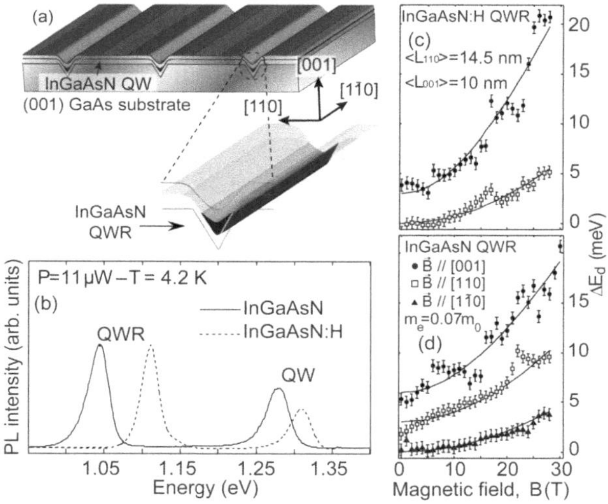

Figure 8.11 (a) Sketch of the V-groove QWR array (pitch = 1 μm, groove width = 300 nm), displaying the three orientations of B employed during the measurements. (Below) Zoom-in of the QWR. (b) Micro-PL spectra of the investigated sample before (continuous line) and after (dashed line) H irradiation. The peak in the high-energy region of the spectrum is due to the QW that forms on the planar part of the patterned substrate. (c) Diamagnetic shift of the hydrogenated InGaAsN QWR peak for B//[001] (black dots) and B//[110] (white-filled squares). (d) Diamagnetic shift of the untreated InGaAsN QWR peak for B//[001] (black dots), B//[110] (white-filled squares), and B//[1−10] (black triangles). In (c and d), different datasets were vertically shifted for clarity, whereas the solid lines are fits obtained modeling the QWR confinement with an anisotropic parabolic potential. By performing these fits for several QWRs, the average QWR thickness (L_{001}) and width (L_{110}) could be estimated, together with the electron effective mass of the InGaAsN alloy (Felici et al., 2012a). Panel (a) was adapted from Ref. (Felici et al., 2012b) and panels (c and d) from Ref. (Felici et al., 2012a).

dilute nitrides by H irradiation (Polimeni et al., 2001a, 2003a; Trotta et al., 2012b) was exploited to progressively "turn off" the effects of N incorporation and to precisely estimate the concentration of N (and In) in the QWRs. Both the H-induced blueshift of the QWR emission (~70 meV, see Fig. 8.11b) and the value of the electron effective mass in the untreated sample (m_e ~0.07 m_0, as estimated from micro-magneto-PL, see Figs. 8.11c,d) are consistent with a N concentration of $\sim$1%. Further, N incorporation is found to enhance the In intake in the QWR, resulting in an additional red shift (the energy of the hydrogenated QWR emission, 1.12 eV, is consistent with an In concentration of $\sim$40%) and suggesting the potential for obtaining QWRs emitting at 1.55 μm and beyond.

The possibility to fine-tune the effective nitrogen concentration in V-groove InGaAsN QWRs by hydrogen irradiation paves the way to the postgrowth engineering of the optical end electronic properties of *single-semiconductor nanostructures.* To achieve this goal, the ability to perform H irradiation in *a spatially controlled way*—as described in the previous sections—will be crucial. As an example of the high potential of this approach, we would like to mention the possibility of depositing an orthogonally oriented grating of metallic wires (Trotta et al., 2010) on an array of V-groove InGaAsN QWRs. Upon hydrogenation, such processing would result in the fabrication of an ordered ensemble of dilute nitride QDs, inherently fit for the integration with an array of coupled PhC nanocavities. Such a photonic structure would be extremely interesting for the investigation of photon-mediated QD–QD coupling (Greentree et al., 2006) and for the realization of ultrafast PhC lasers (Altug et al., 2006) operating at telecom wavelengths.

8.6 "Direct Writing" Methods

As we saw in the previous sections, the lithographic steps required by in-plane BGE are not particularly cumbersome or demanding, and fully within the reach of standard electron beam lithography systems. Nevertheless, the development of methods to modulate the in-plane potential profile of dilute nitrides by *locally dissociating* the N–H complexes responsible for N passivation might allow for greater flexibility in the design and fabrication of nanostructures, laying the foundations for a cheaper, faster approach to nanofabrication. In the following section, we briefly introduce two very promising methods for the *direct writing* of hydrogenated dilute nitride semiconductors, respectively, relying on an *electron beam* (Felici et al., 2006a) and on *focused laser light* (Balakrishnan et al., 2011, 2012) (see also Section 2.5.2) to break the N–H bonds.

8.6.1 *Electron Beam Writing*

A very promising strategy for obtaining a direct, in-plane modulation of the energy gap of hydrogenated dilute nitrides is based upon *electron irradiation*. As widely reported in the literature (Silvestre et al., 2000; Pavesi et al., 2004), the possibility of focusing electrons on a small area (beam diameter $\sim$10÷100 nm) gives an insightful tool for the local investigation of the interaction of hydrogen with point defects. In particular, it has been shown that an energetic (10÷50 keV) electron (e^-) beam can break the bonds that hydrogen forms with silicon impurities in GaAs, through an electronic excitation of the Si–H complexes (Silvestre et al., 2000). Within the context of dilute nitrides, a very similar phenomenon was reported in Ref. (Felici et al., 2006a), wherein the possibility to locally remove H from a $GaAs_{1-x}N_x$ sample by e^- beam irradiation was detailed for the first time.

Figure 8.12a displays two monochromatic CL images ($T = 5$ K, beam current $I_b = 10$ nA) of a $GaAs_{1-x}N_x$ sample ($x = 0.47\%$), acquired at the location of a 30×40 μm^2 rectangular region irradiated with electrons (beam current $I_b = 400$ nA, beam energy $E_b = 5$ keV). A typical irradiation time of 40 seconds was enough to obtain a complete recovery of the $GaAs_{1-x}N_x$ band-gap value (incidentally, several intermediate band-gap values could be obtained by using shorter electron irradiation times or lower doses). The two images displayed in panel (a) were collected with the monochromator energy set at $E_{det} = 1.415$ eV (left image) and 1.465 eV (right image), respectively. Therefore, the two bright areas (one complementary to the other) correspond to photons with energy equal to that of the radiation emitted by the as-grown (left) or hydrogenated (right) sample at $T = 5$ K. A spectral analysis of the light emitted from the electron-irradiated rectangle—as well as from the area outside the rectangle—is shown in Fig. 8.12b. The CL spectrum of the electron-irradiated sample (middle line) nearly coincides with that of pristine $GaAs_{1-x}N_x$ (topmost line) and differs markedly from that of the hydrogenated sample (bottommost line). In other words, the rectangular area swept by the intense electron beam acts in all respects as a potential well for carriers. These findings demonstrate that the e^- beam has broken the bonds of the nitrogen–

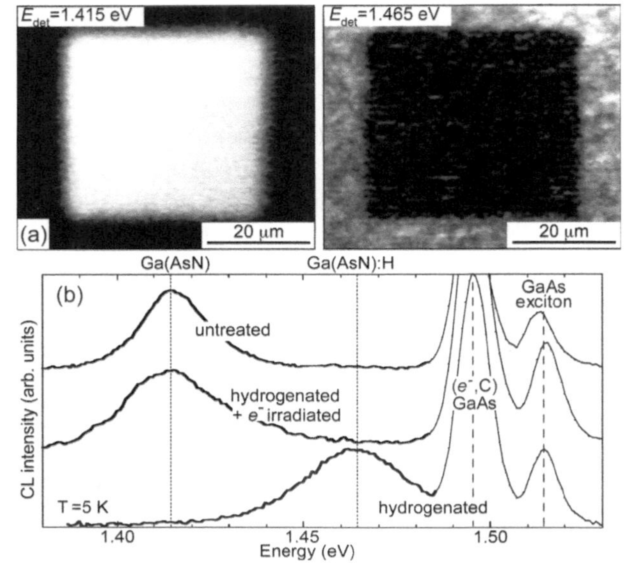

Figure 8.12 In-plane BGE by e^- beam writing. (a) Low-temperature ($T = 5$ K) CL images acquired after sweeping a hydrogenated $GaAs_{1-x}N_x$ sample ($x = 0.47\%$) with an intense electron beam with current I_b = 400 nA and energy E_b = 5 keV (writing operation). The CL images were recorded with a less intense electron beam having I_b = 10 nA and E_b = 20 keV (reading operation). (Left image) The detection energy, E_{det}, is set equal to that of the photons emitted from the GaAsN QW region. (Right image) Same as the left, but E_{det} = 1.465 eV, that is, the energy of the defect band only present in the hydrogenated $GaAs_{1-x}N_x$ barrier region; see panel (b). (b) CL spectra recorded at T = 5 K in the region outside (bottommost line) and inside (middle line) the rectangular area swept by the intense electron beam. The CL spectrum of the $GaAs_{1-x}N_x$ sample before hydrogen irradiation is also shown (topmost line) for comparison purposes. Dotted vertical lines indicate carrier recombination from the $GaAs_{1-x}N_x$ region (either hydrogenated or not). Dashed vertical lines indicate carrier recombination from the GaAs band-gap exciton and free-electron to carbon impurity transition [(e^-, C)]. Adapted from Ref. (Felici et al., 2006a).

dihydrogen complexes responsible for the dramatic changes in the electronic properties of the crystal and—coupled to the very high spatial resolution (<10 nm) achievable by state-of-the-art e^- beam lithography systems—may open the way to the definition of a very effective protocol for the *direct writing* of nanostructures.

8.6.2 *Laser Annealing*

In Ref. (Balakrishnan et al., 2011), Balakrishnan et al. demonstrated that *laser light*—focused on the sample surface with a conventional microscope objective—is also capable of dissociating the N–2H complexes responsible for N passivation in $GaAs_{1-x}N_x$, locally ensuring a *full hydrogen removal* at relatively low power densities (10^5 W/cm^2). As discussed in much greater detail in Section 2.5.2, this finding raised a series of fundamental questions, including whether the physical process driving the laser-induced modification of the N–H bonds in the III–V host crystal is a *thermal* (i.e., due to local heating) or a *photonic* mechanism and whether this "laser annealing" technique could be applied to other technologically relevant dilute nitride materials. These questions were addressed in Ref. (Balakrishnan et al., 2012), where it was established that (*i*) this phenomenon is associated with *photon absorption*, preferentially occurring in resonance with the dissociation energy of the N–H complexes and (*ii*) it is universal to other dilute nitrides, such as $GaP_{1-x}N_x$ and $In_yGa_{1-y}As_{1-x}N_x$.

If properly harnessed, this effect could serve as the basis for the development of a new, inexpensive method for the direct writing of dilute nitride nanostructures, completely devoid of complex lithographic/etching steps. A very promising step in this direction is represented by the recent finding—reported by Intonti et al. in Ref. (Intonti et al., 2011)—that laser annealing can also be performed through the tip of a scanning near-field optical microscope (SNOM). In principle, this may allow for reducing the extent of the illuminated area to sizes compatible with the resolution of SNOM ($<$100 nm)—much below the $\sim$1 μm spot size achievable with a conventional microscope (see, e.g., Fig. 2.12)—and, thus, for the realization of quantum-confined nanostructures with this method.

8.7 Other Technologically Relevant Properties Affected by Hydrogenation

In the context of photonics and nanotechnology, the most relevant effects of hydrogen irradiation on the properties of GaAsN epitaxial layers are surely represented by the full H-induced recovery of the energy gap of pure GaAs and, to a somewhat lower degree, by the consequences of the lattice expansion occurring in hydrogenated samples on the polarization properties of the material (see Section 8.4). As discussed above, indeed, the H-induced reopening of the band gap makes it possible to tailor the potential profile of the material at the nanometer scale, eventually allowing for the realization of site-controlled QDs, ideal for the integration with nanophotonic devices (as detailed in Section 8.3.2).

The formation of N–2H–H complexes in dilute nitrides, however, results in a myriad of other effects, most of which were thoroughly discussed in other chapters of this book. In the following section, we wish to briefly focus on the effects of hydrogen irradiation on the *refractive index* (Geddo et al., 2007; Trotta et al., 2012b) and on the *electrical resistance* (Alvarez et al., 2011) of GaAsN, highlighting their importance for technological applications.

8.7.1 *Refractive Index*

As shown in Fig. 8.13a, H irradiation of a $GaAs_{1-x}N_x$ sample with $x = 2.2\%$ (thickness of N-containing layer = 100 nm) results in the approximate recovery of the *reflectance spectrum* of GaAs. As discussed in greater detail in Refs. (Geddo et al., 2007; Trotta et al., 2012b), this behavior can be qualitatively interpreted as a consequence of the effects of the H-induced reopening of the band gap on the *refractive index*, $n(\lambda)$, of the material. According to semiclassical theories (Adachi, 1999), indeed, in the subgap transparency region of a dielectric material $n(\lambda)$ is expected to continuously decrease as one moves away from the so-called *Sellmeier resonance*, which effectively includes—within a single classical oscillator—all the contributions to absorption processes occurring through the fundamental band gap. In addition, it is interesting to note that at high wavelengths (1200–1650 nm, see Fig. 8.13b), well into the transparency region, the reflectance spectrum of the hydrogenated sample even goes *below* that of GaAs. The reason for such H-induced *overshooting* of the reflectance can probably be ascribed to the presence of a third, weakly bound atom nearby the N–2H complex responsible for N passivation. As already

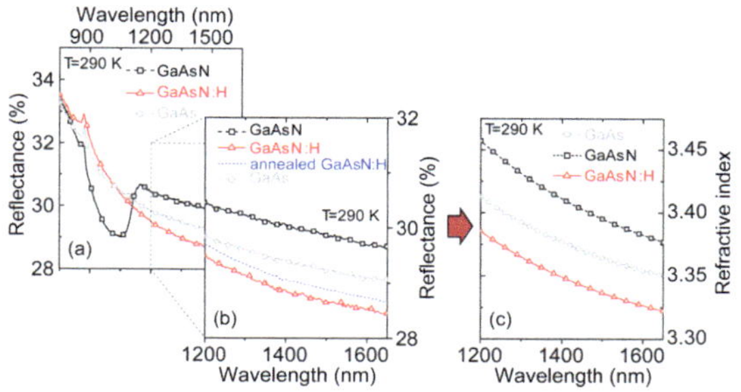

Figure 8.13 (a) Comparison between the reflectance spectra ($T = 290$ K) of a GaAs reference (gray circles) and of a $GaAs_{1-x}N_x$ sample ($x = 2.2\%$, thickness = 100 nm) before (black squares) and after (red triangles) H irradiation. Note the H-induced decrease of the reflectance values in the subgap spectral region. (b) Zoom-in of the reflectance spectra displayed in (a) in the 1200–1650 nm range. The spectrum relative to a hydrogenated $GaAs_{1-x}N_x$ sample after annealing at 225°C is also shown (blue dashed line). The annealing treatment induces a slight increase of the reflectance in the subgap spectral region, thus approaching the values of the GaAs reference. (c) $GaAs_{1-x}N_x$ refractive index spectra before and after H irradiation, as obtained from a best fit to the reflectance spectra in the subgap region for the same samples shown in panel (a). Adapted from Ref. (Trotta et al., 2012b).

discussed in detail elsewhere in this book (see Chapters from 4 to 7, as well as Section 8.2), the presence of this "satellite" H atom has also been associated with the *extra-compressive strain* observed in fully hydrogenated $GaAs_{1-x}N_x$ layers (see also Section 8.4). Similarly to what was already observed for such compressive strain, a five-hour annealing at 225°C of the hydrogenated $GaAs_{1-x}N_x$ sample—sufficient to get rid of the third H atom but not to dissociate the N-2H complex—indeed results in a nearly complete recovery of the reflectance spectrum of pure GaAs.

The reflectance spectra shown in Fig. 8.13b can be used to extract the *refractive index* of the material in the transparency region, through a best-fit procedure performed via the software package WVASE32 by J.A. Woollam Co., Inc., and based on the Levenberg–Marquardt algorithm (Geddo et al., 2007, 2011). The wavelength dependence of $n(\lambda)$ obtained with this procedure is displayed in Fig. 8.13c for an untreated and a hydrogenated $GaAs_{1-x}N_x$ sample, as well as for a GaAs reference. In the 1.3–1.55 μm range, H irradiation results in a relative reduction of the refractive index by ~1.5%–2%. The extent of the H-induced variation of the refractive index falls comfortably within the range required to achieve *photon confinement* in a planar-waveguide geometry (Bhattacharya, 1997), thus suggesting the possibility to realize integrated optical circuits by spatially selective hydrogenation. In particular, optical elements such as planar waveguides and optical couplers working in the telecommunication range could be implemented using the methods presented in this chapter.

8.7.2 *Electrical Resistance*

While the effects of hydrogen irradiation on the optical (Polimeni et al., 2001a; Baldassarri Höger von Högersthal et al., 2001; Polimeni et al., 2003a; Masia et al., 2006; Pettinari et al., 2006; Geddo et al., 2007) and structural (Berti et al., 2007; Ciatto et al., 2005c; Felici et al., 2006b) properties of dilute nitrides (Henini, 2005) have been the object of extensive investigations, the potential influence of the formation of N–2H–H complexes on the *electrical* properties of these materials have been largely overlooked. In Ref. (Alvarez et al., 2011), Alvarez et al. used conductive-probe atomic force microscopy (CP-AFM) to measure the local electrical resistance (R) of several untreated and hydrogenated $GaAs_{1-x}N_x$ samples. It was found that H irradiation leads to a giant increase of R, with ~500 being a typical enhancement factor. While it is reasonable to expect hydrogenation to yield a larger value of R—given the decrease in the intrinsic carrier concentration due to the H-induced recovery of the energy gap of GaAs—the observed enhancement is approximately 1 order of magnitude too large to be explained solely in these terms. Since H-related damages can be ruled out (the resistance of pristine $GaAs_{1-x}N_x$ can be fully restored via thermal annealing), in Ref. (Alvarez et al., 2011) this behavior was ascribed to a tenfold—or more—reduction of the carrier-scattering time, most likely due to the formation of the nitrogen–hydrogen complexes responsible for N passivation.

Regardless of the specific mechanism responsible for this effect, its coupling to the masked-hydrogenation procedure outlined in

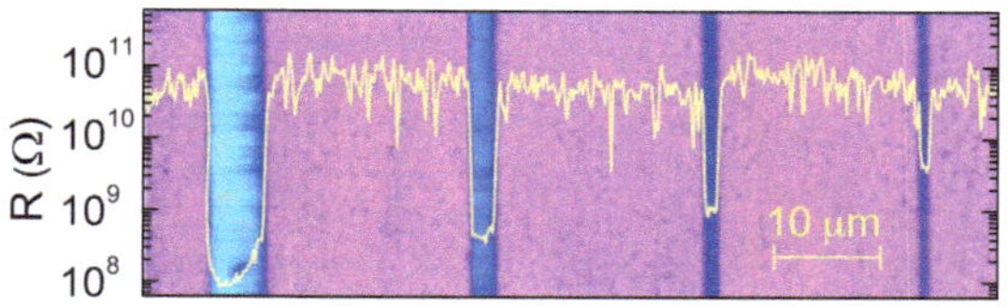

Figure 8.14 Conductive-probe atomic force microscopy (CP-AFM) 2D mapping of the local resistance R of a $GaAs_{1-x}N_x$ sample ($x = 0.9\%$, $t = 200$ nm). The surface of the sample was patterned with H-opaque titanium wires—having widths equal to 5, 2, 1, and 0.5 μm (from left to right)—before H irradiation (hydrogenation temperature = 300°C). The wire-to-wire separation is 20 μm. The yellow line superimposed on the 2D mapping of R provides the value of the local electrical resistance (ordinate axis) along a line perpendicular to the wires' axis. Adapted from Ref. (Alvarez et al., 2011).

this present chapter allows for defining low-resistance $GaAs_{1-x}N_x$ "channels," embedded in a high-resistance $GaAs_{1-x}N_x$:H matrix (see Fig. 8.14). In principle, this ability could be exploited for the realization of microscopic circuits, such as on-chip microscale coils, and of the electrical interconnects of the site-controlled nanostructures fabricated by in-plane BGE.

8.8 Conclusions and Outlook

To summarize, in this chapter we saw how the possibility to finely adjust the amount of hydrogen incorporated in different regions of a dilute nitride sample—either by deposition of H-opaque masks (*in-plane BGE*, see Section 8.2–8.4) or possibly by the direct-writing methods introduced in Section 8.6—can be exploited to tailor the optical, structural, and even electrical (see Section 8.7.2) properties of these materials with nanometric precision. Such precision, due in equal parts to the ultrasharp diffusion profile of H in dilute nitrides (better than 5 nm/decade (Trotta et al., 2009a); see also Section 8.2) and to the impressive level of performance achieved by state-of-the-art electron beam lithography systems ($\leq$10 nm beam and step size (Mohan et al., 2010)), can in principle be exploited for the fabrication of nanometer-sized light emitters with arbitrary *shape* and *position*. In Section 8.3.2, for example, the application of this concept to the

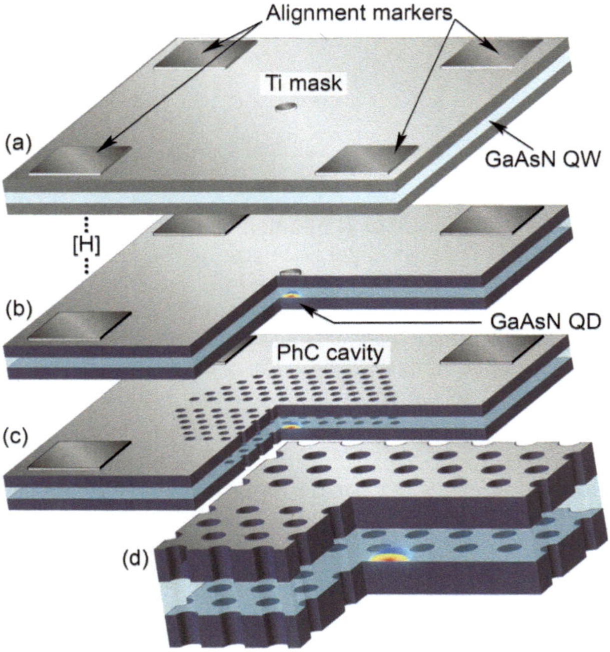

Figure 8.15 Processing steps needed to embed a QD fabricated by in-plane BGE in a PhC microcavity. (a) A Ti mask (dark gray; diameter d ~100 nm; see Section 8.3.2 and Ref. (Trotta et al., 2011)) is deposited on a GaAs membrane (gray), embedding a GaAsN QW (light blue). The mask is precisely positioned (with a positioning error as low as ≤10–20 nm (Mohan et al., 2010)) with respect to a set of alignment markers (dark gray), which will serve as reference points for the following processing steps. (b) A single QD is obtained by hydrogenating the masked sample. (c) A PhC cavity (also aligned to the markers) is fabricated around the QD by etching holes through the fully passivated (GaAs-like) QW. (d) Zoom-in of the QD region, following the fabrication of the PhC cavity.

realization of site-controlled GaAsN/GaAsN:H QDs was discussed (Trotta et al., 2011).

On the one hand, the near-perfect site control that is inherent to in-plane BGE renders the nanoemitters fabricated with this technique particularly well suited for the integration with optical (e.g., PhC) microcavities (see also Section 8.1). The (relatively simple) processing steps required for integrating a site-controlled

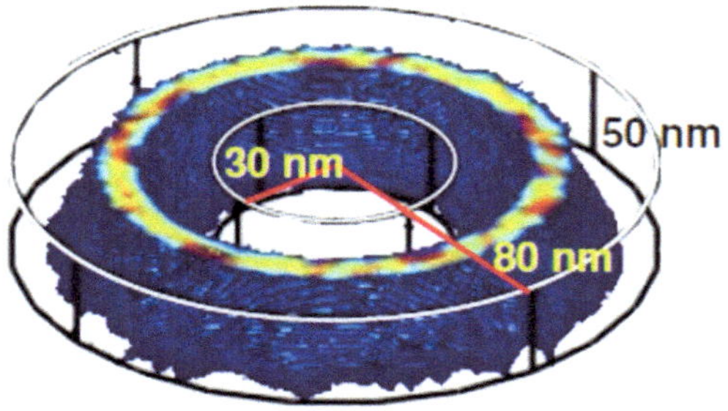

Figure 8.16 Posthydrogenation distribution of electronically active N atoms in a $GaAs_{1-x}N_x$ epilayer ($x = 0.9\%$ and $T = 50$ nm) patterned with a Ti *ring*. The inner (outer) ring diameter is 30 (80) nm. The irradiation temperature used for the calculation was 200°C. Red and blue correspond to maximum and minimum N concentrations, respectively. Adapted from Ref. (Trotta et al., 2012b).

QD fabricated by in-plane BGE with a PhC microcavity are outlined in Fig. 8.15. On the other hand, the unprecedented flexibility afforded by in-plane BGE in the definition of the carrier-confining potential could be exploited in the fabrication of *arbitrarily shaped nanostructures*. As an example, Fig. 8.16 reports the 3D simulation of the formation of a *quantum ring*, obtained by depositing an appropriately shaped mask on a 50 nm thick $GaAs_{1-x}N_x$ epilayer (Trotta et al., 2012b). Although very interesting from a theoretical point of view (Chakraborty and Pietiläinen, 1994), regularly shaped quantum rings are extremely difficult to obtain with conventional self-assembly techniques based on epitaxial growth (Granados et al., 2005), thus greatly increasing the appeal of alternative fabrication methods such as in-plane BGE.

Apart from the undeniably very intriguing possibility to control the shape and position of the fabricated nanoemitters with a precision determined by the electron beam lithography system writing H-opaque masks, however, in-plane BGE presents several additional advantages. Among these advantages, the relative simplicity of the growth process—molecular beam epitaxy (MBE) growth of (In)GaAsN QWs and epilayers is quite well established at this point; see, for example, Ref. (Liu et al., 2003)—and the possibility of erasing/rewriting the QD pattern through multiple annealing/hydrogenation procedures (Polimeni et al., 2001a, 2003a) are surely worth mentioning. Further, the application of this method to samples with extremely dilute N concentrations ([N] $\leq 10^{17}$ cm^{-3} for a 5 nm-thick QW)—such that the average N–N distance is of the order of the size of the masks (as low as 50 nm)—offers the interesting opportunity to isolate individual *N atoms, pairs, or clusters.* In dilute nitrides, indeed, impurity complexes formed by one, two, or more N atoms are widely reported (Thomas and Hopfield, 1966; Francoeur et al., 2004; Merz et al., 1969; Marcet et al., 2010) to give rise to sharp ($<$100 μeV [99]) emission lines associated with the radiative recombination of the excitons—and biexcitons; see Ref. (Merz et al., 1969; Marcet et al., 2010)—bound to the complex. Similar to the well-known case of NV centers in diamond (Barth et al., 2009; Tomljenovic-Hanic et al., 2009; Togan et al., 2010), the coupling of these radiatively active defects to the modes of PhC microcavities might prove very interesting for the realization of single- and entangled-photon sources, potentially useful in quantum computing schemes (it should be noted that single-photon emission has already been demonstrated for these defects; see, e.g., Ref. (Ikezawa et al., 2007)).

In conclusion, the intensive use of nanofabrication techniques based on the *spatially selective hydrogenation* of dilute nitrides will open new avenues for nanophotonics by providing the ability to deterministically place single, arbitrarily shaped nanostructures (and, possibly, N clusters) at specific points of carefully designed (Felici et al., 2010) PhC structures. In turn, this will allow for the realization of complex photonic devices—for example, single- and entangled-photon sources, as well as single-photon routers, switches, and delay lines—and will be instrumental for the practical implementation of quantum computing schemes, thus contributing to the establishment of nanophotonics as a mature technology.

Acknowledgments

We would like to thank A. Miriametro for his invaluable support and technical assistance. We gratefully acknowledge A. Polimeni, M. Capizzi, F. Masia, and G. Pettinari, who greatly contributed to different stages of the research presented here. Together with N. Balakrishnan and A. Patanè (University of Nottingham, UK), G. Pettinari also needs to be acknowledged for the laser annealing experiments reported in Section 8.6.2 (see also Section 2.5.2). We thank F. Martelli (IMM-CNR, Italy) and S. Rubini (TASC-CNR, Italy) for growing most of the samples presented in this chapter and M. Francardi, A. Gerardino, and L. Mariucci (CNR-IFN) for performing the sample processing and electron beam lithography steps required by in-plane BGE. D. Giubertoni (Fondazione Kessler, Trento, Italy) was in charge of SIMS measurements, whereas time-resolved measurements were done at LENS, Firenze, Italy, in collaboration with L. Cavigli, M. Gurioli, and A. Vinattieri. Refractive index measurements were performed by M. Geddo, M. Patrini, M. Galli, and G. Guizzetti (Università di Pavia, Italy). The authors acknowledge J. Alvarez and J.-P. Kleider (LGEP-CNRS, Paris, France) for measurements of the electrical resistance by CP-AFM. We also need to thank R. Carron, D. Fekete, P. Gallo, B. Dwir, A. Rudra, and E. Kapon (Ecole Polytechnique Fédérale de Lausanne, Switzerland) for providing InGaAsN V-groove QWR samples and for fruitful discussions. Finally, the authors thank A. Amore Bonapasta (IMM-CNR, Italy), P. J. Klar (Justus-Liebig-Universität, Giessen, Germany), E. P. O'Reilly (NMRC, National University of Ireland, Cork), M. Berti and G. Bisognin (Università di Padova, Italy), F. Boscherini (Università di Bologna, Italy), G. Ciatto (Soleil Synchrotron, Paris, France), and M. Stavola (Lehigh University, USA) for productive collaborations and discussions. High-magnetic-field measurements were performed at the High Field Magnet Laboratory (Nijmegen, the Netherlands) and supported by EuroMagNET under EU Contract No. 228043. The authors appreciate local support in Nijmegen by J. C. Maan and P. C. M. Christianen. Marco Felici would like to acknowledge funding by the EU under Grant Agreement No. PIEFGA-2011-301363 and by the Italian Ministry for Education, University and Research within the Futuro in Ricerca (FIRB) program (DeLIGHTeD project, Prot. RBFR12RS1W), as well as support from COST Action MP0805.

Bibliography

Adachi, S. (1999). *Optical Constants of Crystalline and Amorphous Semiconductors* (Kluwer Academic, Boston, MA).

Adams, A. R. (1986). Band-structure engineering for low-threshold high-efficiency semiconductorlasers, *Electron. Lett.*, **22**, 249.

Akahane, Y., Asano, T., Song, B., and Noda, S. (2003). Investigation of high-Q channel drop filters using donor-type defects in two-dimensional photonic crystal slabs, *Appl. Phys. Lett.*, **83**, 1512.

Akahane, Y., Asano, T., Song, B., and Noda, S. (2005). Fine-tuned high-Q photonic-crystal nanocavity, *Opt. Express*, **13**, 1202.

Allison, G., Spasov, S., Patanè, A., Eaves, L., Kozlova, N., Freudenberger, J., Hopkinson, M., and Hill, G. (2008). Electron effective mass and Si-donor binding energy in $GaAs_{1-x}N_x$ probed by a high magnetic field, *Phys. Rev. B*, **77**, 125210.

Alt, H. C., Egorov, A. Y., Riechert, H., Wiedemann, B., Meyer, J. D., Michelmann, R. W., and Bethge, K. (2000). Infrared absorption study of nitrogen in N-implanted GaAs and epitaxially grown $GaAs_{1-x}N_x$ layers, *Appl. Phys. Lett.*, **77**, 3331.

Alt, H. C., Egorov, A. Y., Riechert, H., Wiedemann, B., Meyer, J. D., Michelmann, R. W., and Bethge, K. (2001). Local vibrational mode absorption of nitrogen in GaAsN and InGaAsN layers grown by molecular beam epitaxy, *Physica B*, **302–303**, 282.

Alt, H. C., Messerer, P., Köhler, K., and Riechert, H. (2009). H- and D-related mid-infrared absorption bands in $Ga_{1-y}In_yAs_{1-x}N_x$ epitaxial layers, *Phys. Status Solidi B*, **246**, 200.

Altug, H., Englund, D., and Vuckovic, J. (2006). Ultrafast photonic crystal nanocavity laser, *Nat. Phys.*, **2**, 484.

Alvarez, J., Kleider, J.-P., Trotta, R., Polimeni, A., Capizzi, M., Martelli, F., Mariucci, L., and Rubini, S. (2011). Giant and reversible enhancement of the electrical resistance of $GaAs_{1-x}N_x$ by hydrogen irradiation, *Phys. Rev. B*, **84**, 085331.

Amidani, L., Ciatto, G., Bondino, F., and Boscherini, F. (2014). Local electronic structure of dilute elements in alloys, *Phys. Rev. B* (submitted).

Amore Bonapasta, A. (1991). Hydrogen passivation of carbon-doped gallium arsenide, *Phys. Rev. B*, **48**, 8771.

Amore Bonapasta, A., and Capizzi, M. (1998). Hydrogen as a deep impurity in semiconductors and its interaction with deep centers in III-V compounds, *Defect Diffus. Forum*, **157–159**, 133.

Amore Bonapasta, A., and Filippone, F. (2003). Local and lattice relaxations in hydrogenated $GaAs_yN_{1-y}$ alloys, *Phys. Rev. B*, **68**, 073202.

Amore Bonapasta, A., Filippone, F., and Giannozzi, P. (2003). Nitrogen passivation by atomic hydrogen in $GaAs_yN_{1-y}$ and $In_xGa_{1-x}As_yN_{1-y}$ alloys, *Phys. Rev. B*, **68**, 115202.

Amore Bonapasta, A., Filippone, F., and Giannozzi, P. (2004). Structure, electronic properties, and formation mechanisms of hydrogen-nitrogen complexes in GaP_yN_{1-y} alloys, *Phys. Rev. B*, **69**, 115207.

Amore Bonapasta, A., Filippone, F., and Giannozzi, P. (2005). Electron correlation effecs on the hydrogen passivation of $Mn_xGa_{1-x}As$ dilute magnetic semiconductors, *Phys. Rev. B*, **72**, 121202(R).

Amore Bonapasta, A., Filippone, F., Giannozzi, P., Capizzi, M., and Polimeni, A. (2002). Structure and passivation effects of mono- and dihydrogen complexes in $GaAs_{1-y}N_y$ alloys, *Phys. Rev. Lett.*, **89**, 216401.

Amore Bonapasta, A., Filippone, F., and Mattioli, G. (2007). H-induced dangling bonds in H–isoelectronic-impurity complexes formed in $GaAs_{1-y}N_y$ alloys, *Phys. Rev. Lett.*, **98**, 206403.

Amore Bonapasta, A., and Giannozzi, P. (2000). Effects of strain and local charge on the formation of deep defects in III-V ternary alloys, *Phys. Rev. Lett.*, **84**, 3923.

Amore Bonapasta, A., and Pavesi, L. (1996). Hydrogen interaction with shallow and deep centers in GaAs, *Int. J. Quantum Chem.*, **57**, 823.

Amsel, G., and Davies, J. (1969). La microanalyse de l'azote par l'observation directe de réactions nucléaires applications, *Rev. Phys. Appl.*, **4**, 383.

Andreev, A. D., and O'Reilly, E. P. (2004). Theoretical study of Auger recombination in a GaInNAs 1.3 μm quantum well laser structure, *Appl. Phys. Lett.*, **84**, 1826.

Anisimov, V. I., Aryasetiawan, F., and Lichtenstein, A. I. (1997). First-principles calculations of the electronic structure and spectra of strongly correlated systems: the LDA+U method, *J. Phys.: Condens. Matter*, **9**, 767.

Ankudinov, A. L., Ravel, B., Rehr, J. J., and Conradson, S. D. (1998). Real-space multiple-scattering calculation and interpretation of x-ray-absorption near-edge structure, *Phys. Rev. B*, **58**, 7565.

Ashcroft, N. W., and Mermin, N. D. (1976). *Solid State Physics* (Saunders College/Harcourt Brace College).

Atlasov, K. A., Calic, M., Karlsson, K. F., Gallo, P., Rudra, A., Dwir, B. and Kapon, E. (2009). Photonic-crystal microcavity laser with site-controlled quantum-wire active medium, *Opt. Express*, **17**, 18178.

Atlasov, K. A., Felici, M., Karlsson, K. F., Gallo, P., Rudra, A., Dwir, B. and Kapon, E. (2010). 1D photonic band formation in finite-size photonic-crystal waveguides, *Opt. Express*, **18**, 117.

Atlasov, K. A., Karlsson, K. F., Deichsel, E., Rudra, A., Dwir, B., and Kapon, E. (2007). Site-controlled single quantum wire integrated into a photonic-crystal membrane microcavity, *Appl. Phys. Lett.*, **90**, 153107.

Atlasov, K. A., Karlsson, K. F., Rudra, A., Dwir, B., and Kapon, E. (2008). Wavelength and loss splitting in directly coupled photonic-crystal defect microcavities, *Opt. Express*, **16**, 16255.

Atlasov, K. A., Rudra, A., Dwir, B., and Kapon, E. (2011). Large mode splitting and lasing in optimally coupled photonic-crystal microcavities, *Opt. Express*, **19**, 2619.

Averbeck, R., Jaschke, G., Geelhaar, L., and Riechert, H. (2004). Record-low thresholds of InGaAsN/GaAs SQWs lasers, in *Proc. 19th Int. Semicond. Laser Conf.* (Matsue-shi, Japan), doi: 10.1109/ ISLC.2004.1382759.

Averbeck, R., Jaschke, G., Geelhaar, L., and Riechert, H. (2005). Low threshold InGaAsN/GaAs lasers beyond 1500 nm, *J. Cryst. Growth*, **278**, 224.

Baba, T., Fukaya, N., and Yonekura, J. (1999). Observation of light propagation in photonic crystal optical waveguides with bends, *Electron. Lett.*, **35**, 654.

Baier, M., Pelucchi, E., Kapon, E., Varoutsis, S., Gallart, M., Robert-Philip, I., and Abram, I. (2004). Single photon emission from site-controlled pyramidal quantum dots, *Appl. Phys. Lett.*, **84**, 648.

Baier, M. H., Malko, A., Pelucchi, E., Oberli, D. Y., and Kapon, E. (2006). Quantum-dot exciton dynamics probed by photon-correlation spectroscopy, *Phys. Rev. B*, **73**, 205321.

Balakrishnan, N., Patanè, A., Makarovsky, O., Polimeni, A., Capizzi, M., Martelli, F., and Rubini, S. (2011). Laser writing of the electronic activity of N- and H-atoms in GaAs, *Appl. Phys. Lett.*, **99**, 021105.

Balakrishnan, N., Pettinari, G., Makarovsky, O., Turyanska, L., Fay, M. W., De Luca, M., Polimeni, A., Capizzi, M., Martelli, F., Rubini, S. and Patanè, A. (2012). Band-gap profiling by laser writing of hydrogen-containing III-N-Vs, *Phys. Rev. B*, **86**, 155307.

Baldassarri Höger von Högersthal, G., Bissiri, M., Polimeni, A., Capizzi, M., Fischer, M., Reinhardt, M., and Forchel, A. (2001). Hydrogen-induced band gap tuning of (InGa)(AsN)/GaAs single quantum wells, *Appl. Phys. Lett.*, **78**, 3472.

Baldassarri Höger von Högersthal, G., Polimeni, A., Masia, F., Bissiri, M., Capizzi, M., Gollub, D., Fischer, M., and Forchel, A. (2003). Magnetophotoluminescence studies of (InGa)(AsN)/GaAs heterostructures, *Phys. Rev. B*, **67**, 233304.

Bank, S. R. (2006). *High-Performance 1.55 μm GaInNAsSb Lasers Grown on GaAs*, PhD thesis, Stanford University.

Bank, S. R., Bae, H. P., Yuen, H. B., Wistey, M. A., Goddard, L. L., and Harris, J. S. (2006). Room-temperature continuous-wave 1.55 μm GaInNAsSb laser on GaAs, *Electron. Lett.*, **42**, 156.

Bank, S. R., Wistey, M. A., Goddard, L. L., Yuen, H. B., Lordi, V., and Harris, J. S. (2004). Low-threshold continuous-wave 1.5-μm GaInNAsSb lasers grown on GaAs, *IEEE J. Quantum Electron.*, **40**, 656.

Bank, S. R., Yuen, H. B., Wistey, M. A., Lordi, V., Bae, H. P., and Harris, J. S. (2005). Effects of growth temperature on the structural and optical properties of 1.55 μm GaInNAsSb quantum wells grown on GaAs, *Appl. Phys. Lett.*, **87**, 21908.

Barrau, J., Amand, T., Brousseau, M., Simes, R. J., and Goldstein, L. (1992). Induced electrostatic confinement of the electron gas in tensile strained InGaAs/InGaAsP quantum well lasers, *J. Appl. Phys.*, **71**, 5768.

Barth, M., Nuesse, N., Loechel, B., and Benson, O. (2009). Controlled coupling of a single-diamond nanocrystal to a photonic crystal cavity, *Opt. Lett.*, **34**, 1108.

Bastard, G. (1990). *Wave Mechanics Applied to Semiconductor Heterostructures* (Editions de Physique, Paris).

Bayer, M., Gutbrod, T., Reithmaier, J. P., Forchel, A., Reinecke, T. L., Knipp, P. A., Dremin, A. A., and Kulakovskii, V. D. (1998). Optical modes in photonic molecules, *Phys. Rev. Lett.*, **81**, 2582.

Becke, A. D. (1988). Density-functional exchange-energy approximation with correct asymptotic behavior, *Phys. Rev. A*, **38**, 309.

Bellaiche, L., Wei, S., and Zunger, A. (1996). Localization and percolation in semiconductor alloys: GaAsN vs GaAsP, *Phys. Rev. B*, **54**, 17568.

Benfatto, M., and Felici, R. (2001). Resonant atomic scattering factor theory: a multiple scattering approach, *Phys. Rev. B*, **64**, 115410.

Bergman, K., Stavola, M., Pearton, S. J., and Lopata, J. (1998). Donor-hydrogen complexes in passivated silicon, *Phys. Rev. B*, **37**, 2770.

Berti, M., Bisognin, G., De Salvador, D., and Napolitani, E. (2008). *Beam Injection Based Nanocharacterization of Advanced Materials* (Research Signpost Trivandrum, Kerala, India).

Berti, M., Bisognin, G., De Salvador, D., Napolitani, E., Vangelista, S., Polimeni, A., Capizzi, M., Boscherini, F., Ciatto, G., Rubini, S., Martelli, F., and Franciosi, A. (2007). Formation and dissolution of D-N complexes in dilute nitrides, *Phys. Rev. B*, **76**, 205323.

Bhattacharya, P. (1997). *Semiconductor Optoelectronic Devices* (Prentice Hall, Englewood Cliffs, NJ).

Biasiol, G., and Kapon, E. (1998). Mechanisms of self-ordering of quantum nanostructures grown on nonplanar surfaces, *Phys. Rev. Lett.*, **81**, 2962.

Bimberg, D. (1978). Anomaly of the linear and quadratic Zeeman effect of an effective-mass acceptor: C in GaAs, *Phys. Rev. B*, **18**, 1794.

Bimberg, D., Grundmann, M., and Ledentsov, N. N. (1999). *Quantum Dot Heterostructures* (Wiley, Chichester).

Bisognin, G., De Salvador, D., Drigo, A. V., Napolitani, E., Sambo, A., Berti, M., Polimeni, A., Felici, M., Capizzi, M., Güngerich, M., Klar, P. J., Bais, G., Jabeen, F., Piccin, M., Rubini, S., Martelli, F. and Franciosi, A. (2006). Hydrogen-nitrogen complexes in dilute nitride alloys: origin of the compressive lattice strain, *Appl. Phys. Lett.*, **89**, 061904.

Bisognin, G., De Salvador, D., Mattevi, C., Berti, M., Drigo, A. V., Ciatto, G., Grenouillet, L., Duvaut, P., Gilet, P., and Mariette, H. (2004). Determination of lattice parameter and of N lattice location in $In_xGa_{1-x}N_yAs_{1-y}$/GaAs and GaN_yAs_{1-y}/GaAs epilayers, *J. Appl. Phys.*, **95**, 48.

Bisognin, G., De Salvador, D., Napolitani, E., Berti, M., Polimeni, A., Capizzi, M., Rubini, S., Martelli, F., and Franciosi, A. (2008). High-resolution x-ray diffraction in situ study of very small complexes: the case of hydrogenated dilute nitrides, *J. Appl. Cryst.*, **41**, 366.

Bisognin, G., De Salvador, D., Napolitani, E., Berti, M., Polimeni, A., Felici, M., Capizzi, M., Bais, G., Jabeen, F., Piccin, M., Rubini, S., Martelli, F., and Franciosi, A. (2007). Thermal evolution of small N–D complexes in deuterated dilute nitrides revealed by in-situ high resolution x-ray diffraction, *Phys. Status Solidi A*, **204**, 2766.

Bissiri, M., Baldassarri Höger von Högersthal, G., Polimeni, A., Capizzi, M., Gollub, D., Fischer, M., Reinhardt, M., and Forchel, A. (2002a). Role of N clusters in $In_xGa_{1-x}As_{1-y}N_y$ band-gap reduction, *Phys. Rev. B*, **66**, 033311.

Bissiri, M., Baldassarri Höger von Högersthal, G., Polimeni, A., Gaspari, V., Ranalli, F., Capizzi, M., Amore Bonapasta, A., Jiang, F., Stavola, M., Gollub, D., Fischer, M., Reinhardt, M., and Forchel, A. (2002b). Hydrogen-induced passivation of nitrogen in $GaAs_{1-y}N_y$, *Phys. Rev. B*, **65**, 235210.

Borenstein, J. T., Corbett, J. W., and Pearton, S. J. (1993). Kinetic-model for hydrogen reactions in boron-doped silicon, *J. Appl. Phys.*, **73**, 2751.

Bosch, H. S., and Hale, G. M. (1992). Improved formulas for fusion cross-sections and thermal reactivities, *Nucl. Fusion*, **32**, 611.

Boscherini, F. (2013). X-ray absorption fine structure in the study of semiconductor heterostructures and nanostructures, in *Characterization of Semiconductor Heterostructures and Nanostructures*, 2nd ed. (Elsevier, Amsterdam).

Bransden, B. H., and Joachain, C. J. (2003). *Physics of Atoms and Molecules* (Pearson Education, Prentice Hall, USA).

Broderick, C. A., Usman, M., Sweeney, S. J., and O'Reilly, E. P. (2012). Band engineering in dilute nitride and bismide semiconductor lasers, *Semicond. Sci. Technol.*, **27**, 094011.

Buda, F., Chiarotti, G. L., Car, R., and Parrinello, M. (1989). Proton diffusion in crystalline silicon, *Phys. Rev. Lett.*, **63**, 294.

Bunker, G. (2010). *Introduction to XAFS: A Practical Guide to X-Ray Absorption Fine Structure Spectroscopy* (Cambridge University Press, UK).

Burt, M. G. (1999). Fundamentals of envelope function theory for electronic states and photonic modes in nanostructures, *J. Phys. Cond. Matt.*, **11**, R53.

Buyanova, I. A., and Chen, W. M. (2004). *Physics and Applications of Dilute Nitrides* (Taylor & Francis, New York).

Buyanova, I. A., Chen, W. M., Izadifard, M., Pearton, S. J., Bihler, C., Brandt, M. S., Hong, Y. G., and Tu, C. W. (2007). Hydrogen passivation of nitrogen in GaNAs and GaNP alloys: how many H atoms are required for each N atom? *Appl. Phys. Lett.*, **90**, 021920.

Buyanova, I. A., Chen, W. M., Pozina, G., Bergman, J. P., Monemar, B., Xin, H. P., and Tu, C. W. (1999). Mechanism for low-temperature photoluminescence in GaNAs/GaAs structures grown by molecular-beam epitaxy, *Appl. Phys. Lett.*, **75**, 124429.

Buyanova, I. A., Chen, W. M., Pozina, G., Monemar, B., and Ni, W. X. (1997). Mechanism for thermal quenching of luminescence in SiGe/Si structures grown by molecular beam epitaxy: role of nonradiative defects, *Appl. Phys. Lett.*, **71**, 3676–3678.

Buyanova, I. A., Chen, W. M., and Tu, C. W. (2003a). Recombination processes in N-containing III-V ternary alloys, *Solid State Electron.*, **47**, 467–475.

Buyanova, I. A., Ferreira, A. C., Holtz, P. O., Monemar, B., Campman, K., Merz, J. L., and Gossard, A. C. (1996). Effect of hydrogen passivation on Be-doped AlGaAs/GaAs quantum wells, *Appl. Phys. Lett.*, **68**, 1365–1367.

Buyanova, I. A., Izadifard, M., Chen, W. M., Polimeni, A., Capizzi, M., Xin, H. P., and Tu, C. W. (2003b). Hydrogen-induced improvements in optical quality of GaNAs alloys, *Appl. Phys. Lett.*, **82**, 3662.

Buyanova, I. A., Izadifard, M., Ivanov, I. G., Birch, J., Chen, W. M., Felici, M., Polimeni, A., Capizzi, M., Hong, Y. G., Xin, H. P., and Tu, C. W. (2004). Direct experimental evidence for unusual effects of hydrogen on the electronic and vibrational properties of GaN_xP_{1-x} alloys: a proof for a general property of dilute nitrides, *Phys. Rev. B*, **70**, 245215.

Cabib, D., Fabri, E., and Fiorio, G. (1971). The ground state of the exciton in a magnetic field, *Solid State Commun.*, **9**, 1517.

Cabib, D., Fabri, E., and Fiorio, G. (1972). Ground and first excited states of excitons in a magnetic field, *Il Nuovo Cimento B*, **10**, 185.

Cai, Y., and Thorpe, M. F. (1992). Length mismatch in random semiconductor alloys. I general theory for quaternaries, *Phys. Rev. B*, **46**, 15872.

Calic, M., Gallo, P., Felici, M., Atlasov, K. A., Dwir, B., Rudra, A., Biasiol, G., Sorba, L., Tarel, G., Savona, V., and Kapon, E. (2011). Phonon-mediated coupling of InGaAs/GaAs quantum-dot excitons to photonic crystal cavities, *Phys. Rev. Lett.*, **106**, 227402.

Capizzi, M., Coluzza, C., Emiliani, V., Frankl, P., Frova, A., Sarto, F., Amore Bonapasta, A., Sobiesierski, Z., and Sacks, R. N. (1992). Hydrogen activated radiative states in GaAs/GaAlAs heterostructures and InGaAs/GaAs multiquantum wells, *J. Appl. Phys.*, **72**, 1454–1459.

Capizzi, M., Emiliani, V., Frova, A., and Sarto, F. (1993). Effect of ion-gun hydrogenation on the photoluminescence of degenerate n-type GaAs:Si, *Phys. Rev. B*, **47**, 4301–4306.

Caplan, P. J., Poindexter, E. H., Deal, B. E., and Razouk, R. R. (1979). ESR centers, interface states, and oxide fixed charge in thermally oxidized silicon wafers, *J. Appl. Phys.*, **50**, 5847–5854.

Carrère, H., Marie, X., Lombez, L., and Amand, T. (2006). Optical gain of InGaAsN/InP quantum wells for laser applications, *Appl. Phys. Lett.*, **89**, 181115.

Carron, R., Fekete, D., Gallo, P., Dwir, B., Rudra, A., Felici, M., Bartova, B., Cantoni, M., and Kapon, E. (2011a). Dilute nitride InGaAsN/GaAs V-groove quantum wires emitting at 1.3 μm wavelength at room temperature, *Appl. Phys. Lett.*, **99**, 101107.

Carron, R., Gallo, P., Dwir, B., Rudra, A., and Kapon, E. (2011b). Dilute-nitride GaInAsN/GaAs site-controlled pyramidal quantum dots, *Appl. Phys. Lett.*, **99**, 181113.

Cauchois, Y. (1956). Distribution spectrale observée dans une région d'absorption propre de divers cristaux, in *Compte Rendus de l'Académie des Sciences (Paris)*, **242**, 100–102.

Chadderton, L. T. (1965). *Radiation Damage in Crystals* (Methuen, London).

Chakraborty, T., and Pietiläinen, P. (1994). Electron-electron interaction and the persistent current in a quantum ring, *Phys. Rev. B*, **50**, 8460.

Chen, T. R., Chang, B., Chiu, L. C., Yu, K. L., Margalit, S., and Yariv, A. (1983). Carrier leakage and temperature dependence of InGaAsP lasers, *Appl. Phys. Lett.*, **43**, 217.

Chen, W. M. (2000). Applications of optically detected magnetic resonance in semiconductor layered structures, *Thin Solid Films*, **364**, 45–52.

Chen, W. M., Buyanova, I. A., Buyanov, A. V., Lundstrom, T., Bi, W. G., and Tu, C. W. (1996). Intrinsic doping: a new approach for n-type modulation doping in InP-based heterostructures, *Phys. Rev. Lett.*, **77**, 2734–2737.

Chevallier, J., Dautremont-Smith, W. C., Tu, C. W., and Pearton, S. J. (1985). Donor neutralization in GaAs(Si) by atomic hydrogen, *Appl. Phys. Lett.*, **47**, 108.

Ciatto, G. (2014). Complexes and clusters in crystalline semiconductors, in *X-Ray Absorption Spectroscopy of Semiconductors* (Springer, Berlin).

Ciatto, G., Boscherini, F., Amore Bonapasta, A., Filippone, F., Polimeni, A. and Capizzi, M. (2005a). Nitrogen-hydrogen complex in $GaAs_xN_{1-x}$ revealed by x-ray absorption spectroscopy, *Phys. Rev. B*, **71**, 201301(R).

Ciatto, G., Boscherini, F., Amore Bonapasta, A., Filippone, F., Polimeni, A., Capizzi, M., Berti, M., Bisognin, G., De Salvador, D., Floreano, L., Martelli, F., Rubini, S., and Grenouillet, L. (2009). Local structure of nitrogen-hydrogen complexes in dilute nitrides, *Phys. Rev. B*, **79**, 165205.

Ciatto, G., d'Acapito, F., Boscherini, F., and Mobilio, S. (2004). Treatment of EXAFS data taken in the fluorescence mode in non-linear conditions, *J. Synchrotron Radiat.*, **11**, 278.

Ciatto, G., D'Acapito, F., Sanna, S., Fiorentini, V., Polimeni, A., Capizzi, M., Mobilio, S., and Boscherini, F. (2005b). Comparison between experimental and theoretical determination of the local structure of the $GaAs_{1-y}N_y$ dilute nitride alloy, *Phys. Rev. B*, **71**, 115210.

Ciatto, G., Harmand, J.-C., Glas, F., Largeau, L., Le Du, M., Boscherini, F., Malvestuto, M., Floreano, L., Glatzel, P., and Alonso Mori, R. (2007). Anions relative location in the group-V sublattice of GaAsSbN/GaAs epilayers: XAFS measurements and simulations, *Phys. Rev. B*, **75**, 245212.

Ciatto, G., Renevier, H., Proietti, M. G., Polimeni, A., Capizzi, M., Mobilio, S., and Boscherini, F. (2005c). Effects of hydrogenation on the local structure of $In_xGa_{1-x}As_{1-y}N_y$ quantum wells and $GaAs_{1-y}N_y$ epilayers, *Phys. Rev. B*, **72**, 085322.

Clerjaud, B., Côte, D., Hahn, W.-S., Lebkiri, A., Ulrici, W., and Wasik, D. (1996). Nitrogen-dihydrogen complex in GaP, *Phys. Rev. Lett.*, **77**, 4930.

Cococcioni, M., and de Gironcoli, S. (2005). Linear response approach to the calculation of the effective interaction parameters in the LDA+U method, *Phys. Rev. B*, **71**, 035105.

Cohen, A. J., Mori-Sànchez, P., and Yang, W. (2008). Insights into current limitations of density functional theory, *Science*, **321**, 792.

Cohen-Tannoudji, C., Diu, B., and Laloe, F. (1973). *Mécanique Quantique*, Vol. II (Hermann).

Coleman, J. (1993). Chapter 8, in *Quantum Well Lasers* (Academic Press, New York).

Cotton, F. A., and Wilkinson, G. (1988). *Advanced Inorganic Chemistry* (John Wiley, New York).

CPMD (2012). (Copyright IBM Corp 1990-2008, Copyright MPI fuer Festkoerperforschung, Stuttgart 1997-2001, http://www.cpmd.org).

D. Marx and J. Hutter (2000). *Ab initio Molecular Dynamics: Theory and Implementation, NIC*, Vol. 1 (John von Neumann Institute for Computing, Jülich), ISBN 3-00-005618, 301.

D'Acapito, F. (2011). Advanced methods for the analysis of x-ray absorption spectroscopy data applied to semiconductors, *Semic. Science Technol.*, **26**, 064004.

Dagnelund, D., Tu, C. W., Polimeni, A., Capizzi, M., Chen, W. M., and Buyanova, I. A. (2013). Effect of thermal annealing on defects in post-growth hydrogenated GaNP, *Phys. Status Solidi C*, **10**, 561.

Dagnelund, D., Vorona, I., Nosenko, G., Wang, X. J., Tu, C. W., Yonezu, H., Polimeni, A., Capizzi, M., Chen, W. M., and Buyanova, I. A. (2012). Effects of hydrogenation on non-radiative defects in GaNP and GaNAs alloys:

an optically detected magnetic resonance study. *J. Appl. Phys.*, **111**, 023501.

Dagnelund, D., Wang, X. J., Tu, C. W., Polimeni, A., Capizzi, M., Chen, W. M. and Buyanova, I. A. (2011). Effect of postgrowth hydrogen treatment on defects in GaNP, *Appl. Phys. Lett.*, **98**, 141920.

Dannecker, T., Jin, Y., Cheng, H., Gorman, C., Buckeridge, J., Uher, C., Fahy, S., Kurdak, C., and Goldman, R. S. (2010). Nitrogen composition dependence of electron effective mass in $GaAs_{1-x}N_x$, *Phys. Rev. B*, **82**, 125203.

Davies, G., Lightowlers, E. C., Stavola, M., Bergman, K., and Svensson, B. (1987). The 3942-cm^{-1} optical band in irradiated silicon, *Phys. Rev. B*, **35**, 2755.

De Luca, M., Pettinari, G., Ciatto, G., Amidani, L., Filippone, F., Polimeni, A., Fonda, E., Boscherini, F., Amore Bonapasta, A., Giubertoni, D., Knübel, A., Lebedev, V., and Capizzi, M. (2012). Identification of four-hydrogen complexes in In-rich $In_xGa_{1-x}N$ ($x>0.4$) alloys using photoluminescence, x-ray absorption, and density functional theory, *Phys. Rev. B*, **86**, 201202(R).

Derbyshire, G., Cheung, K. C., Sangsingkeow, P., and Hasnain, S. (1999). A low-profile monolithic multi-element Ge detector for x-ray fluorescence applications, *J. Synchrotron Radiat.*, **6**, 62.

Diehl, W., Brick, P., Kunert, B., Reinhard, S., Volz, K., and Stolz, W. (2007). Low threshold 1260 nm (GaIn)(NAs) semiconductor disk laser, *Appl. Phys. Lett.*, **91**, 071103.

Dieumegard, D., Dubreuil, D., and Amsel, G. (1979). Analysis and depth profiling of deuterium with the D(^{3}He, p)^{4}He reaction by detecting the protons at backward angles, *Nucl. Instrum. Methods*, **166**, 431.

Dixon, P., Richardson, D., Jones, R., Latham, C. D., Öberg, S., Torres, V. J. B., and Briddon, P. R. (1998). Nitrogen-hydrogen defects in GaP, *Phys. Status Solidi B*, **210**, 321.

Dovesi, R., Saunders, V. R., Roetti, C., Orlando, R., Zicovich-Wilson, C. M., Pascale, F., Civalleri, B., Doll, K., Harrison, N. M., Bush, I. J., D'Arco, P., and Llunell, M. (2006). *Crystal06 User's Manual*, University of Torino, Torino.

Du, M.-H., Limpijumnong, S., and Zhang, S. B. (2005). Hydrogen pairs and local vibrational frequencies in H-irradiated $GaAs_{1-y}N_y$, *Phys. Rev. B*, **72**, 073202.

Dutta, N. K., and Nelson, R. J. (1981). Temperature dependence of threshold of InGaAsP/InP double heterostructure lasers and Auger recombination, *Appl. Phys. Lett.*, **38**, 407.

Elias, D. C., Nair, R. R., Mohiuddin, T. M. G., Morozov, S. V., Blake, P., Halsall, M. P., Ferrari, A. C., Boukhvalov, D. W., Katsnelson, M. I., Geim, A. K., and Novoselov, K. (2009). Control of graphene's properties by reversible hydrogenation: evidence for graphane, *Science*, **323**, 610.

Endicott, J., Patanè, A., Ibánez, J., Eaves, L., Bissiri, M., Hopkinson, M., Airey, R., and Hill, G. (2003). Magnetotunneling spectroscopy of dilute Ga(AsN) quantum wells, *Phys. Rev. Lett.*, **91**, 126802.

Englund, D., Fushman, I., and Vuckovic, J. (2005). General recipe for designing photonic crystal cavities, *Opt. Express*, **13**, 5961.

Englund, D., Shields, B., Rivoire, K., Hatami, F., Vuckovic, J., Park, H. and Lukin, M. D. (2010). Deterministic coupling of a single nitrogen vacancy center to a photonic crystal cavity, *Nano Lett.*, **10**, 3922.

Erbil, A., Cargill III, G. S., Frahm, R., and Boehme, R. F. (1988). Total-electron-yield current measurements for near-surface extended x-ray-absorption fine structure, *Phys. Rev. B*, **37**, 2450.

Erol, A. (2008). *Dilute III-V Nitride Semiconductors and Material Systems* (Springer, Berlin).

Fahy, S., Lindsay, A., Ouerdane, H., and O'Reilly, E. P. (2006). Alloy scattering of n-type carriers in GaN_xAs_{1-x}, *Phys. Rev. B*, **74**, 035203.

Fahy, S., and O'Reilly, E. P. (2003). Intrinsic limits on electron mobility in dilute nitride semiconductors, *Appl. Phys. Lett.*, **83**, 3731.

Fehse, R., Tomić, S., Adams, A. R., Sweeney, S. J., O'Reilly, E. P., Andreev, A. D., and Riechert, H. (2002). A quantitative study of radiative, Auger, and defect related recombination processes in 1.3-μm GaInNAs-based quantum-well lasers, *IEEE J. Sel. Top. Quantum Electron.*, **8**, 801.

Felici, M., Atlasov, K. A., Surrente, A., and Kapon, E. (2010). Semianalytical approach to the design of photonic crystal cavities, *Phys. Rev. B*, **82**, 115118.

Felici, M., Gallo, P., Mohan, A., Dwir, B., Rudra, A., and Kapon, E. (2009). Site-controlled InGaAs quantum dots with tunable emission energy, *Small*, **5**, 938.

Felici, M., Pettinari, G., Carron, R., Lavenuta, G., Tartaglini, E., Polimeni, A., Fekete, D., Gallo, P., Dwir, B., Rudra, A., Christianen, P. C. M., Maan, J. C., Capizzi, M., and Kapon, E. (2012a). Magneto-optical properties of single site-controlled InGaAsN quantum wires grown on prepatterned GaAs substrates, *Phys. Rev. B*, **85**, 155319.

Felici, M., Polimeni, A., Miriametro, A., Capizzi, M., Xin, H. P., and Tu, C. W. (2005). Free carrier and/or exciton trapping by nitrogen pairs in dilute $GaP_{1-x}N_x$, *Phys. Rev. B*, **71**, 045209.

Felici, M., Polimeni, A., Salviati, G., Lazzarini, L., Armani, N., Masia, F., Capizzi, M., Martelli, F., Lazzarino, M., Bais, G., Piccin, M., Rubini, S. and Franciosi, A. (2006a). In-plane bandgap engineering by modulated hydrogenation of dilute nitride semiconductors, *Adv. Mater.*, **18**, 1993.

Felici, M., Polimeni, A., Tartaglini, E., Notargiacomo, A., De Luca, M., Carron, R., Fekete, D., Dwir, B., Rudra, A., Capizzi, M., and Kapon, E. (2012b). Reduced temperature sensitivity of the polarization properties of hydrogenated InGaAsN V-groove quantum wires, *Appl. Phys. Lett.*, **101**, 151114.

Felici, M., Trotta, R., Masia, F., Polimeni, A., Miriametro, A., Capizzi, M., Klar, P. J., and Stolz, W. (2006b). Compositional disorder in $GaAs_{1-x}N_x$:H investigated by photoluminescence, *Phys. Rev. B*, **74**, 085203.

Filippone, F., Mattioli, G., Alippi, P., and Amore Bonapasta, A. (2011). Clusters and Magnetic Anchoring Points in (Ga,Fe)N Condesed Magnetic Semiconductors, *Phys. Rev. Lett.*, **107**, 196401.

Filippone, F., and Parrinello, M. (2001). Vibrational analysis from linear response theory, *Chem. Phys. Lett.*, **345**, 179.

Fischer, M., Gollub, D., Reinhardt, M., Kamp, M., and Forchel, A. (2003). GaInNAs for GaAs based lasers for the 1.3 to 1.5 μm range, *J. Cryst. Growth*, **251**, 353.

Fischer, M., Reinhardt, M., and Forchel, A. (2000). GaInAsN/GaAs laser diodes operating at 1.52 μm, *Electron. Lett.*, **36**, 1208.

Fisher, M. A., Adams, A. R., O'Reilly, E. P., and Harris, J. J. (1987). Resonant electron scattering due to the central cell of impurities observed in AlGaAs using hydrostatic pressure, *Phys. Rev. Lett.*, **59**, 2341.

Floreano, L., Naletto, G., Cvetko, D., Gotter, R., Malvezzi, M., Marassi, L., Morgante, A., Santaniello, A., Verdini, A., and Tommasini, F. (1999). Performance of the grating-crystal monochromator of the ALOISA beamline at the Elettra Synchrotron, *Rev. Sci. Instrum.*, **70**, 3855.

Fowler, B., Martin, K. R., Washer, K., and Stavola, M. (2005). Structure and vibrational properties of N–H_2 complexes in GaAs:N, *Phys. Rev. B*, **72**, 035208.

Francoeur, S., Klem, J. F., and Mascarenhas, A. (2004). Optical spectroscopy of single impurity centers in semiconductors, *Phys. Rev. Lett.*, **93**, 067403.

Gallo, P., Felici, M., Dwir, B., Atlasov, K. A., Karlsson, K. F., Rudra, A., Mohan, A., Biasiol, G., Sorba, L., and Kapon, E. (2008). Integration of site-controlled pyramidal quantum dots and photonic crystal membrane cavities, *Appl. Phys. Lett.*, **92**, 263101.

Geddo, M., Ciabattoni, T., Guizzetti, G., Galli, M., Patrini, M., Polimeni, A., Trotta, R., Capizzi, M., Bais, G., Piccin, M., Rubini, S., Martelli, F. and Franciosi, A. (2007). Photoreflectance and reflectance investigation of deuterium-irradiated GaAsN, *Appl. Phys. Lett.*, **90**, 091907.

Geddo, M., Patrini, M., Guizzetti, G., Galli, M., Trotta, R., Polimeni, A., Capizzi, M., Martelli, F., and Rubini, S. (2011). Optical study of hydrogen-irradiated GaAsN/GaAs heterostructures, *J. Appl. Phys.*, **109**, 123511.

Ghiringhelli, G., Tagliaferri, A., Braicovich, L., and Brookes, N. (1998). A simple spherical grating by-pass monochromator dedicated to soft x-ray emission spectroscopy, *Rev. Sci. Instrum.*, **69**, 1610.

Giannozzi, P., Baroni, S., Bonini, N., Calandra, M., Car, R., Cavazzoni, C., Ceresoli, D., Chiarotti, G. L., Cococcioni, M., Dabo, I., Corso, A. D., Fabris, S., Fratesi, G., de Gironcoli, S., Gebauer, R., Gerstmann, U., Gougoussis, C., Kokalj, A., Lazzeri, M., Martin-Samos, L., Marzari, N., Mauri, F., Mazzarello, R., Paolini, S., Pasquarello, A., Paulatto, L., Sbraccia, C., Scandolo, S., Sclauzero, G., Seitsonen, A. P., Smogunov, A., Umari, P., and Wentzcovitch, R. M. (2009). Quantum Espresso: a modular and open-source software project for quantum simulations of materials, http://www.quantum-espresso.org.

Glatzel, P., and Bergmann, U. (2005). High resolution 1s core hole x-ray spectroscopy in 3d transition metal complexes—electronic and structural information. *Coord. Chem. Rev.*, **249**, 65.

Goedecker, S., Huetter, J., and Teter, M. (1996). Separable dual-space Gaussian pseudopotentials, *Phys. Rev. B*, **54**, 1703.

Gonze, X., Stumpf, R., and Scheffler, M. (1991). Analysis of separable potentials, *Phys. Rev. B*, **44**, 8503.

Granados, D., Garcia, J. M., Ben, T., and Molina, S. I. (2005). Vertical order in stacked layers of self-assembled In(Ga)As quantum rings on GaAs (001), *Appl. Phys. Lett.*, **86**, 071918.

Greentree, A., Tahan, C., Cole, J., and Hollenberg, L. (2006). Quantum phase transitions of light, *Nat. Phys.*, **2**, 856.

Groot, F. D., and Kotani, A. (2008). *Core Level Spectroscopy of Solids* (CRC Press, Boca Raton, USA).

Hague, C. F., Underwood, J. H., Avila, A., Delaunay, R., Ringuenet, H., Marsi, M., and Sacchi, M. (2005). Plane-grating flat-field soft x-ray spectrometer, *Rev. Sci. Instrum*, **76**, 023110.

Hahn, W. S., Clerjaud, B., Côte, D., Gendron, F., Porte, C., Ulrici, W., Wasik, D., and Wilkening, W. (1994). *Defects in Semiconductors 17: Nitrogen-Hydrogen Complexes in GaP and GaAs* (Trans Tech, Aedermannsdorf).

Hai, P. N., Chen, W. M., Buyanova, I. A., Xin, H. P., and Tu, C. W. (2000). Direct determination of electron effective mass in GaNAs/GaAs quantum wells, *Appl. Phys. Lett.*, **77**, 1843.

Haller, E. E., Hansen, W. L., and Goulding, F. S. (1981). Physics of ultra-pure germanium, *Adv. Phys.*, **30**, 93–138.

Harris, C., Lindsay, A., and O'Reilly, E. P. (2008). Evolution of N defect states and optical transitions in ordered and disordered $GaP_{1-x}N_x$ alloys, *J. Phys.: Condens. Matter*, **20**, 295211.

Harrison, J., and Hauser, J. R. (1976). Alloy scattering in ternary III-V compounds, *Phys. Rev. B*, **13**, 5347.

Hartmann, A., Ducommun, Y., Kapon, E., Hohenester, U., and Molinari, E. (2000). Few-particle effects in semiconductor quantum dots: observation of multicharged excitons, *Phys. Rev. Lett.*, **84**, 5648.

Hartwigsen, C., Goedecker, S., and Hutter, J. (1998). Relativistic separable dual-space gaussian pseudopotentials from H to Rn, *Phys. Rev. B*, **58**, 3641.

Henini, M. (2005). *Dilute Nitride Semiconductors* (Elsevier, Amsterdam).

Hennessy, K., Badolato, A., Winger, M., Gerace, D., Atatüre, M., Gulde, S., Fält, S., Hu, E., and Imamoğlu, A. (2007). Quantum nature of a strongly coupled single quantum dot-cavity system, *Nature*, **445**, 896.

Higashi, T., Sweeney, S. J., Phillips, A. F., Adams, A. R., O'Reilly, E. P., Uchida, T., and Fujii, T. (1999). Experimental analysis of temperature dependence in 1.3-μm AlGaInAs-InP strained MQW lasers, *IEEE J. Sel. Top. Quantum Electron.*, **5**, 413.

Hjalmarson, H. P., Vogl, P., Wolford, D., and Dow, J. (1980). Theory of substitutional deep traps in covalent semiconductors, *Phys. Rev. Lett.*, **44**, 810.

Hofmann, M., Wagner, A., Ellmers, C., Schlichenmeier, C., Schafer, S., Höhnsdorf, F., Koch, J., Stolz, W., Koch, S. W., Rühle, W. W., Hader, J., Moloney, J. V., O'Reilly, E. P., Borchert, B., Egorov, A. Y. and Riechert, H. (2001). Gain spectra of (GaIn)(NAs) laser diodes for the 1.3-μm-wavelength regime, *Appl. Phys. Lett.*, **78**, 3009.

Hofmann, M. R., Gerhardt, N., Wagner, A. M., Ellmers, C., Höhnsdorf, F., Koch, J., Stolz, W., Koch, S. W., Rühle, W. W., Hader, J., Moloney, J. V., O'Reilly, E. P., Borchert, B., Egorov, A. Y., H. Riechert, H. C. S. and Chow, W. W. (2002). Emission dynamics and optical gain of 1.3-μm (GaIn)(NAs)/GaAs lasers. Quantum electronics, *IEEE J. Quantum Electron.*, **38**, 213.

Hohenberg, P., and Kohn, W. (1964). Inhomogeneous electron gas, *Phys. Rev.*, **136**, B864.

Holý, V., Pietsch, U., and Baumbach, T. (1999). *High-Resolution X-Ray Scattering from Thin Films and Multilayers* (Springer-Verlag).

Hudson, A. J., Stevenson, R. M., Bennett, A. J., Young, R. J., Nicoll, C. A., Atkinson, P., Cooper, K., Ritchie, D. A., and Shields, A. J. (2007). Coherence of an entangled exciton-photon state, *Phys. Rev. Lett.*, **99**, 266802.

Huggenberger, A., Heckelmann, S., Schneider, C., Hoefling, S., Reitzenstein, S., Worschech, L., Kamp, M., and Forchel, A. (2011). Narrow spectral linewidth from single site-controlled In(Ga)As quantum dots with high uniformity, *Appl. Phys. Lett.*, **98**, 131104.

Hughes, S., Yao, P., Milde, F., Knorr, A., Dalacu, D., Mnaymneh, K., Sazonova, V., Poole, P., Aers, G., Lapointe, J., Cheriton, R., and Williams, R. (2011). Influence of electron-acoustic phonon scattering on off-resonant cavity feeding within a strongly coupled quantum-dot cavity system, *Phys. Rev. B*, **83**, 165313.

Ignatov, A., Patanè, A., Makarovsky, O., and Eaves, L. (2006). Terahertz response of hot electrons in dilute nitride Ga(AsN) alloys, *Appl. Phys. Lett.*, **88**, 032107.

Ikezawa, M., Sakuma, Y., and Masumoto, Y. (2007). Single photon emission from individual nitrogen pairs in GaP, *Jpn. J. Appl. Phys. II*, **46**, L871.

Intonti, F., Matarazzo, V., Nasir, A., Makarovsky, O., Campion, R., Patanè, A., Kumar, S., Rastelli, A., Schmidt, O. G., and Gurioli, M. (2011). Nano-sized light emitting diodes by near field laser exposure, *Appl. Phys. Lett.*, **98**, 183102.

Ivchenko, E. L., Kalevich, V. K., Shiryaev, A. Y., Afanasiev, M. M. and Masumoto, Y. (2010). Optical orientation and spin-dependent recombination in GaAsN alloys under continuous-wave pumping, *J. Phys.: Condens. Matter*, **22**, 465804.

Jaklevic, J., Kirby, J. A., Klein, M. P., Robertson, A. S., Brown, G. S. and Eisenberger, P. (1977). Fluorescence detection of exafs: sensitivity enhancement for dilute species and thin films, *Sol. State. Commun.*, **23**, 679.

Janesko, B. G., Henderson, T. M., and Scuseria, G. E. (2009). Screened hybrid density functionals for solid-state chemistry and physics, *Phys. Chem. Chem. Phys.*, **11**, 443.

Janotti, A., Lyons, J. L., and Van de Walle, C. G. (2012). Hybrid functional calculations of native point defects in InN, *Phys. Status Solidi A*, **209**, 65.

Janotti, A., Wei, S. H., Zhang, S. B., Kurtz, S., and Van de Walle, C. G. (2003). Interactions between nitrogen, hydrogen, and gallium vacancies in $GaAs_{1-x}N_x$ alloys, *Phys. Rev. B*, **67**, 161201.

Janotti, A., Zhang, S. B., and Wei, S.-H. (2002a). Hydrogen vibrational modes in GaP:N: the pivotal role of nitrogen in stabilizing the H_2^* complex, *Phys. Rev. Lett.*, **88**, 125506.

Janotti, A., Zhang, S. B., Wei, S.-H., and Van de Walle, C. G. (2002b). Effects of hydrogen on the electronic properties of dilute GaAsN alloys, *Phys. Rev. Lett.*, **89**, 086403.

Jiang, F., Stavola, M., Capizzi, M., Polimeni, A., Amore Bonapasta, A. and Filippone, F. (2004). Vibrational spectroscopy of hydrogenated $GaAs_{1-y}N_y$: a structure-sensitive test of an H_2^*(N) model, *Phys. Rev. B*, **69**, 041309(R).

Jimenez-Mier, J., van Ek, J., Ederer, D. L., Callcott, T. A., Jia, J. J., Carlisle, J., Terminello, L., Asfawi, A., and Perera, R. C. (1999). Dynamical behavior of x-ray absorption and scattering at the L edge of titanium compounds: experiment and theory, *Phys. Rev. B*, **59**, 2649.

Jin, Y. (2010). *Influence of N incorporation on the electronic properties of dilute nitride (In)GaAsN alloys*, Ph.D. thesis, University of Michigan.

Johnson, N. M., Herring, C., and Chadi, D. J. (1986). Interstitial hydrogen and neutralization of shallow-donor impurities in single-crystal silicon, *Phys. Rev. Lett.*, **56**, 769–772.

Joly, Y. (2001). X-ray absorption near-edge structure calculations beyond the muffin-tin approximation, *Phys. Rev. B*, **63**, 125120.

Joly, Y. (2003). Calculating x-ray absorption near-edge structure at very low energy, *J. Synchrotron Radiat.*, **10**, 58.

Jones, R. E., Yuen, H. B., Liu, T., and Misra, P. (2011). Lattice matchable alloy for solar cells, US Patent Application 20110232730.

Kane, E. O. (1957). Band structure of indium antimonide, *J. Phys. Chem. Solids*, **1**, 249.

Kaplyanskii, A. A. (1963). Noncubic centers in cubic crystals and their piezospectroscopic investigation, *Opt. Spectrosc.*, **16**, 329.

Kapon, E. (1999). *Semiconductor Lasers: I and II* (Academic Press, New York).

Kapon, E., Hwang, D. M., and Bhat, R. (1989). Stimulated-emission in semiconductor quantum wire heterostructures, *Phys. Rev. Lett.*, **63**, 430.

Karaiskaj, D., Mascarenhas, A., Klem, J. F., Volz, K., Stolz, W., Adamcyk, M. and Tiedje, T. (2007). Excitons bound to nitrogen pairs in GaAs as seen by photoluminescence of high spectral and spatial resolution, *Phys. Rev. B*, **76**, 12509.

Kennedy, D. P., and O'Brien, R. R. (1965). Analysis of the impurity atom distribution near the diffusion mask for a planar p-n junction, *IBM J. Res. Dev.*, **9**, 179.

Kent, P. R. C., Bellaiche, L., and Zunger, A. (2002). Pseudopotential theory of dilute III-V nitrides, *Semicond. Sci. Technol.*, **17**, 851.

Kent, P. R. C., and Zunger, A. (2001a). Evolution of III-V nitride alloy electronic structure: the localized to delocalized transition, *Phys. Rev. Lett.*, **86**, 2613.

Kent, P. R. C., and Zunger, A. (2001b). Theory of electronic structure evolution in GaAsN and GaPN alloys, *Phys. Rev. B*, **64**, 115208.

Khitrova, G., Gibbs, H. M., Kira, M., Koch, S. W., and Scherer, A. (2006). Vacuum Rabi splitting in semiconductors, *Nat. Phys.*, **2**, 81.

Kilic, C., and Zunger, A. (2002). *n*-type doping of oxides by hydrogen, *Appl. Phys. Lett.*, **81**, 73.

Kim, K. S., Yang, G. M., Shim, H. W., Lim, K., Suh, E.-K., and Lee, H. J. (1997). Photoluminescence characterization of biaxial tensile strained GaAs, *J. Appl. Phys.*, **82**, 5103.

Kim, Y.-S., and Chang, K. J. (2002). Nitrogen-monohydride versus nitrogen-dihydride complexes in GaAs and $GaAs_{1-x}N_x$ alloys, *Phys. Rev. B*, **66**, 073313.

Kiravittaya, S., Rastelli, A., and Schmidt, O. (2006). Photoluminescence from seeded three-dimensional InAs/GaAs quantum-dot crystals, *Appl. Phys. Lett.*, **88**, 043112.

Klar, P. J., Grüning, H., Güngerich, M., Heimbrodt, W., Koch, J., Torunski, T., Stolz, W., Polimeni, A., and Capizzi, M. (2003). Global changes of the band structure and the crystal lattice of Ga(N,As) due to hydrogenation, *Phys. Rev. B*, **67**, 121206(R).

Klar, P. J., Grüning, H., Heimbrodt, W., Koch, J., Höhnsdorf, F., Stolz, W., Vicente, P. M. A., and Camassel, J. (2000). From N isoelectronic impurities to N-induced bands in the GaN_xAs_{1-x} alloy, *Appl. Phys. Lett.*, **76**, 3439.

Klar, P. J., Grüning, H., Heimbrodt, W., Weiser, G., Koch, J., Volz, K., Stolz, W., Koch, S. W., Tomić, S., Choulis, S. A., Hosea, T., O'Reilly, E. P., Hofmann, M., Hader, J., and Moloney, J. V. (2002). Interband transitions of quantum wells and device structures containing Ga(N,As) and (Ga,In)(N, As), *Semicond. Sci. Technol.*, **17**, 830.

Kleekajai, S., Jiang, F., Colon, K., Stavola, M., Fowler, W. B., Martin, K. R., Polimeni, A., Capizzi, M., Hong, Y. G., Xin, H. P., Tu, C. W., Gais, G., Rubini, S., and Martelli, F. (2008). Vibrational properties of the H–N–H complex in dilute III-N-V alloys: infrared spectroscopy and density functional theory, *Phys. Rev. B*, **77**, 085213.

Klinger, M. I., Emtsev, V. V., Mashovets, T. V., Ryvkin, S. M., and Vitovskii, N. A. (1981). Ionization mechanism of the defect production in semiconductors. *Radiat. Eff. Defect S.*, **56**, 229–239.

Kobayashi, N., and Kobayashi, Y. (1991). As and P desorption from III-V semiconductor surface in metalorganic chemical vapor deposition studied by surface photo-absorption, *Jpn. J. Appl. Phys.*, **30**, L1699.

Kolmogorov, A., Petrovsky, I., and Piscounov, N. (1937). Etude de l' équation de la diffusion avec croissance de la quantité de matière et son application á un probléme biologique, *Moscow Univ. Bull. Math*, **1**, 1–25.

Kondow, M., Kitatani, T., Nakatsuka, S., Larson, M. C., Hakahara, K., Yazawa, Y., Okai, M., and Uomi, K. (1997). GaInNAs: a novel material for long-wavelength semiconductor lasers, *IEEE J. Sel. Top. Quantum Electron.*, **3**, 719.

Kondow, M., Uomi, K., Niwa, A., Kitatani, T., Watahiki, S., and Yazawa, Y. (1996). Gas-source molecular beam epitaxy of GaN_xAs_{1-x} using a N radical as the N source, *Jpn. J. Appl. Phys.*, **33**, L1056.

Koningsberger, D. C., and Prins, R. (1988). *Principles, Applications and Techniques of EXAFS, SEXAFS and XANES* (Wiley, New York).

Kozlova, N. V., Pettinari, G., Makarovsky, O., Mori, N., Polimeni, A., Capizzi, M., Zhuang, Q. D., Krier, A., and Patanè, A. (2013). Non-resonant hydrogen dopants in In(AsN): a route to high electron concentrations and mobilities, *Phys. Rev. B*, **87**, 165207.

Kulik, H. J., Cococcioni, M., Scherlis, D. A., and Marzari, N. (2006). Density functional theory in transition-metal chemistry: a self-consistent Hubbard *U* approach, *Phys. Rev. Lett.*, **97**, 103001.

Kunold, A., Balocchi, A., Zhao, F., Amand, T., Ben Abdallah, N., Harmand, J. and Marie, X. (2011). Giant spin-dependent photo-conductivity in GaAsN dilute nitride semiconductor, *Phys. Rev. B*, **83**, 165202.

Kuramochi, E., Notomi, M., Mitsugi, S., Shinya, A., Tanabe, T., and Watanabe, T. (2006). Ultrahigh-Q photonic crystal nanocavities realized by the local width modulation of a line defect, *Appl. Phys. Lett*, **88**, 041112.

Kurtz, S., Webb, J., Gedvilas, L., Friedman, D., Geisz, J., Olson, J., King, R., Joslin, D., and Karam, N. (2001). Structural changes during annealing of GaInAsN, *Appl. Phys. Lett.*, **78**, 748.

Landau, L. D., and Lifshitz, E. M. (1986). *Theory of Elasticity* (Pergamon Press).

Lee, C., Yang, W., and Parr, R. G. (1988). Development of the Colle-Salvetti correlation-energy formula into a functional of the electron density, *Phys. Rev. B*, **37**, 785.

Lee, P. A., Citrin, P. H., Eisenberger, P., and Kincaid, B. M. (1981). Extended x-ray absorption fine structure: its strengths and limitations as a structural tool, *Rev. Mod. Phys.*, **53**, 769.

Leifer, K., Pelucchi, E., Watanabe, S., Michelini, F., Dwir, B., and Kapon, E. (2007). Narrow ($\approx$ 4 mev) inhomogeneous broadening and its correlation with confinement potential of pyramidal quantum dot arrays, *Appl. Phys. Lett.*, **91**, 081106.

Li, L. H., Sallet, V., Patriarche, G., Largeau, L., Bouchoule, S., Travers, L. and Harmand, J. C. (2003). Investigations on GaInNAsSb quinary alloy for 1.5 μm laser emission on GaAs, *Appl. Phys. Lett.*, **83**, 1298.

Li, W., Pessa, M., and Likonen, J. (2001). Lattice parameter in GaNAs epilayers on GaAs: deviation from Vegard's law, *Appl. Phys. Lett.*, **78**, 2864.

Liebich, S., Zimbrich, M., Beyer, A., Lange, C., Franzbach, D. J., Chaterjee, S., Hossain, N., Sweeney, S. J., Volz, K., and Stolz, W. (2011). Laser operation of Ga(NAsP) lattice-matched to (001) silicon substrate, *Appl. Phys. Lett.*, **99**, 071109.

Lindsay, A., and O'Reilly, E. P. (1999). Theory of enhanced bandgap non-parabolicity in GaN_xAs_{1-x} and related alloys, *Solid State Commun.*, **112**, 443.

Lindsay, A., and O'Reilly, E. P. (2001). Influence of nitrogen resonant states on the electronic structure of GaN_xAs_{1-x}, *Solid State Commun.*, **118**, 313.

Lindsay, A., and O'Reilly, E. P. (2003). A universal model for trends in A_1-type defect states in zincblende and diamond semiconductor structures, *Physica B*, **340-342**, 434.

Lindsay, A., and O'Reilly, E. P. (2004a). A tight-binding-based analysis of the band anti-crossing model in GaN_xAs_{1-x}, *Physica E*, **21**, 901.

Lindsay, A., and O'Reilly, E. P. (2004b). Unification of the band anticrossing and cluster-state models of dilute nitride semiconductor alloys, *Phys. Rev. Lett.*, **93**, 196402.

Lindsay, A., O'Reilly, E. P., Andreev, A. D., and Ashley, T. (2008). Theory of conduction band structure of InN_xSb_{1-x} and GaN_xSb_{1-x} dilute nitride alloys, *Phys. Rev. B*, **77**, 165205.

Liu, H. Y., Hopkinson, M., Navaretti, P., Gutierrez, M., Ng, J. S., and David, J. P. R. (2003). Improving optical properties of 1.55 μm GaInNAs/GaAs multiple quantum wells with Ga(In)NAs barrier and space layer, *Appl. Phys. Lett.*, **83**, 4951.

Liu, X., Pistol, M.-E., and Samuelson, L. (1990a). Excitons bound to nitrogen pairs in GaAs, *Phys. Rev. B*, **42**, 7504.

Liu, X., Pistol, M.-E., Samuelson, L., Schwetlick, S., and Seifert, W. (1990b). Nitrogen pair luminescence in GaAs, *Appl. Phys. Lett.*, **56**, 1451.

Longa, S. D., Arcovito, A., Girasole, M., Hazemann, J. L., and Benfatto, M. (2001). Quantitative analysis of x-ray absorption near edge structure data by a full multiple scattering procedure: the Fe-CO geometry in photolyzed carbonmonoxy-myoglobin single crystal, *Phys. Rev. Lett.*, **87**, 155501.

Loridant-Bernard, D., Meziere, S., Constant, M., Dupuy, N., Sombret, S. and Chevalier, J. (1998). Infrared study of light-induced reactivation of neutralized dopants in hydrogenated n-type GaAs doped with silicon, *Appl. Phys. Lett.*, **73**, 644.

Luo, X. D., Xu, Z. Y., Ge, W. K., Pan, Z., Li, L. H., and Lin, Y. W. (2001). Photoluminescence properties of a $GaN_{0.015}As_{0.985}$/GaAs single quantum well under short pulse excitation, *Appl. Phys. Lett.*, **79**, 958.

Ma, B. S., Su, F. H., Ding, K., Li, G. H., Zhang, Y., Mascarenhas, A., Xin, H. P., and Tu, C. W. (2005). Pressure behavior of the alloy band edge and nitrogen-related centers in $GaAs_{0.999}N_{0.001}$, *Phys. Rev. B*, **71**, 045213.

Ma, Y. (1994). X-ray absorption, emission, and resonant inelastic scattering in solids, *Phys. Rev. B*, **49**, 5799.

Marcet, S., Ouellet-Plamondon, C., Éthier Majcher, G., Saint-Jean, P., André, R., Klem, J., and Francoeur, S. (2010). Charged excitons and biexcitons bound to isoelectronic centers, *Phys. Rev. B*, **82**, 235311.

Masia, F., Pettinari, G., Polimeni, A., Felici, M., Miriametro, A., Capizzi, M., Lindsay, A., Healy, S. B., O'Reilly, E. P., Cristofoli, A., Bais, G., Piccin, M., Rubini, S., Martelli, F., Franciosi, A., Klar, P. J., Volz, K. and Stolz, W. (2006). Interaction between conduction band edge and nitrogen states probed by carrier effective-mass measurements in $GaAs_{1-x}N_x$, *Phys. Rev. B*, **73**, 073201.

Masia, F., Polimeni, A., Baldassarri Höger von Högersthal, G., M.Bissiri, Capizzi, M., Klar, P. J., and Stolz, W. (2003). Early manifestation of localization effects in diluted Ga(AsN), *Appl. Phys. Lett.*, **82**, 4474.

Massies, J., and Grandjean, N. (1993). Surfactant effect on the surface diffusion length in epitaxial growth, *Phys. Rev. B*, **48**, 8502.

Matsubara, M., Godet, J., and Pizzagalli, L. (2010). Theoretical study of hydrogen stability and aggregation in dislocation cores in silicon, *Phys. Rev. B*, **82**, 024107.

Mattila, T., Wei, S. H., and Zunger, A. (1999). Localization and anticrossing of electron levels in $GaAs_{1-x}N_x$ alloys, *Phys. Rev. B*, **60**, R11245.

Mattioli, G., Alippi, P., Filippone, F., Caminiti, R., and Amore Bonapasta, A. (2010). Deep versus shallow behavior of intinsic defects in rutile and anatase TiO_2 polymorphs, *J. Phys. Chem. C*, **114**, 21694.

McCluskey, M. D. (2000). Local vibrational modes of impurities in semiconductors, *J. Appl. Phys.*, **87**, 3593.

McCluskey, M. D., and Haller, E. E. (1999). Hydrogen in III-V and II-VI semiconductors, in *Hydrogen in Semiconductors II* (Academic Press).

Merz, J., Faulkner, R., and Dean, P. (1969). Excitonic molecule bound to the isoelectronic nitrogen trap in GaP, *Phys. Rev.*, **188**, 1228.

Mohan, A., Gallo, P., Felici, M., Dwir, B., Rudra, A., Faist, J., and Kapon, E. (2010). Record-low inhomogeneous broadening of site-controlled quantum dots for nanophotonics, *Small*, **6**, 1268.

Moller, W., and Besembacher, F. (1980). A note on the He+D nuclear-reaction cross section, *Nucl. Instrum. Methods*, **168**, 111.

Morrow, R. A. (1989). Modeling the diffusion of hydrogen in GaAs, *J. Appl. Phys.*, **66**, 2973.

Naber, J. A., and James, H. M. (1961). *Bull. Am. Phys. Soc.*, **6**, 303.

Nagamune, Y., Arakawa, Y., Tsukamoto, S., Nishioka, M., Sasaki, S., and Miura, N. (1992). Photoluminescence spectra and anisotropic energy shift of GaAs quantum wires in high magnetic-fields, *Phys. Rev. Lett.*, **69**, 2963.

Newville, M., Ravel, B., Haskel, D., Rehr, J. J., Stern, E. A., and Yacoby, Y. (1995). Analysis of multiple-scattering XAFS data using theoretical standards, *Physica B*, **208&209**, 154.

Noda, S. (2006). Seeking the ultimate nanolaser, *Science*, **314**, 260.

Nordgren, J., Bray, G., Gamn, S., Nyholm, R., Rubensson, J., and Wassdahl, N. (1989). Soft x-ray emission spectroscopy using monochromatized synchrotron radiation, *Rev. Sci. Instrum.*, **60**, 1690.

Northrup, J. E. (2002). Theory of intrinsic and H-passivated screw dislocations in GaN, *Phys. Rev. B*, **66**, 045204.

Oestreich, M., and Rühle, W. W. (1995). Temperature dependence of the electron landé g factor in GaAs, *Phys. Rev. Lett.*, **74**, 2315.

O'Reilly, E. P. (1989). Valence band engineering in strained-layer structures, *Semicond. Sci. Technol.*, **4**, 121.

O'Reilly, E. P., and Lindsay, A. (2010). Tight-binding approach to calculation of localised perturbations in semiconductors, *J. Phys.: Conf. Ser.*, **242**, 012002.

O'Reilly, E. P., Lindsay, A., and Fahy, S. (2004). Theory of the electronic structure of dilute nitride alloys: beyond the band-anti-crossing model, *J. Phys.: Condens. Matter*, **16**, S3257.

O'Reilly, E. P., Lindsay, A., Klar, P. J., Polimeni, A., and Capizzi, M. (2009). Trends in the electronic structure of dilute nitride alloys, *Semicond. Sci. Technol.*, **24**, 033001.

O'Reilly, E. P., Lindsay, A., Tomić, S., and Kamal-Saadi, M. (2002). Tight-binding and k·p models for the electronic structure of Ga(In)NAs and related alloys, *Semicond. Sci. Technol.*, **17**, 870.

Oyanagi, H., Martini, M., and Saito, M. (1998). Nineteen-element high-purity Ge solid-state detector array for fluorescence x-ray absorption fine structure studies, *Nucl. Instrum. Meth. A*, **403**, 58.

Pan, J. W., and Chyi, J. I. (1996). Theoretical study of the temperature dependence of 1.3-μm AlGaInAs-InP multiple-quantum-well lasers, *IEEE J. Quantum Electron.*, **32**, 2133.

Pankove, J. I., Carlson, D. E., Berkeyheiser, J. E., and Wance, R. O. (1983). Neutralization of shallow acceptor levels in silicon by atomic-hydrogen, *Phys. Rev. Lett.*, **51**, 2224–2225.

Pankove, J. I., and Johnson, N. M. (1991). *Hydrogen in Semiconductors, in Semiconductors and Semimetals*, Vol. 34 (Academic Press, Boston).

Pauling, L. (1967). *The Nature of the Chemical Bond* (Cornell University Press, Ithaca).

Pavesi, L., and Giannozzi, P. (1991). H passivation of Si impurities in GaAs, *Phys. Rev. B*, **43**, 2446.

Pavesi, L., and Giannozzi, P. (1992). Atomic and molecular hydrogen in gallium arsenide: a theoretical study, *Phys. Rev. B*, **46**, 4621.

Pavesi, L., and Guzzi, M. (1994). Photoluminescence of $Al_xGa_{1-x}As$ alloys, *J. Appl. Phys.*, **75**, 4779.

Pavesi, M., Manfredi, M., Salviati, G., Armani, N., Rossi, F., Meneghesso, G., Levada, S., Zanoni, E., Du, S., and Eliashevich, I. (2004). Optical evidence of an electrothermal degradation of InGaN-based light-emitting diodes during electrical stress, *Appl. Phys. Lett.*, **84**, 3403.

Pearton, S. J. (1994). Hydrogen in crystalline semiconductors: part II; III-V compounds, *Int. J. Mod. Phys. B*, **8**, 1247.

Pearton, S. J., Corbett, J. W., and Stavola, M. (1992). *Hydrogen in Crystalline Semiconductors* (Springer-Verlag, Berlin).

Pearton, S. J., Dautremont-Smith, W. C., Chevallier, J., Tu, C. W. and Cummings, K. D. (1986). Hydrogenation of shallow-donor levels in GaAs, *J. Appl. Phys.*, **59**, 2821–2827.

People, R., and Bean, J. C. (1985). Calculation of critical layer thickness versus lattice mismatch for Ge_xSi_{1-x}/Si strained-layer heterostructures, *Appl. Phys. Lett.*, **47**, 322.

Perkins, J. D., Mascarenhas, A., Zhang, Y., Geisz, J. F., Friedman, D. J., Olson, J. M., and Kurtz, S. R. (1999). Nitrogen-activated transitions, level repulsion, and band gap reduction in $GaAs_{1-x}N_x$ with $x < 0.03$, *Phys. Rev. Lett.*, **82**, 3312.

Peter, E., Senellart, P., Martrou, D., Lemaître, A., Hours, J., Gérard, J., and Bloch, J. (2005). Exciton-photon strong-coupling regime for a single quantum dot embedded in a microcavity, *Phys. Rev. Lett.*, **95**, 067401.

Pettinari, G., Masia, F., Capizzi, M., Polimeni, A., Losurdo, M., Bruno, G., Kim, T. H., Choi, S., Brown, A., Lebedev, V., Cimalla, V., and Ambacher, O. (2008). Experimental evidence of different hydrogen donors in n-type InN, *Phys. Rev. B*, **77**, 125207.

Pettinari, G., Masia, F., Polimeni, A., Felici, M., Frova, A., Capizzi, M., Lindsay, A., O'Reilly, E. P., Klar, P. J., Stolz, W., Bais, G., Piccin, M., Rubini, S., Martelli, F., and Franciosi, A. (2006). Influence of nitrogen-cluster states on the gyromagnetic factor of electrons in $GaAs_{1-x}N_x$, *Phys. Rev. B*, **74**, 245202.

Pettinari, G., Patanè, A., Polimeni, A., Capizzi, M., Lu, X., and Tiedje, T. (2012a). Effects of hydrogen on the electronic properties of Ga(AsBi) alloys, *Appl. Phys. Lett.*, **101**, 222103.

Pettinari, G., Polimeni, A., and Capizzi, M. (2012b). *Photoluminescence: a tool for investigating optical, electronic, and structural properties of semiconductors, Chapter 5 in Semiconductor Research* (Spring-Verlag, Berlin, Heidelberg).

Pettinari, G., Polimeni, A., Masia, F., Trotta, R., Felici, M., Capizzi, M., Niebling, T., Stolz, W., and Klar, P. J. (2007). Electron mass in dilute nitrides and its anomalous dependence on hydrostatic pressure, *Phys. Rev. Lett.*, **98**, 146402.

Phillips, A. F., Sweeney, S. J., Adams, A. R., and Thijs, P. J. A. (1999). The temperature dependence of 1.3- and 1.5-μm compressively strained InGaAs(P) MQW semiconductor lasers, *IEEE J. Sel. Top. Quantum Electron.*, **5**, 401.

Phillips, J. C. (1973). *Bonds and Bands in Semiconductors* (Academic Press, New York).

Polimeni, A. (2007). private communication.

Polimeni, A., Baldassarri Höger von Högersthal, G., Bissiri, M., Capizzi, M., Fischer, M., Reinhardt, M., and Forchel, A. (2001a). Effect of hydrogen on the electronic properties of $In_xGa_{1-x}As_{1-y}N_y/GaAs$ quantum wells, *Phys. Rev. B*, **63**, 201304R.

Polimeni, A., Baldassarri Höger von Högersthal, G., Bissiri, M., Capizzi, M., Frova, A., Fischer, M., Reinhardt, M., and Forchel, A. (2002). Role of hydrogen in III-N-V compound semiconductors, *Semicond. Sci. Technol.*, **17**, 797.

Polimeni, A., Baldassarri Höger von Högersthal, G., Masia, F., Frova, A., Capizzi, M., Sanna, S., Fiorentini, V., Klar, P. J., and Stolz, W. (2004). Tunable variation of the electron effective mass and exciton radius in hydrogenated $GaAs_{1-x}N_x$, *Phys. Rev. B*, **69**, 041201(R).

Polimeni, A., Bissiri, M., Felici, M., Capizzi, M., Buyanova, I. A., Chen, W. M., Xin, H. P., and Tu, C. W. (2003a). Nitrogen passivation induced by atomic hydrogen: the $GaP_{1-y}N_y$ case, *Phys. Rev. B*, **67**, 201303.

Polimeni, A., Capizzi, M., Geddo, M., Fischer, M., Reinhardt, M., and Forchel, A. (2001b). Effect of nitrogen on the temperature dependence of the energy gap in $In_xGa_{1-x}As_{1-y}N_y$/GaAs single quantum wells, *Phys. Rev. B*, **63**, 195320.

Polimeni, A., Ciatto, G., Ortega, L., Jiang, F., Boscherini, F., Filippone, F., Amore Bonapasta, A., Stavola, M., and Capizzi, M. (2003b). Lattice relaxation by atomic hydrogen irradiation of III-N-V semiconductor alloys, *Phys. Rev. B*, **68**, 085204.

Polimeni, A., Marangio, D., Capizzi, M., Frova, A., and Martelli, F. (1994). Giant photoluminescence enhancement in deuterated highly strained InAs/GaAs quantum wells, *Appl. Phys. Lett.*, **65**, 1254.

Polimeni, A., Masia, F., Pettinari, G., Trotta, R., Felici, M., Capizzi, M., Lindsay, A., O'Reilly, E. P., Niebling, T., Stolz, W., and Klar, P. J. (2008a). Role of strain and properties of N clusters at the onset of the alloy limit in $GaAs_{1-x}N_x$, *Phys. Rev. B*, **77**, 155213.

Polimeni, A., Pettinari, G., Trotta, R., Masia, F., Felici, M., Capizzi, M., Lindsay, A., O'Reilly, E. P., Niebling, T., Stolz, W., Klar, P. J., Martelli, F., and Rubini, S. (2008b). Photoluminescence under magnetic field and hydrostatic pressure for probing the electronic properties of GaAsN, *Phys. Status Solidi A*, **205**, 107.

Pollak, F., and Cardona, M. (1968). Piezo-electroreflectance in Ge, GaAs, and Si, *Phys. Rev.*, **172**, 816.

Proietti, M. G., Renevier, H., Hodeau, J. L., Garcia, J., Berar, J. F. and Wolfers, P. (1999). Diffraction-anomalous-fine-structure spectroscopy applied to the study of III-V strained semiconductors, *Phys. Rev. B*, **59**, 5479.

Prokofyeva, T., Sauncy, T., Seon, M., Holtz, M., Qiu, Y., Nikishin, S. and Temkin, H. (1998). Raman studies of nitrogen incorporation in $GaAs_{1-x}N_x$, *Appl. Phys. Lett.*, **73**, 1409.

Putrino, A., Sebastiani, D., and Parrinello, M. (2000). Generalized variational density functional perturbation theory, *J. Chem. Phys.*, **113**, 7102.

Raisanen, J., Keinonen, J., and Karttunen, V. (1988). Redistribution of implanted H in annealings of n-type GaAs, *J. Appl. Phys.*, **64**, 2334.

Rammohan, K., Tang, Y., Rich, D. H., Goldman, R. S., Wieder, H. H. and Kavanagh, K. L. (1995). Relaxation-induced polarized luminescence from $In_xGa_{1-x}As$ films grown on GaAs(001), *Phys. Rev. B*, **51**, 5033.

Reithmaier, J. P., Sęk, G., Löffler, A., Hofmann, C., Kuhn, S., Reitzenstein, S., Keldysh, L. V., Kulakovskii, V. D., Reinecke, T. L. and Forchel, A. (2004). Strong coupling in a single quantum dot-semiconductor microcavity system, *Nature*, **432**, 197.

Renevier, H., Grenier, S., Arnaud, S., Berar, J. F., Caillot, B., Hodeau, J. L., Letoublon, A., Proietti, M. G., and Ravel, B. (2003). Diffraction anomalous fine-structure spectroscopy at beamline BM2 at the european synchrotron radiation facility, *J. Synchrotron Rad.*, **10**, 435.

Riechert, H., Ramakrishnan, A., and Steinle, G. (2002). Development of InGaAsN-based 1.3 μm VCSELs, *Semicond. Sci. Technol.*, **17**, 892.

Rossi, J., Wolfe, C., and Dimmock, J. (1970). Acceptor luminescence in high-purity n-type GaAs, *Phys. Rev. Lett.*, **25**, 1614.

Roth, L. M., Lax, B., and Zwerdling, S. (1959). Theory of optical magneto-absorption effects in semiconductors, *Phys. Rev.*, **114**, 90.

Rühle, W., and Göbel, E. (1976). New aspects of the magnetoluminescence of a band to acceptor transition in GaAs, *Phys. Status Solidi B*, **78**, 311.

Rutherford, E. (1911). The scattering of α and β particles by matter and the structure of the atom, *Philos. Mag.*, **6**, 21.

Sah, C. T., Sun, J. Y., and Tzou, J. J. (1983). Deactivation of the boron acceptor in silicon by hydrogen, *Appl. Phys. Lett.*, **43**, 204–206.

Sakurai, J. J. (1967). *Advanced Quantum Mechanics* (Benjamin Cummings).

Salter, C. L., Stevenson, R. M., Farrer, I., Nicoll, C. A., Ritchie, D. A. and Shields, A. J. (2010). An entangled-light-emitting diode, *Nature*, **465**, 594.

Sankey, O. F., Dow, J. D., and Hess, K. (1982). Theory of resonant scattering in semiconductors due to impurity central-cell potentials, *Appl. Phys. Lett.*, **41**, 664.

Sanna, S., and Fiorentini, V. (2004). Lattice constant, effective mass, and gap recovery in hydrogenated $GaAs_{1-x}N_x$, *Phys. Rev. B*, **69**, 125208.

Santos, P. V., Johnson, N. M., and Street, R. A. (1991). Light-enhanced hydrogen motion in a-Si:H, *Phys. Rev. Lett.*, **67**, 2686.

Sayers, D. E., Stern, E. A., and Lytle, F. W. (1971). New technique for investigating noncrystalline structures: Fourier analysis of the extended x-ray-absorption fine structure, *Phys. Rev. Lett.*, **27**, 1204.

Schaich, W. L. (1973). Comment on the theory of extended x-ray-absorption fine structure, *Phys. Rev. B*, **8**, 4028.

Schnohr, C., and Ridgway, M. (2014). *X-Ray Absorption Spectroscopy of Semiconductors* (Springer, Berlin).

Seki, S., Yamanaka, T., Lui, W., Yoshikuni, Y., and Yokoyama, K. (1994). Theoretical-analysis of pure effects of strain and quantum confinement on differential gain in InGaAsP/lnP strained-layer quantum-well lasers, *IEEE J. Quantum Electron.*, **30**, 500.

Shafi, M., Mari, R. H., Khatab, A., Henini, M., Polimeni, A., Capizzi, M. and Hopkinson, M. (2011). Deep levels in H-irradiated $GaAs_{1-x}N_x$ ($x < 0.01$) grown by molecular beam epitaxy, *J. Appl. Phys.*, **110**, 124508.

Shan, W., Walukiewicz, W., Ager III, J. W., Haller, E. E., Geisz, J. F., Friedman, D. J., Olson, J. M., and Kurtz, S. R. (1999a). Band anticrossing in GaInNAs alloys, *Phys. Rev. Lett.*, **82**, 1221.

Shan, W., Walukiewicz, W., Ager III, J. W., Haller, E. E., Geisz, J. F., Friedman, D. J., Olson, J. M., and Kurtz, S. R. (1999b). Effect of nitrogen on the band structure of GaInNAs alloys, *J. Appl. Phys*, **86**, 2349.

Shantharama, L. G., Adams, A. R., Ahmad, C. N., and Nicholas, R. J. (1984). The $k \cdot p$ interaction in InP and GaAs from the band-gap dependence of the effective mass, *J. Phys. C*, **17**, 4429.

Silvestre, S., Bernard-Loridant, D., Constant, E., Constant, M., and Chevallier, J. (2000). Electron-beam-induced reactivation of Si dopants in hydrogenated GaAs: a minority carrier generation effect or an energetic electron excitation effect? *Appl. Phys. Lett.*, **77**, 3206.

Skiba-Szymanska, J., Jamil, A., Farrer, I., Ward, M. B., Nicoll, C. A., Ellis, D. J. P., Griffiths, J. P., Anderson, D., Jones, G. A. C., Ritchie, D. A. and Shields, A. J. (2011). Narrow emission linewidths of positioned InAs quantum dots grown on pre-patterned GaAs(100) substrates, *Nanotechnology*, **22**, 065302.

Skierbiszewski, C. (2002). Experimental studies of the conduction band structure of GaInNAs alloys, *Semicond. Sci. Technol.*, **17**, 803.

Song, B. S., Noda, S., Asano, T., and Akahane, Y. (2005). Ultra-high-Q photonic double-heterostructure nanocavity, *Nat. Mater.*, **4**, 207.

Spasov, S., Allison, G., Patanè, A., Eaves, L., Tretyakov, M. Y., Ignatov, A., Hopkinson, M., and Hill, G. (2008). High field electron dynamics in dilute nitride Ga(AsN), *Appl. Phys. Lett.*, **93**, 022111.

Srinivasan, K., and Painter, O. (2003). Fourier space design of high-Q cavities in standard and compressed hexagonal lattice photonic crystals, *Opt. Express*, **11**, 579.

Stavola, M. (1998). Vibrational spectroscopy of light element impurities in semiconductors, in *Semiconductors and Semimetals 51B* (Academic Press, Boston).

Stragier, H., Cross, J. O., Rehr, J. J., Sorensen, L. B., Bouldin, C. E. and Woicik, J. C. (1992). Diffraction anomalous fine structure: a new x-ray structural technique, *Phys. Rev. Lett.*, **69**, 3064.

Strocov, V. N., Nilsson, P. O., Schmitt, T., Augustsson, A., Gridneva, L., Debowska-Nilsson, D., Claessen, R., Egorov, A. Y., Ustinov, V. M. and Alferov, Z. I. (2004). Nitrogen local electronic structure in Ga(In)AsN alloys by soft-x-ray absorption and emission: implications for optical properties, *Phys. Rev. B*, **69**, 035206.

Suemune, I., Uesugi, K., and Walukiewicz, W. (2000). Role of nitrogen in the reduced temperature dependence of band-gap energy in GaNAs, *Appl. Phys. Lett.*, **77**, 3021.

Surrente, A., Gallo, P., Felici, M., Dwir, B., Rudra, A., and Kapon, E. (2009). Dense arrays of ordered pyramidal quantum dots with narrow linewidth photoluminescence spectra, *Nanotechnology*, **20**, 415205.

Tansu, N., Kirsch, N. J., and Mawst, L. J. (2002). Low-threshold-current-density 1300-nm dilute-nitride quantum well lasers, *Physica E*, **81**, 2523.

Tansu, N., and Mawst, L. J. (2002). Temperature sensitivity of 1300-nm InGaAsN quantum-well lasers, *IEEE Photon. Technol. Lett.*, **14**, 1052.

Teo, B. K., and Joy, D. C. (1981). *EXAFS Spectroscopy: Techniques and Applications* (Plenum, New York).

Tesmer, J. R., and Nastasi, M. (1995). *Handbook of Modern Ion Beam Materials Analysis* (Materials Research Society, Pittsburgh).

Teukam, Z., Chevallier, J., Saguy, C., Kalish, R., Ballutaud, D., Barbé, M., Jomard, F., Tromson-Carli, A., Cytermann, C., Butler, J. E., Bernard, M., Baron, C., and Deneuville, A. (2003). Shallow donors with high n-type electrical conductivity in homoepitaxial deuterated boron-doped diamond layers, *Nat. Mater.*, **2**, 482.

Thinh, N. Q., Buyanova, I. A., Hai, P. N., Chen, W. M., Xin, H. P., and Tu, C. W. (2001). Signature of an intrinsic point defect in GaN_xAs_{1-x}, *Phys. Rev. B*, **63**, 033203.

Thomas, D. G., and Hopfield, J. J. (1966). Isoelectronic traps due to nitrogen in gallium phosphide, *Phys. Rev.*, **150**, 680.

Tisch, U., Finkman, E., and Salzman, J. (2002). The anomalous bandgap bowing in GaAsN, *Appl. Phys. Lett*, **81**, 463.

Togan, E., Chu, Y., Trifonov, A. S., Jiang, L., Maze, J., Childress, L., Dutt, M. V. G., Sørensen, A. S., Hemmer, P. R., Zibrov, A. S., and Lukin, M. D. (2010). Quantum entanglement between an optical photon and a solid-state spin qubit, *Nature*, **466**, 730.

Tomić, S., and O'Reilly, E. P. (2002). Gain characteristics of ideal dilute nitride quantum well lasers, *Physica E*, **12**, 1102.

Tomić, S., and O'Reilly, E. P. (2003). Optimization of material parameters in 1.3-μm InGaAsN-GaAs lasers, *IEEE Photon. Technol. Lett.*, **15**, 6.

Tomić, S., O'Reilly, E. P., Fehse, R., Sweeney, S. J., Adams, A. R., Andreev, A. D., Choulis, S. A., Hosea, T. J. C., and Riechert, H. (2003). Theoretical and experimental analysis of 1.3-μm InGaAsN/GaAs lasers, *IEEE J. Sel. Top. Quantum Electron.*, **9**, 1228.

Tomić, S., O'Reilly, E. P., Klar, P. J., Grüning, H., Heimbrodt, W., Chen, W. M., and Buyanova, I. A. (2004). Influence of conduction-band nonparabolicity on electron confinement and effective mass in GaN_xAs_{1-x}/ GaAs quantum wells, *Phys. Rev. B*, **69**, 245305.

Tomljenovic-Hanic, S., Greentree, A. D., De Sterke, C. M., and Prawer, S. (2009). Flexible design of ultrahigh-Q microcavities in diamond-based photonic crystal slabs, *Opt. Express*, **17**, 6465.

Tormen, M., De Salvador, D., Natali, M., Drigo, A., Romanato, F., Rossetto, G., Boscherini, F., and Mobilio, S. (1999). Bond length variation in $In_{0.25}Ga_{0.75}As$/InP epitaxial layers thicker than the critical thickness, *J. Appl. Phys.*, **86**, 2533.

Trotta, R., Atkinson, P., Plumhof, J. D., Zallo, E., Rezaev, R. O., Kumar, S., Baunack, S., Schroter, J. R., Rastelli, A., and Schmidt, O. G. (2012a). Nanomembrane quantum-light-emitting diodes integrated onto piezo-electric actuators, *Adv. Mater.*, **24**, 2668.

Trotta, R., Cavigli, L., Felisari, L., Polimeni, A., Vinattieri, A., Gurioli, M., Capizzi, M., Martelli, F., Rubini, S., Mariucci, L., Francardi, M. and Gerardino, A. (2010). Quantum confinement effects in hydrogen-intercalated $Ga_{1-x}As_xN_x$-$GaAs_{1-x}N_x$: H planar heterostructures investigated by photoluminescence spectroscopy, *Phys. Rev. B*, **81**, 235327.

Trotta, R., Giubertoni, D., Polimeni, A., Bersani, M., Capizzi, M., Martelli, F., Rubini, S., Bisognin, G., and Berti, M. (2009a). Hydrogen diffusion in $GaAs_{1-x}N_x$, *Phys. Rev. B*, **80**, 195206.

Trotta, R., Polimeni, A., and Capizzi, M. (2012b). Hydrogen incorporation in III-N-V semiconductors: from macroscopic to nanometre control of the material physical properties, *Adv. Funct. Mater.*, **22**, 1782.

Trotta, R., Polimeni, A., Capizzi, M., Giubertoni, D., Bersani, M., Bisognin, G., Berti, M., Rubini, S., Martelli, F., Mariucci, L., Francardi, M. and Gerardino, A. (2008). Effect of hydrogen incorporation temperature in in-plane engineered GaAsN/GaAsN:H heterostructure, *Appl. Phys. Lett.*, **92**, 221901.

Trotta, R., Polimeni, A., Capizzi, M., Martelli, F., Rubini, S., Francardi, M., Gerardino, A., and Mariucci, L. (2009b). Light polarization control in strain-engineered GaAsN/GaAsN:H heterostructures, *Appl. Phys. Lett*, **94**, 261905.

Trotta, R., Polimeni, A., Martelli, F., Pettinari, G., Capizzi, M., Felisari, L., Rubini, S., Francardi, M., Gerardino, A., Christianen, P. C. M., and Maan, J. C. (2011). Fabrication of site-controlled quantum dots by spatially selective incorporation of hydrogen in Ga(AsN)/GaAs heterostructures, *Adv. Mater.*, **23**, 2706.

Trotta, R., Zallo, E., Ortix, C., Atkinson, P., Plumhof, J. D., den Brink, J. V., Rastelli, A., and Schmidt, O. G. (2012c). Universal recovery of the energy-level degeneracy of bright excitons in InGaAs quantum dots without a structure symmetry, *Phys. Rev. Lett.*, **109**, 147401.

Troullier, N., and Martins, J. (1991). Efficient pseudopotentials for plane-wave calculations, *Phys. Rev. B*, **43**, 1993.

Ugliengo, P. (2006). *MOLDRAW Program to display and manipulate molecular and crystal structures*, University of Torino, Torino, available at http://www.moldraw.unito.it.

Ugliengo, P., Viterbo, D., and Chiari, G. (1993). MOLDRAW: molecular graphics on a personal computer, *Z. Kristallogr.*, **207**, 9.

Ungaro, G., Roux, G. L., Teissier, R., and Harmand, J. C. (1999). GaAsSbN: a new low-bandgap material for GaAs substrates, *Electron. Lett.*, **35**, 1246.

Van de Walle, C. G., Denteneer, P. J. H., Bar-Yam, Y., and Pantelides, S. T. (1989). Theory of hydrogen diffusion and reactions in crystalline silicon, *Phys. Rev. B*, **39**, 10791.

Van de Walle, C. G., Limpijumnong, S., and Neugebauer, J. (2001). First-principles studies of beryllium doping of GaN, *Phys. Rev. B*, **63**, 245205.

Van de Walle, C. G., Lyons, J. L., and Janotti, A. (2010). Controlling the conductivity of InN, *Phys. Status Solidi A*, **207**, 1024.

Van de Walle, C. G., and Neugebauer, J. (2003). Universal alignment of hydrogen levels in semiconductors, insulators and solutions, *Nature*, **423**, 626.

Van de Walle, C. G., and Neugebauer, J. (2006). Hydrogen in semiconductors, *Annu. Rev. Mater. Sci.*, **36**, 179–201.

Vanderbilt, D. (1990). Soft self-consistent pseudopotentials in a generalized eigenvalue formalism, *Phys. Rev. B*, **41**, 7892.

Vaughan, M. P., and Ridley, B. K. (2007a). Effect of non-parabolicity on the density of states for high-field mobility calculations in dilute nitrides, *Phys. Status Solidi C*, **4**, 686.

Vaughan, M. P., and Ridley, B. K. (2007b). Electron-nitrogen scattering in dilute nitrides, *Phys. Rev. B*, **75**, 195205.

Vavilov, V. S., Kiv, A. E., and Niyazova, O. (1975). The subthreshold radiation effects in semiconductors, *Phys. Status Solidi A*, **32**, 11–33.

Vegard, L. (1921). Die konstitution der mischkristalle und die raumfüllung der atome, *Z. Phys.*, **5**, 17.

Virkkala, V., Havu, V., Tuomisto, F., and Puska, M. J. (2012). Hybrid functional study of band structures of $GaAs_{1-x}N_x$ and $GaSb_{1-x}N_x$ alloys, *Phys. Rev. B*, **85**, 085134.

Vogl, P. (1984). Predictions of deep impurity level energies in semiconductors, *Adv. Electron. Electron Phys.*, **62**, 101.

Volz, K., Koch, J., Kunert, B., and Stolz, W. (2003). Doping behaviour of Si, Te, Zn and Mg in lattice-matched (GaIn)(NAs)/GaAs bulk films, *J. Cryst. Growth*, **248**, 451.

Volz, K., Torunski, T., Lackner, D., Rubel, O., Stolz, W., Baur, C., Mueller, S., Dimroth, F., and Bett, A. W. (2007). Material development for improved 1 eV (GaIn)(NAs) solar cell structures, *J. Sol. Energy Eng.*, **129**, 266.

Vorona, I. P., McHedlidze, T., Dagnelund, D., Buyanova, I. A., Chen, W. M. and Kohler, K. (2006). Optically detected magnetic resonance studies of point defects in Ga(Al)NAs, *Phys. Rev. B*, **73**, 125204.

Vouilloz, F., Oberli, D. Y., Dupertuis, M. A., Gustafsson, A., Reinhardt, F. and Kapon, E. (1997). Polarization anisotropy and valence band mixing in semiconductor quantum wires, *Phys. Rev. Lett.*, **78**, 1580.

Wang, X. J., Buyanova, I. A., Zhao, F., Lagarde, D., Balocchi, A., Marie, X., Tu, C. W., Harmand, J. C., and Chen, W. M. (2009a). Room-temperature defect-engineered spin filter based on a non-magnetic semiconductor, *Nat. Mater.*, **8**, 198.

Wang, X. J., Puttisong, Y., Tu, C. W., Ptak, A. J., Kalevich, V. K., Egorov, A. Y., Geelhaar, L., Riechert, H., Chen, W. M., and Buyanova, I. A. (2009b).

Dominant recombination centers in Ga(In)NAs alloys: Ga interstitials, *Appl. Phys. Lett.*, **95**, 241904.

Warren, B. E. (1990). *X-ray Diffraction* (Dover, New York).

Wei, S. H., and Zunger, A. (1996). Giant and composition-dependent optical bowing coefficient in GaAsN alloys, *Phys. Rev. Lett.*, **76**, 664.

Weinstein, B. A., Stambach, S. R., Ritter, T. M., Maclean, J. O., and Wallis, D. J. (2003). Evidence for selective delocalization of N-pair states in dilute $GaAs_{1-x}N_x$, *Phys. Rev. B*, **68**, 035336.

Weiringen, A. V., and Warmholtz, N. (1956). On the permeation of hydrogen and helium in single crystal silicon and germanium at elevated temperatures, *Physica*, **22**, 849–865.

Wen, L., Bekisli, F., Stavola, M., Fowler, W. B., Trotta, R., Polimeni, A., Capizzi, M., Rubini, S., and Martelli, F. (2010). Detailed structure of the H–N–H center in $GaAs_{1-y}N_y$ revealed by vibrational spectroscopy under uniaxial stress, *Phys. Rev. B*, **81**, 233201.

Wen, L., Stavola, M., Fowler, W. B., Trotta, R., Polimeni, A., Capizzi, M., Bisognin, G., Berti, M., Rubini, S., and Martelli, F. (2012). Microscopic origin of compressive strain in hydrogen-irradiated dilute $GaAs_{1-y}N_y$ alloys: role of $N–H_n$ centers with $n>2$ and their thermal stability, *Phys. Rev. B*, **86**, 085206.

Woicik, J. C., Pellegrino, J. G., Steiner, B., Miyano, K. E., Bompadre, S. G., Sorensen, L. B., Lee, T. L., and Khalid, S. (1997). Bond-length distortions in strained semiconductor alloys, *Phys. Rev. Lett.*, **79**, 5026.

Wolford, D. J., Bradley, J. A., Fry, K., and Thompson, J. (1984). The nitrogen isoelectronic trap in GaAs, in *Proceedings of the 17th International Conference on the Physics of Semiconductors* (Springer, New York), 627.

Wormington, M., Panaccione, C., Matney, K. M., and Bowen, K. (1999). Characterisation of structures from x-ray scattering data using genetic algorithms, *Philos. Trans. R. Soc. London, A*, **357**, 2827.

Xin, H. P., and Tu, C. W. (2000). Photoluminescence properties of GaNP/GaP multiple quantum wells grown by gas source molecular beam epitaxy, *Appl. Phys. Lett.*, **77**, 2180.

Xin, H. P., Tu, C. W., and Geva, M. (1999). Annealing behavior of p-type $Ga_{0.892}In_{0.108}N_xAs_{1-x}$ ($0 \leq x \leq 0.024$) grown by gas-source molecular beam epitaxy, *Appl. Phys. Lett.*, **75**, 1416.

Yang, X., Heroux, J. B., Mei, L. F., and Wang, W. I. (2001). InGaAsNSb/GaAs quantum wells for 1.55 μm lasers grown by molecular-beam epitaxy, *Appl. Phys. Lett.*, **78**, 4068.

Yang, X., Jurlovic, M. J., Heroux, J. B., and Wang, W. I. (1999). Molecular beam epitaxial growth of InGaAsN:Sb/GaAs quantum wells for long-wavelength semiconductor lasers, *Appl. Phys. Lett.*, **75**, 178.

Yoshie, T., Scherer, A., Hendrickson, J., Khitrova, G., Gibbs, H. M., Rupper, G., Ell, C., Shchekin, O. B., and Deppe, D. G. (2004). Vacuum Rabi splitting with a single quantum dot in a photonic crystal nanocavity, *Nature*, **432**, 200.

Yuan, J., Wang, H., Veldhoven, R. P. J. V., Wang, J., Vries, T. D., Smalbrugge, B., Jin, C. Y., Nouwens, P., Geluk, E. J., Silov, A. Y., and Noetzel, R. (2011). Controlling polarization anisotropy of site-controlled InAs/InP (100) quantum dots, *Appl. Phys. Lett.*, **98**, 201904.

Zangrando, M., Finazzi, M., Zacchigna, M., Cocco, D., Rochow, R. and Parmigiani, F. (2004). Polarized high-brilliance and high-resolution soft x-ray source at ELETTRA: the performance of beamline BACH, *Rev. Sci. Instrum.*, **75**, 31.

Zavada, J. M., Jenkinson, H. A., Sarkis, R. G., and Wilson, R. G. (1985). Hydrogen depth profiles and optical characterization of annealed, proton-implanted n-type GaAs, *J. Appl. Phys.*, **58**, 3731.

Zhang, Y., Fluegel, B., Mascarenhas, A., Xin, H. P., and Tu, C. W. (2000). Optical transitions in the isoelectronically doped semiconductor GaP:N: an evolution from isolated centers, pairs, and clusters to an impurity band, *Phys. Rev. B*, **62**, 4493.

Zhang, Y., and Mascarenhas, A. (2000). Isoelectronic impurity states in GaAs:N, *Phys. Rev. B*, **61**, 15562.

Zory, P. S. J. (1993). *Quantum Well Lasers* (Academic Press, New York).

Zwiller, V., Pistol, M., Hessman, D., Cederstrom, R., Seifert, W. and Samuelson, L. (1999). Time-resolved studies of single semiconductor quantum dots, *Phys. Rev. B*, **59**, 5021.

Index